MULTIVARIATE ANALYSIS FOR THE BIOBEHAVIORAL AND SOCIAL SCIENCES

MULTIVARIATE ANALYSIS FOR THE BIOBEHAVIORAL AND SOCIAL SCIENCES
A Graphical Approach

BRUCE L. BROWN
SUZANNE B. HENDRIX
DAWSON W. HEDGES
TIMOTHY B. SMITH

A JOHN WILEY & SONS, INC., PUBLICATION

Published by John Wiley & Sons, Inc., Hoboken, New Jersey.
Published simultaneously in Canada.

For general information on our other products and services or for technical support, please contact our Customer Care Department within the United States at (800) 762-2974, outside the United States at (317) 572-3993 or fax (317) 572-4002.

Wiley also publishes its books in a variety of electronic formats. Some content that appears in print may not be available in electronic formats. For more information about Wiley products, visit our web site at www.wiley.com.

Library of Congress Cataloging-in-Publication Data:

Brown, Bruce (Bruce L.)
  Multivariate analysis for the biobehavioral and social sciences / Bruce L. Brown, Suzanne B. Hendrix, Dawson W. Hedges, Timothy B. Smith.
    p. cm.
  Includes bibliographical references and index.
  ISBN 978-0-470-53756-5
  1. Social sciences–Statistical methods. 2. Multivariate analysis. I. Title.
  HA29.B82857 2012
  300.1'519535–dc23

                                                                2011018420

Printed in the United States of America.

10 9 8 7 6 5 4 3 2 1

CONTENTS

PREFACE

The plan of this book is unusual. It is unusual in that we have elevated graphics to the status of an equal partner in the data analysis process. Our intent is to demonstrate the centrality of good graphics to the scientific process, to provide a graphical concomitant for each of the classical multivariate statistical methods presented, and to demonstrate the superiority of graphical expressions in clarifying and laying bare the meaning in the data.

The plan of the book is also unusual in the pedagogical approach taken. The first three chapters are all preparatory: giving an overview of multivariate methods (Chapter 1), reviewing the fundamental principles of elementary statistics as "habits" that are necessary preparation for understanding multivariate methods (Chapter 2), and then introducing matrix algebra (Chapter 3).

The most unusual aspect of the book, however, is the six methods chapters (4, 5, 6, 7, 8, and 9)—the core of the book. We introduce each with a published paper that is in some way exemplary as a "research-publication case study." We first showcase the method in a strong piece of published work to demonstrate to the student the practical value of that method. The next section answers the question "How do you do that?" It is our intent to answer that question fully with a complete but simplified demonstration of the mathematics and the concepts of the method. This is a unique feature of the book. There are books that are fully enabling mathematically, and there are also books that are highly accessible to the beginning student, but this book is unique in combining these two characteristics by the use of simplest case demonstrations.

The next step in each chapter is to demonstrate how the analysis of the data is accomplished using one of the commonly used statistical packages, such as Stata®, SAS®, or SPSS®. One of the major tasks in demonstrating statistical packages is that of instructing the student in the reading of output. The simplest case demonstration of the full computational process is an effective way to deal with that aspect. After carrying out the full analysis with simplest case data, and then using Stata, SAS, or SPSS to analyze the same simple set of data, the meaning of each of the parts in the computer output becomes obvious and clear.

Although this book covers many of the commonly used multivariate methods and builds upon them graphically, this approach is also applicable to a wide variety of additional methods, both modern regression methods and also extensions of multivariate methods. Factor analysis has grown into structural equations modeling, MANOVA and ANOVA have grown into multilevel, hierarchical, and mixed models, and general linear models have grown into generalized linear models that can deal with a broad variety of data types, including categorical. All of these can be supplemented and improved upon with graphics.

The focus of this book on the application of graphics to the classical methods is, we believe, the appropriate beginning given the relative simplicity of the fundamental multivariate methods. The principles the student learns here can then more easily be expanded in future texts to the full power of advanced derivative methods. It is curious that the rather simple multiple regression model is the foundation of many if not most of the higher-level developments. We have closed the book with a simple presentation of multiple regression in Chapter 9, both as a look backward and also as a look forward. It is a basic example of the application of matrix methods to multiple variables, but also as a prelude to the higher-level methods.

We are grateful to those exemplary researchers and quantitative methods scholars whose work we have built upon. We are grateful to our students from whom we have learned, many of whom appear in this book. Most of all, we are grateful to our families who have been supportive and patient through this process.

BRUCE L. BROWN
SUZANNE HENDRIX
DAWSON W. HEDGES
TIMOTHY B. SMITH

CHAPTER ONE

OVERVIEW OF MULTIVARIATE AND REGRESSION METHODS

1.1 INTRODUCTION

More information about human functioning has accrued in the past five decades than in the preceding five millennia, and many of those recent gains can be attributed to the application of multivariate and regression statistics. The scientific experimentation that proliferated during the 19th century was a remarkable advance over previous centuries, but the advent of the computer in the mid-20th century opened the way for the widespread use of complex analytic methods that exponentially increased the pace of discovery. Multivariate and regression methods of data analysis have completely transformed the bio-behavioral and social sciences.

Multivariate and regression statistics provide several essential tools for scientific inquiry. They allow for detailed descriptions of data, and they identify patterns impossible to discern otherwise. They allow for empirical testing of complex theoretical propositions. They enable enhanced prediction of events, from disease onset to likelihood of remission. Stated simply, multivariate statistics can be applied to a broad variety of research questions about the human condition.

Given the widespread application and utility of multivariate and regression methods, this book covers many of the statistical methods commonly used in a broad range of bio-behavioral and social sciences, such as psychology,

Multivariate Analysis for the Biobehavioral and Social Sciences: A Graphical Approach,
First Edition. Bruce L. Brown, Suzanne B. Hendrix, Dawson W. Hedges, Timothy B. Smith.
© 2012 John Wiley & Sons, Inc. Published 2012 by John Wiley & Sons, Inc.

business, biology, medicine, education, and sociology. In these disciplines, mathematics is not typically a student's primary focus. Thus, the approach of the book is *conceptual*. This does not mean that the mathematical account of the methods is compromised, just that the mathematical developments are employed in the service of the conceptual basis for each method. The math is presented in an accessible form, called *simplest case*. The idea is that we seek a demonstration for each method that uses the *simplest case* we can find that has all the key attributes of the full-blown cases of actual practice. We provide exercises that will enable students to learn the simplified case thoroughly, after which the focus is expanded to more realistic cases.

We have learned that it is possible to make these complex mathematical concepts accessible and enjoyable, even to those who may see themselves as nonmathematical. It is possible with this *simplest-case* approach to teach the underlying conceptual basis so thoroughly that some students can perform many multivariate and regression analyses on simple "student-accommodating" data sets from memory, without referring to written formulas. This kind of deep conceptual acquaintance brings the method up close for the student, so that the meaning of the analytical results becomes clearer.

This first chapter defines *multivariate data analysis methods* and introduces the fundamental concepts. It also outlines and explains the structure of the remaining chapters in the book. All analysis method chapters follow a common format. The main body of each chapter starts with an example of the method, usually from an article in a prominent journal. It then explains the rationale for each method and gives complete but simplified numerical demonstrations of the various expressions of each method using *simplest-case data*. At the end of each chapter is the section entitled *Study Questions*, which consists of three types: *essay questions, calculation questions*, and *data-analysis questions*. There is a complete set of answers to all of these questions available electronically on the website at https://mvgraphics.byu.edu.

1.2 MULTIVARIATE METHODS AS AN EXTENSION OF FAMILIAR UNIVARIATE METHODS

The term *multivariate* denotes the analysis of multiple dependent variables. If the data set has only one dependent variable, it is called *univariate*. In elementary statistics, you were probably introduced to the two-way analysis of variance (ANOVA) and learned that any ANOVA that is two-way or higher is referred to as a *factorial* model. *Factorial* in this instance means having multiple *independent* variables or factors. The advantage of a factorial ANOVA is that it enables one to examine the interaction between the independent variables in the effects they exert upon the dependent variable.

Multivariate models have a similar advantage, but applied to the multiple dependent variables rather than independent variables. Multivariate methods enable one to deal with the *covariance* among the dependent variables in a

Table 1.1 Overview of Univariate and Multivariate Statistical Methods

Description and Number of Predictor (Independent) Variables	Univariate Method	Multivariate Method
	One quantitative outcome (dependent) variable	Multiple quantitative outcome (dependent) variables
No predictor variable	—	Factor analysis
		Principal component analysis
		Cluster analysis
One categorical predictor variable, two levels	t tests	Hotelling's T^2 tests
	z tests	Profile analysis using Hotelling's T^2
One categorical predictor, variable, three or more levels	ANOVA, one-way models	MANOVA, one-way models
Two or more categorical predictor variables	ANOVA, factorial models	MANOVA, factorial models
Categorical predictor(s) with one or more quantitative control variables	ANCOVA, one-way or factorial models	MANCOVA, one-way or factorial models
One quantitative predictor variable	Bivariate regression	Multivariate regression
Two or more quantitative predictor variables	Multiple regression	Multivariate multiple regression Canonical correlation*

way that is analogous to the way factorial ANOVA enables one to deal with interaction.

Fortunately, many of the multivariate methods are straightforward extensions of the corresponding univariate methods (Table 1.1). This means that your considerable investment up to this point in understanding univariate statistics will go a long way toward helping you to understand multivariate statistics. (This is particularly true of Chapters 7, 8, and 9, where the t-tests are extended to multivariate t-tests, and various ANOVA models are extended to corresponding multiple ANOVA [MANOVA] models.) Indeed, one can think of multivariate statistics in a simplified way as just the same univariate methods that you already know (t-test, ANOVA, correlation/regression, etc.) rewritten in *matrix algebra* with the matrices extended to include multiple dependent variables.

Matrix algebra is a tool for more efficiently working with data matrices. Many of the formulas you learned in elementary statistics (variance, covariance, correlation coefficients, ANOVA, etc.) can be expressed much more compactly and more efficiently with matrix algebra. Matrix multiplication in particular is closely connected to the calculation of variances and covariances in that it directly produces sums of squares and sums of products of input vectors. It is as if matrix algebra were invented specifically for the calculation of covariance structures. Chapter 3 provides an introduction to the fundamentals of matrix algebra. Readers unfamiliar with matrix algebra should therefore carefully read Chapter 3 prior to the other chapters that follow, since all are based upon it.

The second prerequisite for understanding this book is a *knowledge of elementary statistical methods:* the normal distribution, the binomial distribution, confidence intervals, *t*-tests, ANOVA, correlation coefficients, and regression. It is assumed that you begin this course with a fairly good grasp of basic statistics. Chapter 2 provides a review of the fundamental principles of elementary statistics, expressed in matrix notation where applicable.

1.3   MEASUREMENT SCALES AND DATA TYPES

Choosing an appropriate statistical method requires an accurate categorization of the data to be analyzed. The four kinds of measurement scales identified by S. Smith Stevens (1946) are nominal, ordinal, interval, and ratio. However, there are almost no examples of interval data that are not also ratio, so we often refer to the two collectively as an interval/ratio scale. So, effectively, we have only three kinds of data: those that are categorical (nominal), those that are ordinal (ordered categorical), and those that are fully quantitative (interval/ratio). As we investigate the methods of this book, we will discover that ordinal is not a particularly meaningful category of data for multivariate methods. Therefore, from the standpoint of data, the major distinction will be between those methods that apply to fully quantitative data (interval/ratio), those that apply to categorical data, and those that apply to data sets that have both quantitative and categorical data in them.

Factor analysis (Chapter 4) is an example of a method that has only quantitative variables, as is multiple regression. Log-linear models (Chapter 9) are an example of a method that deals with data that are completely categorical. MANOVA (Chapter 8) is an example of an analysis that requires both quantitative and categorical data; it has categorical independent variables and quantitative dependent variables.

Another important issue with respect to data types is the distinction between discrete and continuous data. Discrete data are whole numbers, such as the number of persons voting for a proposition, or the number voting against it. Continuous data are decimal numbers that have an infinite number of possible points between any two points. In measuring cut lengths of wire, it is possible in principal to identify an infinitude of lengths that lie between any two points, for example, between 23 and 24 inches. The number possible, in practical terms, depends on the accuracy of one's measuring instrument. Measured length is therefore continuous. By extension, variables measured in biomedical and social sciences that have multiple possible values along a continuum, such as oxytocin levels or scores on a measure of personality traits, are treated as continuous data.

All categorical data are by definition discrete. It is not possible for data to be both categorical and also continuous. Quantitative data, on the other hand, can be either continuous or discrete. Most measured quantities, such as height, width, length, and weight, are both continuous and also fully quantitative (interval/ratio). There are also, however, many other examples of data that are fully quantitative and yet discrete. For example, the count of the number of persons in a room is discrete, because it can only be a whole number, but it is also fully quantitative, with interval/ratio properties. If there are 12 persons in one room and twenty-four in another, it makes sense to say that there are twice as many persons in the second room. Counts of number of persons therefore have interval/ratio properties.[1]

When all the variables are measured on the same scale, we refer to them as *commensurate*. When the variables are measured with different scales, they are *noncommensurate*. An example of commensurate data would be width, length, and height of a box, each one measured in inches. An example of non-commensurate would be if the width of the box and its length were measured in inches, but the height was measured in centimeters. (Of course, one could make them commensurate by transforming all to inches or all to centimeters.) Another example of noncommensurate variables would be IQ scores and blood lead levels. Variables that are not commensurate can always be made so by standardizing them (transforming them into Z-scores or percentiles). A few multivariate methods, such as profile analysis (associated with Chapter 7 in connection with Hotelling's T^2), or principal component analysis of a covariance matrix (Chapter 4) require that variables be commensurate, but most of the multivariate methods do not require this.

1.4   FOUR BASIC DATA SET STRUCTURES FOR MULTIVARIATE ANALYSIS

Multivariate and regression data analysis methods can be creatively applied to a wide variety of types of data set structures. However, four basic types of data set structures include most of the multivariate and regression data sets that will be encountered. These four basic types of data fit almost all of the statistical methods introduced in this book.

---

[1] See Chapter 2, Section 2.1, for a review of the properties of a ratio scale and also of the other three types of scales.

FOUR BASIC TYPES OF DATA SET STRUCTURE

Type 1: Single sample with multiple variables measured on each sampling unit.
Possible methods include factor analysis, principal component analysis, cluster analysis, and confirmatory factor analysis.
Type 2: Single sample with two sets of multiple variables (an X set and a Y set) measured on each sampling unit.
Possible methods include canonical correlation, multivariate multiple regression, and structural equations modeling.
Type 3: Two samples with multiple variables measured on each sampling unit.
Possible methods include Hotelling's T^2 test, discriminant analysis, and some varieties of classification analysis.
Type 4 More than two samples with multiple variables measured on each sampling unit.
Possible methods include MANOVA, multiple discriminant analysis, and some varieties of classification analysis.

The first type of data set structure is *a single sample with multiple variables measured on each sampling unit*. An example of this kind of data set would be the scores of 300 people on seven psychological tests. Multivariate methods that apply to this kind of data are discussed in Chapter 4 and include *principal component analysis*, *factor analysis*, and *confirmatory factor analysis*. These methods provide answers to the question, "What is the covariance structure of this set of multiple variables?"

The second type of data set structure is *a single sample with two sets of multiple variables (an X set and a Y set) measured on each unit*. An example of data of this kind would be a linked data set of mental health inpatients' records, with the X set of variables consisting of several indicators of physical health (e.g., blood serum levels), and the Y set of variables consisting of several indicators of neurological functioning (e.g., results of testing). Multivariate methods that can be applied to this kind of data include *canonical correlation* (Chapter 6) and *multivariate multiple regression* (Chapter 9). These methods provide answers to the question, "What are the linear combinations of variables in the X set and in the Y set that are maximally predictive of the other set?" Another method that can be used with a single sample with two sets of multiple variables would be SEM, *structural equations modeling*. However, SEM can also be applied when there are *more than two* sets of multiple variables. In fact, it can handle any number of sets of multiple variables. It is the general case of which these other methods are special cases, and as such it has a great deal of potential analytical power.

The third type of data set structure is *two samples with multiple variables measured on each unit*. An example would be a simple experiment with an experimental group and a control group, and with two or more dependent variables measured on each observation unit. For example, the effects of a certain medication could be assessed by applying it to 12 patients selected at random (the experimental group) and not applying it to the other 12 patients (the control group), using multiple dependent variable measurements (such as scores on several tests of patient functioning). Multivariate methods that can be applied to this kind of data are *Hotelling's* T^2 *test* (Chapter 7), *profile analysis, discriminant analysis* (Chapter 7), and some varieties of *classification analysis*. The Hotelling's T^2 test is the multivariate analogue of the ordinary *t*-test, which applies to two-sample data when there is only one dependent variable. The Hotelling's T^2 test extends the logic of the t test to compare two groups and analyze statistical significance holistically for the combined set of multiple dependent variables. The T^2 test answers the question, "Are the vectors of means for these two samples significantly different from one another?" Discriminant analysis and other classification methods can be used to find the optimal linear combination of the multiple dependent variables to best separate the two groups from one another.

The fourth type of data set structure is similar to the third but extended to three or more samples (with multiple dependent variables measured on each of the units of observation). For example, the same test of the effects of medication on hospitalized patients could be done with two types of medication plus the control group, making three groups to be compared simultaneously and multivariately. The major method here is *MANOVA*, or *multivariate ANOVA* (Chapter 8), which is the multivariate analog of ANOVA. In fact, for every ANOVA model (two-way, three-way, repeated measures, etc.), there exists a corresponding MANOVA model. MANOVA models answer all the same questions that ANOVA models do (significance of main effects and interactions), but holistically within multivariate spaces rather than just for a single dependent variable. *Multiple discriminant analysis* and *classification analysis* methods can also be applied to multivariate data having three or more groups, to provide a spatial representation that optimally separates the groups.

1.5   PICTORIAL OVERVIEW OF MULTIVARIATE METHODS

Diagrammatic representations can help explain and differentiate among the various multivariate statistical methods. Several such methods are described pictorially in this section, starting with factor analysis (Chapter 4), a method that applies to the simplest of the four data set structures just described, *a single sample with multiple variables measured on each sampling unit or unit of observation.* Principal component analysis (Chapter 4) also applies to this simple data set structure. The ways in which these two methods differ will be more fully explained in Chapter 4, but one difference can be seen from the schematic diagram of each method given below. The bottom part of each figure shows the matrix organization of the input data, with rows representing observations and columns representing variables, and the two methods are seen to be identical in this aspect.

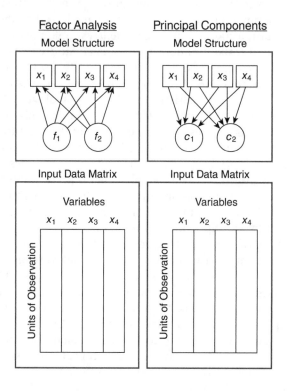

The top part of each figure shows the structure of the model, how the observed variables (x_1 through x_4 for this example) are related to the underlying *latent* variables, which are the factors (f_1 and f_2) for factor analysis, and the components (c_1 and c_2) for principal component analysis. As can be seen by the direction of the arrows, principal components are defined as *linear combinations* (which can be thought of as weighted sums) of the observed variables. However, in factor analysis, the direction is reversed. The observed variables are expressed as linear combinations of the factors. Another difference is that in principal component analysis, we seek to explain a large part of the total variance in the observed variables with the components, but in factor analysis, we seek to account for the covariances or correlations among the variables. (Note that latent variables are represented with circles, and manifest/observed variables are represented with squares, consistent with structural equation modeling notation.)

Multiple regression, also referred to as OLS or "ordinary least-squares regression," is probably the simplest of the methods presented in this book, but in its many variations, it is also the most ubiquitous. It is the foundation for understanding a number of the other methods, as it is the basis for the general linear model. ANOVA is a special case of multiple regression (multiple regression with categorical dummy variables as the predictor variables, the X variables in the diagram below), and when data are unbalanced (unequal cell sizes), multiple regression is by far the most efficient way to analyze the data (as will be demonstrated in Chapter 9). Logistic regression and the generalized linear model (Chapter 9) are adaptations of multiple regression to deal with a wide variety of data types, categorical as well as quantitative. Multilevel linear models, mixed models, and hierarchical linear models are high-level derivatives of regression. The simple data set structure of OLS regression consists of merely several independent variables (also referred to as "predictor variables") being used to predict one dependent variable (also referred to as the "criterion variable").

Multiple Regression

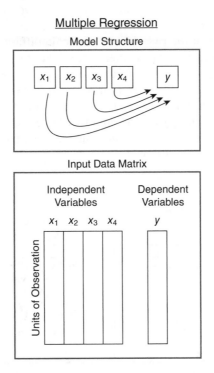

Canonical correlation is similar to multiple regression (and the multiple correlation coefficient on which multiple regression is based), but it deals with two sets of multiple variables rather than one. As such, it fits the second type of data set structure explained above, *a single sample with two sets of multiple variables (an* X *set and a* Y *set) measured on each unit.* Multiple regression gives the correlation coefficient between the best possible linear combination of a group of X variables and a single Y variable. Canonical correlation, by extension, gives the correlation coefficient between two linear combinations, one on the X set of multiple variables and one on the Y set of multiple variables. In other words, latent variables are extracted from both the X set of variables and the Y set of variables to fit the criterion that the correlation between the corresponding latent variables in the X set and the Y set is maximal. It is like a double multiple regression that is recursive, where the best possible linear combination of X variables for predicting Y variables is obtained, and also vice versa. This is shown in the diagram on the left below. To return to the example given above for this kind of linked multivariate data set, the canonical correlation of the mental health inpatient data set described would give the best possible linear combination of blood serum levels for predicting neurological functioning, but since it is recursive (bidirectional), it also gives the best possible combination of neurological functioning for predicting blood serum levels.

A slight change in the way the analysis is conceived and the calculations are performed turns canonical correlation into a double factor analysis, as shown in the diagram at the right below. The main difference here is theoretical, in how the latent variables (the linear combinations of observed variables) are interpreted. In the application of canonical correlation as a double factor analysis shown below, the interpretation is that the observed variables are in fact combinatorial expressions of the underlying latent variables, labeled here with the Greek letters chi (χ), for the latent variables for the X set, and nu (η) for the latent variables for the Y set. The concepts and mathematics for canonical correlation are presented in Chapter 6.

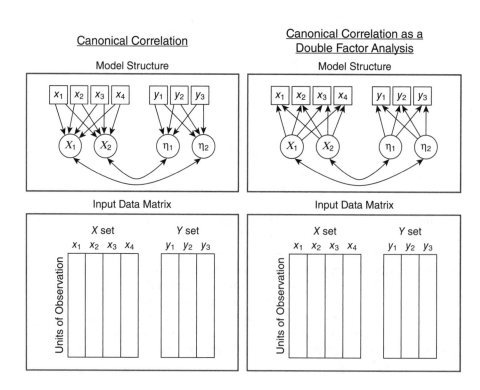

Another method closely related to canonical correlation and multiple regression is multivariate multiple regression, as shown in the diagram. This is essentially the same computational machinery as canonical correlation, except that the latent variables are not recursive. That is, the X set is thought of as being predictive of the Y set, but not vice versa. This is shown in the diagram by the arrows only going one way. An example of this would be predicting a Y set of mutual-fund performance variables from an X set of market index variables. The X set of variables on the left are combined together into the left-hand latent variables labeled as χ_1 and χ_2. These are the linear combinations of market indices that are most predictive of performance on the entire set of mutual funds as a whole, but this is mediated through the right-hand latent variables η_1 and η_2, which are combined together to predict the performance on each of the mutual funds, the Y variables. This is analogous to the way that simple bivariate correlation is recursive (the Pearson product moment correlation coefficient between X and Y is the same as that between Y and X), but simple bivariate regression is not. The regression equation is used for predicting Y from X but not usually for predicting X from Y.

Multivariate Multiple Regression

Model Structure

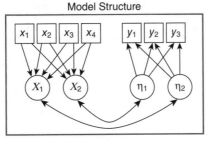

Input Data Matrix

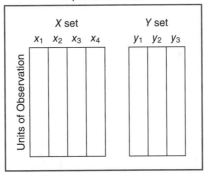

Another way to think of multivariate multiple regression is as the multivariate extension of multiple regression. Instead of predicting one dependent variable from a linear combination of independent variables, one predicts a set of multiple dependent variables from linear combinations of independent variables.

None of the methods discussed so far are specifically intended for data from true experimental designs. In fact, most multivariate methods are for correlational rather than experimental methods. However, a number of the multivariate methods are specifically designed to deal with truly experimental data having multiple dependent variables. These are the T^2, MANOVA, ANCOVA, and MANCOVA methods presented in Chapters 7, 8, and 9 (one half of the methods chapters in this book). These methods fit the third and fourth types of data set structure discussed above, *two samples with multiple variables measured on each unit*, and *three or more samples with multiple variables measured on each unit*. These are illustrated in the two diagrams below. The two-sample type of data set can be analyzed with Hotelling's T^2 (Chapter 7), as shown in the diagram on the left, and data sets with three or more treatment groups require MANOVA (Chapter 8), as shown in the diagram on the right. In the same way that the t-test is a special case of ANOVA, the case restricted to two treatment groups, and the F-ratio of ANOVA is just the square of the corresponding t-value, Hotelling's T^2 is also a special case of MANOVA, and when there are only two groups, the same results will be obtained by using either method.

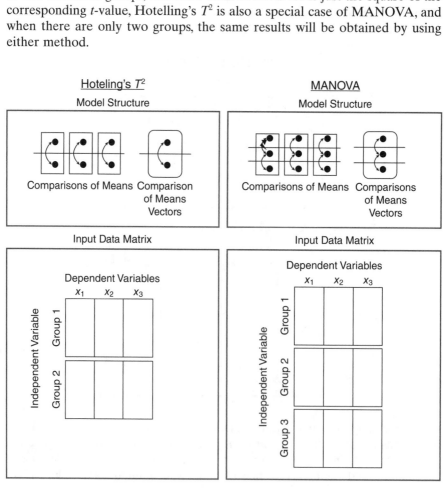

The simplest way to think of analysis of covariance (ANCOVA) is as an ANOVA calculated on the residuals from a regression analysis. That is, ANCOVA, like ANOVA provides tests of whether treatment effects are significant, but with the effects of one or more covariates "regressed out," as shown in the diagram at the left below.

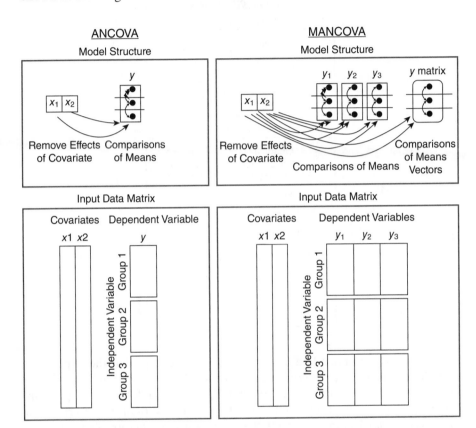

Strictly speaking ANCOVA is not a multivariate method, since there is only one dependent variable. The multivariate version of ANCOVA is *multivariate analysis of covariance* (MANCOVA), in which one essentially calculates a MANOVA with the effects of one or more covariates statistically controlled, as shown in the diagram at the right above. Both of these methods are presented in Chapter 9.

For the final diagram, we again draw upon the third and fourth types of data set structure, *two samples with multiple variables measured on each unit*, and *three or more samples with multiple variables measured on each unit*. These are the types for which Hotelling's T^2 analysis and MANOVA are appropriate. However, the methods of discriminant analysis (Chapter 7) and classification analysis can also be used to good advantage with this kind of data structure, even when it involves data from a true experimental design. The diagram for these methods applied to this kind of data structure is given below.

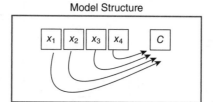

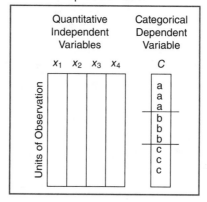

This looks very much like multiple regression, except that the dependent variable is categorical. Discriminant analysis asks the question "what is the best linear combination of a set of quantitative variables (X) to optimally separate categorical groups (C)?" In the usual way of using discriminant analysis and classification methods, the quantitative predictors of group membership would be thought of as independent variables, and the categories of group membership would be thought of as the dependent variable (which is how the diagram above is labeled). However, if one had MANOVA data from a true experiment, with the categories being treatment groups (the independent variable), then discriminant analysis could answer the question "what combination of the dependent variables best accounts for the significant multivariate effects of my experimental and control treatments?" This would reverse what is considered independent and what dependent variables.

The forgoing pictorial overview of methods includes most of the methods presented in this book. Notably absent are the methods of Chapter 5 (cluster analysis, multidimensional scaling, and multivariate graphics). These methods have much in common with factor analysis and principal components, and the diagram for cluster analysis would be similar to principal components.

However, cluster analysis creates taxonomic groupings rather than "factor loadings" (correlations between factors and observed variables), and it can be used to cluster both the variables and also the units of observation. Like factor analysis and principal component analysis, it can be applied to the first kind of data set structure, *a single sample with multiple variables measured on each sampling unit.*

Multidimensional scaling also applies to this kind of data set structure, but in a very unusual way, going directly to the latent variables without having any observed variables. It does this by inferring the latent variables from measured distances among the units of observation. Chapter 5, multivariate graphics, demonstrates how illuminating graphs can be constructed based upon the quantitative methods that apply to the first type of data set structure—factor analysis, principal component analysis, cluster analysis, and multidimensional scaling.

The log-linear methods presented in Chapter 9 are simple, involving two-way contingency tables like those analyzed with chi square. However, log-linear also can be used to deal with higher-order three-way, four-way, and in general "multiway" contingency tables in a more efficient manner. As such, it deals with data that are entirely categorical. However, the use of logarithms converts multiplicative relationships (the multiplication rule of probabilities) into additive relationships, and makes possible the full power of linear models (of the kind used in ANOVA and MANOVA) with categorical data. Generalized linear models use logarithmic (and other) linking functions to render categorical data amenable to linear models analysis. The second half of Chapter 9 introduces logistic regression and other generalized linear models, which can be used with any mix of categorical and quantitative variables. Diagrams for these methods would be very similar to those shown for various types of multiple and multivariate multiple regression.

Structural equations modeling would be difficult to diagram, since it is a general and very malleable set of methods that can be applied in one way or another to most of the types of data set structure presented. The results of virtually all of the other data analysis methods can be obtained from an adaptation of structural equations modeling.

1.6   CORRELATIONAL VERSUS EXPERIMENTAL METHODS

Experimental and correlational studies differ both in the research designs employed and also in the kind of statistics that are used to analyze the data. They also differ in the kinds of questions that can be answered, and in the way they use random processes. Correlational research designs are usually based on **random selection** of subjects, often in a naturalistic setting where there is little control over the variables. Experimental research designs, on the other hand, usually involve tight experimental controls and **random assignment** of subjects to treatment groups to ensure comparability. The critical distinction between experimental and nonexperimental designs is that in true experimental designs, the experimenter manipulates the independent variable by randomly assigning subjects to treatment groups and the control group. Experimental designs enable the researcher to make more definitive conclusions and to attribute causality, whereas the inferences in correlational research are more tenuous. The multivariate methods introduced in Chapter 7 (Hotelling's T-Squared), 8 (MANOVA), 9 (ANCOVA, MANCOVA, repeated measures MANOVA, logistic regression models, etc.) are applicable to the data obtained from true experimental designs. The methods in the remainder of the chapters are used primarily with data from correlational studies and therefore provide less definitive conclusions.

Many seem to believe that the only real disadvantage of nonexperimental studies is that one cannot attribute causality with a high degree of confidence. While this is indeed a serious problem with nonexperimental designs, there are other issues. On page 162 of his 1971 book, Winer makes the very important point that ANOVA was originally developed to be used within the setting of a true experiment where one has control over extraneous variables and subjects are assigned to treatment groups at random. The logic underlying the significance tests and the determination of probabilities of the Type I error is based upon the assumption that treatment effects and error effects are independent of one another. The only assurance one can have that the two are indeed independent is that subjects are assigned at random to treatment groups. In other words, when ANOVA is used to analyze data from a nonexperimental design, and subjects are not assigned at random to treatment groups, there is no assurance that treatment and error effects are independent of one another, and the logic underlying the determination of the probability of the Type I error breaks down. One is, in this case, using the P-values from an ANOVA metaphorically. Winer (1971, 162) concludes this section with the words "hence the importance of randomization in design problems."

1.7   OLD VERSUS NEW METHODS

Factor analysis is a very old multivariate method. It dates back to a turn-of-the-century paper by Spearman (1904) and an earlier one by Pearson (1901). As such, it can be thought of as the fundamental and quintessential correlational multivariate method. Similarly, MANOVA is the fundamental multivariate method for dealing with data from true experimental designs (where variables are manipulated under controlled conditions). MANOVA was a comparatively late development, with its advent in the mid-20th century. However, both of these can be thought of as the "old" multivariate methods, one for correlational data and one for experimental data. These were the major multivariate methods used and taught in the research methodology classes a generation ago.

The young up-and-coming generation of researchers is much more excited about later developments in methods for dealing with both experimental data and also correlational data. By far, the most influential new method for dealing with correlational data is *structural equations modeling* (SEM). It has an updated approach to factor analysis referred to as *confirmatory factor analysis,* but confirmatory factor analysis is only one of a wide variety of SEM models, and one of the simpler ones at that. SEM is general and highly malleable, such that virtually all of the other correlational multivariate methods can be thought of as special cases of it.

SEM concepts are also foundational for the second "new" method, *hierarchical linear models* (HLM). HLM can be used to deal with data from true experimental designs, but it can also be used with correlational studies. At this point, it is not yet strictly speaking a multivariate method, in that the major data analytic packages (SAS, Stata, and SPSS) do not have procedures for a multivariate HLM (although there are ways to finesse the packages to accomplish multivariate goals with HLM). Also, the method is new enough that some of the mathematics for multivariate applications for HLM have yet to be worked out. It does not, however, by any means replace MANOVA (as some have erroneously thought). It is applicable to univariate data involving repeated measures, where one has a "mixed" model, involving a combination of fixed (treatment) and random (repeated measures) variables.

1.8 SUMMARY

Data are only as useful as the analyses performed with them. Increasingly, scientists have recognized that there are extremely few cases in which a single variable exerts sole influence on an isolated outcome. Typically, many factors influence one another, often in complex sequences. The world is multivariate.

Given the multivariate nature of biomedical and social sciences, advanced multivariate and regression methods are becoming increasingly utilized. This book covers those methods most commonly used in the research literature. Factor analysis, principal component analysis, and cluster analysis pertain to a single data set with multiple variables. Canonical correlation and multivariate multiple regression pertain to multivariate data broken down into distinct sets (i.e., classes of combinable variables in the case of canonical correlation, and predictors versus outcomes for multivariate multiple regression models). MANOVA, MANCOVA, and HLM involve continuous data distinguished by categories, and as such, these methods are essential to experimentation wherein groups or conditions are compared. Other statistics cover special cases, such as categorical outcomes (logistic and log-linear models). The common thread through all of these methods for quantitative and categorical outcomes is the general linear model.

The common features but also the clear differences between several multivariate methods have been represented diagrammatically in this chapter. These differences involve distinct configurations of the data and data types, with single-sample data sets of multiple variables being the simplest (as in factor analysis). Multiple-sample data sets (as in MANOVA) and prediction across two or more sets of multiple variables (as in canonical correlation or SEM) are some of the more complex configurations. The principles underlying all of these analyses are quite similar. Once the foundations are learned, different building blocks can be arranged to suit specific analytical purposes. It is also true that after the overall architectural design has been mastered, the student can more easily re-arrange building blocks as needed. The next chapter of this book describes the foundational building blocks of univariate statistics. Thereafter, the chapters progress systematically up through the specific to the most general cases of multivariate statistical methods, ending with methods for categorical outcomes. From simple to elegantly complex, multivariate methods provide the quantitative foundation for contemporary research in the biomedical and social sciences.

STUDY QUESTIONS

A. Essay Questions

1. Explain the difference between univariate statistical methods and multivariate statistical methods.

2. Explain the difference between factorial statistical methods and multivariate statistical methods. Can statistical methods be both factorial and also multivariate? Explain.

3. Discuss the statement that "most multivariate techniques were developed for use in nonexperimental research."

4. Summarize the major kinds of data that are possible using the "four kinds of measurement scale" hypothesized by Stevens.

5. Explain the distinction between continuous and discrete data. Can data be both discrete and also interval/ratio? Explain. Can data be both continuous and also categorical? Explain.

6. There is a major difference between experimental and correlational research. Explain how research designs differ for these two. How do the statistical methods differ? How is randomization applied in each kind of research?

7. Evaluate the concept that although ANOVA methods were developed for experimental research, they can be applied to correlational data, that "the statistical methods 'work' whether or not the researcher manipulated the

independent variable." You may wish to bring Winer's (1971, 162) point about the assumption of independence of treatment effects and error into the discussion.

8. Chapter 1 states that the mathematical prerequisite for understanding this book is matrix algebra. Why is matrix algebra crucial to multivariate statistics?

9. Discuss the taxonomy of the four basic data set structures that are amenable to multivariate analysis; give examples of each and of multivariate methods that can be applied to each.

REFERENCES

Pearson, K. 1901. On lines and planes of closest fit to systems of points in space. *Philosophy Magazine*, 2(6), 559–572.

Spearman, C. 1904. General intelligence, objectively determined and measured. *American Journal of Psychology*, 15, 201–293.

Stevens, S. S. 1946. On the theory of scales of measurement. *Science, 103*(2684), 677–680.

Winer, B. J. 1971. *Statistical Principles in Experimental Design, Second Edition.* New York: McGraw-Hill.

CHAPTER TWO

THE SEVEN HABITS OF HIGHLY EFFECTIVE QUANTS: A REVIEW OF ELEMENTARY STATISTICS USING MATRIX ALGEBRA

2.1 INTRODUCTION

Years ago Stephen R. Covey[1] (1989) published a best-selling book, *The Seven Habits of Highly Effective People*. The book has sold over 15 million copies in 38 languages and has had a worldwide impact in business and in the everyday lives of many people. Many of its concepts, such as "abundance mindset" versus "scarcity mindset," have become a part of common parlance. Each of the seven habits ("be proactive," "begin with the end in mind," "put first things first," "think win-win," "seek first to understand, then to be understood," "synergize," and "sharpen the saw") are not merely rules of how to succeed in business—they help to build an abundant life.

The concept of *habit* is an intriguing one that can be applied to anything one undertakes, including multivariate methods. It is not enough to understand

---

[1] Dr. Covey has an MBA from Harvard and spent most of his professional career at Brigham Young University, where he was a professor of business management and organizational behavior in the Marriott School of Management. He is the first to say that he did not invent the seven habits—he is merely calling attention to several universal principles of positive human interactions.

Multivariate Analysis for the Biobehavioral and Social Sciences: A Graphical Approach, First Edition. Bruce L. Brown, Suzanne B. Hendrix, Dawson W. Hedges, Timothy B. Smith. © 2012 John Wiley & Sons, Inc. Published 2012 by John Wiley & Sons, Inc.

or even give positive assent to the principles. Rather, the aim is to make them well-practiced responses so that one can act effectively, consistently.

The key to mastering the advanced methods that are the subject matter of this book is to have a strong and habitual grasp of the fundamental principles of statistics, the ones that are taught and hopefully learned in the introductory class. William James referred to habit as the "enormous flywheel of society," and it is in this sense that the fundamental principles upon which statistical theory and method are based must be so well practiced, so well assimilated, that they give continuity and confidence in building upon them and extending them into advanced methods.

The purpose of this chapter is twofold. First, it is a review of the principles and methods of elementary statistics, particularly those that will be most important for understanding multivariate analysis and modern regression methods. Second, it provides instruction in how to apply matrix algebra to some of these elementary statistical methods. There are two reasons for taking a matrix algebra approach: (1) the matrix approach is usually more efficient in its own right, and (2) the systematic simplification that comes from the matrix approach will be absolutely necessary in the higher level methods of multivariate statistics. Once one has learned to calculate known methods, such as analysis of variance, by the matrix algebra method, it is a reasonably straightforward process to extend the matrix structures to accommodate multiple dependent variables, thus turning analysis of variance (ANOVA) into multivariate analysis of variance (MANOVA). This chapter will review univariate statistics, but the intent is to provide clarity and to remove confusion about the elementary principles of statistics before venturing into the more complex ones. In each case, we will be looking for the simplest case that captures the key attributes of the method while being readily understandable.

This chapter is structured around seven brief modules or demonstrations intended to capture the essence of the major concepts of elementary statistics. They could be considered "the seven habits of highly effective quants (quantitative researchers)." When these have become, in fact, well-developed habits with a certain amount of automaticity, the process of learning advanced quantitative methods is simplified substantially. They are: (1) the meaning of measurement scales, (2) the meaning of the three measures of central tendency, (3) variance, (4) covariance/correlation, (5) probability, (6) t and z significance tests, and (7) analysis of variance.

The most central and fundamental principles upon which everything else is built are variance, covariance, and probability. From these three, most of the commonly used methods, both of data description and also of statistical significance testing, can be derived. Before reviewing variance, covariance, and the correlation coefficient (which is standardized covariance), and showing how they are simplified and systematized by matrix algebra, we will first summarize two even more basic aspects of elementary statistics encountered in the first three chapters of an elementary statistics book—the four measurement scales and measures of central tendency.

2.2   THE MEANING OF MEASUREMENT SCALES

The four measurement scales were discussed in connection with Chapter 1, Section 1.3. More will be said here about the basic properties of each of the four kinds of data. The table below summarizes the defining properties of each of the scales, that is, the properties of numbers that are meaningful for data of each of the four types. As can be seen from the table, the scales are cumulative from left to right, such that each new scale to the right has the properties of the scale at its left plus some new property. A nominal scale (such as football jersey numbers) signifies only one thing—identity—equal or unequal. It is merely categorical. Indeed some nominal scales, such as religious affiliation, are verbal labels (such as Buddhist, Christian, Muslim, Hindu) rather than numerical.

The ordinal scale also has this identity (equal or unequal) property, but combined with the property of order, greater than or less than. It is still categorical, but now ordered categorical. Even when the labels are verbal, they clearly denote a rank order, such as "first," "second." The interval scale has the properties of the two lower scales (equal/unequal and greater than/less than) plus one additional property. This property is the most subtle and difficult to understand of the four properties, but briefly it is that the quantitative values of intervals are comparable.

properties		nominal scale	ordinal scale	Interval scale	ratio scale
$=, \neq$	equal or unequal	X	X	X	X
$>, <$	greater than or less than		X	X	X
quantitatively comparable intervals, such that ratio comparisons of the deltas are meaningful				X	X
absolute zero point, such that ratio comparisons of the measurement values are meaningful					X

The ratio scale has all of the properties of the three scales below it, plus the property of an absolute zero point. Most interval scale data could also be considered ratio scales. The usual example of an interval scale that is not also a ratio scale is temperature. A Celsius scale and a Fahrenheit scale both clearly have interval properties. If two measurements on a Celsius scale, for example, are two degrees apart, that has the same quantitative meaning anywhere on the scale (the difference between 43° and 41°, compared with the difference between 25° and 23°, etc.). These differences (deltas) between measured values are comparable at any location on the scale. In fact, these *deltas* have ratio properties: a 6° difference between two measurements is three times as much as a 2° difference between two measurements. For these two temperature scales, however, the ratio properties do not apply to the measurements themselves. Seventy-two degrees Fahrenheit does not have twice as much heat as 36°F.

On a true ratio scale, these ratio properties apply to the measurements themselves as well as to the deltas (intervals). That is, on a true ratio scale, where there is a meaningful zero point, the measurements themselves are meaningfully related by ratios. For example, in a Kelvin scale (where 0 K is meaningful in the sense of representing zero molecule movement), a measurement of 120 K does in fact represent twice as much heat (movement) as a measurement of 60 K. Actually, most quantitative, physical measurements are of this kind—a person who weighs 208 lbs. is twice as heavy as one who weighs 104 lbs., a 6-m rod is three times as long as a 3-m rod, and so on. Interval scale data that does not also fit a ratio scale is quite rare, which is why Fahrenheit temperature versus Kelvin temperature is usually the example that is given.

2.3 THE MEANING OF MEASURES OF CENTRAL TENDENCY

The chapter on central tendency in an elementary statistics book usually includes the three measures of mean, median, and mode. It is instructive to consider these three as physical quantities to make each of the three and their relative properties more memorable and understandable. Suppose that one has 56 wooden blocks that each measure precisely 1 in on each side (height, width, and length). These are all stacked on a strong ruler of negligible weight, organized into a skewed distribution as shown in the first figure below. This distribution is clearly right skewed, positively skewed, in that its tail points to the right in a positive direction. If one records the data with a frequency count for each of the seventeen values along the base, the *mode* for these data is found to be 3 (which has 7 blocks/observations), which is seen here to be *the scale value that has the highest stack of blocks*, the most frequent value.

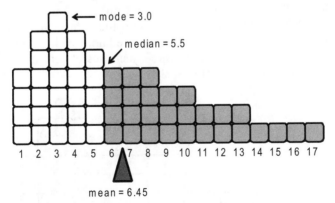

The median is the point that has as many scores above it as below it, and when it is calculated for this distribution, it is found to be 5.5, as shown in the figure. There are 28 gray-shaded blocks from 6 up to 17 (above the median of 5.5), and 28 white-shaded blocks from 1 to 5 (below the median of 5.5). Since each block is one cubic inch in size, the face of each is one square inch. The area of the face of this distribution is therefore 56 square inches, with 28 square inches below the median and 28 square inches above the median. It could be said that *the median is the point that bisects the area*. Notice that the median is not sensitive to shifts in the shape of the distribution, as long as the 28 white blocks remain below 5.5, and the 28 gray blocks remain above 5.5. It can therefore be thought of as having a kind of robustness with regard to fluctuations that may be random and not meaningful, but it also has a kind of insensitivity to the actual properties of the distribution.

The mean as a physical quantity is the balance point. The calculated mean for the data set from this distribution is 6.45. If one were to place a fulcrum under the distribution at the location 6.45, it will balance (assuming the weight of the ruler bar at the bottom is truly negligible). As the balance point, the mean is affected by every score, sometimes profoundly. Suppose for example that the two highest data blocks located at 16 and 17 were moved out on the bar to a location of 35 and 36. This would increase the mean from 6.45 to 7.13, while the median and mode would stay the same. There might also be cases in research when the mean could give misleading values. It is common to use medians in reaction time studies. Suppose that typical reaction times to a cognitive task are 150 to 160 ms, but a lapse in attention produces a reaction time of 762 ms. The mean reaction time would be misleading, drawn away by this single extreme value, but the median would be fairly representative. Similarly, if one were trying to characterize the socioeconomic level of a neighborhood, and most households had an income of $60,000 to $70,000 per year, but one multimillionaire also lived in the neighborhood, the mean household income might be a highly misleading statistic to use.

Notice that in the distribution just presented, the mean is higher than the median, which is higher than the mode. For positively skewed data, it will usually be the case that the mean (balance points) is higher than the mode, and that the median will fall between them. Consider the negatively skewed distribution below in which the frequencies for the 1 through 17 locations are reversed. Here the mean is lower than the median, which is lower than the mode. The relative position of these three measures of central tendency can be used in this way as an indicator of skew.

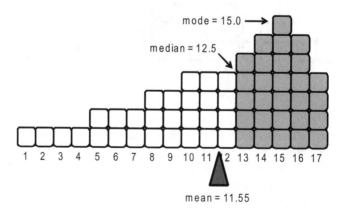

The calculation of the median is rendered somewhat more demanding by adding four more data observations to the right of this negatively skewed distribution as shown in the histogram below. Now there are 60 observations rather than 56, and the median will have 30 observations below it and 30 observations above it, as shown in the histogram. Notice that this time the median is *within* the interval (between 12.5 and 13.5) rather than at the boundary, which means that it will be necessary to interpolate.[2] There are five observations within this interval, but only two of them are to be below the median, so the median will be two-fifths (0.40) of the distance from 12.5 to 13.5, which is added to 12.5, the lower limit of the interval, to yield a median of 12.9. This can also be done by formula. In fact, the median is equivalent to the 50th percentile point, so the percentile point formula can be used to obtain the median. The formula is:

$$\text{Median} = X_{50} = L + \left[\frac{P(N) - cfb}{fw} \right].$$

where P is the proportion corresponding to the percentile (0.50 for the 50th percentile), N is the sample size, L stands for "lower limit" of the interval, cfb stands for "cumulative frequency below the lower limit," fw stands for "frequency within." Sample size, N, is equal to 60. Entering these into the formula, the calculated median agrees with the interpolated value given above.

$$\text{Median} = X_{50} = L + \left[\frac{P(N) - cfb}{fw} \right] = 12.5 + \left[\frac{(0.50)(60) - 28}{5} \right] = 12.5 + \left(\frac{2}{5} \right) = 12.9.$$

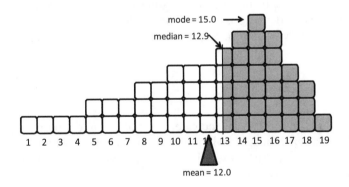

mode = 15.0 →
median = 12.9

1 2 3 4 5 6 7 8 9 10 11 12 13 14 15 16 17 18 19

mean = 12.0

[2] There is a simpler way to obtain the median, and that is to consider any median that lies between 12.5 and 13.5 (for this example) as being a median of 13. This is in fact the method used by Microsoft's Excel spreadsheet. The more precise and complex median calculation method shown here is referred to as the *interpolated median*. As demonstrated in the computational questions at the end of this chapter, there is often good reason to prefer the interpolated median over the simple median.

There is some relationship between the four kinds of measurement scale and these three measures of central tendency. The mode, like the nominal scale, is primarily based upon an identity relationship (equal or unequal), with the mode being defined as the point along the numerical scale that has the greatest number of observations that are equal to one another in value. The median can be thought of as having some properties in common with an ordinal scale. It marks the spot that divides the data in half ordinally, with half of the scores higher than the median and half lower. It would be possible to change interval properties of the data, for example, moving the single value of 17 in the right-skewed distribution above out to 20. The mean would change, because it is reflective of the interval/ratio properties of the data, but the median would not change. The mean thus is dependent upon and reflective of the numerical value of every observation in the data set, whereas the mode and median are not. This is also true of variance and covariance. They are like the mean in that they reflect the numerical properties of every observation in the data set. Most of the methods introduced in this book will be based upon fully quantitative data, and will therefore be based upon means, variances, and covariances.

2.4   VARIANCE AND MATRIX ALGEBRA

After an elementary statistics book introduces methods for quantifying the central tendency of distributions (the mean, the median, and the mode), the next topic is methods for quantifying the spread of a distribution (variance, standard deviation, semi-interquartile range, etc.) The focus here will be on variance, and also on standard deviation, the square root of the variance.

The simple definition of variance is the "mean of the squared deviation scores," or more cryptically the "mean square." The variance is equal to the Sum of Squares (of deviations) divided by the degrees of freedom, represented mnemonically as MS = SS/df. Represented algebraically, the simple formula for variance is

$$S_u^2 = \frac{\sum x_i^2}{N-1},$$ (2.1)

where the "u" subscript on S^2 indicates "unbiased," and where the symbol x_i in the formula represents a deviation score, the deviation of a raw score from its mean: $x_i = X_i - \overline{X}$. Since there is an unbiased variance, there must also be a biased one, and it is given as

$$S_b^2 = \frac{\sum x_i^2}{N}$$ (2.2)

where the sum of squares is divided by N rather than by the degrees of freedom, $N-1$. Almost always the unbiased variance is the one that will be of interest, since the biased one systematically underestimates the true population variance by an amount of $1/N^{th}$.

The mean of any set of values is obtained by taking the sum of those values divided by the number of values. Therefore, to get the mean of the squared deviation scores, those squared deviation scores (distances of scores from their mean) are summed, as shown in the two formulas above, and the sum of squared deviations is divided by N or $N-1$.

Consider the simple little distribution of 8, 6, 5, 5, 4, and 2. The mean of this distribution is five:

$$\overline{X} = \frac{\sum X_i}{N} = \frac{30}{6} = 5$$

The variance is calculated according to Equation 2.1 by first obtaining the deviations of each of the six scores from 5, squaring each deviation score, summing these squares, and then dividing this sum of squares by the degrees of freedom:

$$S^2 = \frac{\sum x_i^2}{N-1} = \frac{(8-5)^2 + (6-5)^2 + (5-5)^2 + (5-5)^2 + (4-5)^2 + (2-5)^2}{6-1}$$

$$= \frac{3^2 + 1^2 + 0^2 + 0^2 + (-1)^2 + (-3)^2}{5} = \frac{9+1+0+0+1+9}{5} = \frac{20}{5} = 4$$

The standard deviation, which is the square root of the variance, for this small and simple distribution is two: $S = \sqrt{S^2} = \sqrt{4} = 2$

The biased variance, obtained by dividing the SS value of 20 by 6 rather than 5, comes out to 3.33, smaller by one sixth than it should be to be unbiased.

Occasionally, on the way to introducing variance, a simpler measure of spread is discussed, the *mean deviation*. The mean deviation statistic is simply the average *absolute* distance of scores from their mean. It is similar to the variance, but simpler. Instead of being the average of the *squared* distances of scores from their mean (as the variance is), it is the average *absolute* distance of scores from their mean. This can be easily demonstrated with a simple histogram of the data just given with each datapoint block having its deviation score written upon it:

$$\overline{X} = 5$$

The mean of the absolute values of these six deviation scores is:

$$MD = \frac{\sum |x_i|}{N} = \frac{|8-5| + |6-5| + |5-5| + |5-5| + |4-5| + |2-5|}{6}$$

$$= \frac{3+1+0+0+1+3}{6} = \frac{8}{6} = 1.33.$$

This mean deviation has a very direct and simple interpretation. It is just the *mean or expected value of deviations away from the mean*. Some scores are three away from the mean, some are one unit away, and some are zero, but the average amount scores deviate from the mean is 1.33. It helps to remember that the variance is only one conceptual step away from this—the biased variance is merely the *mean* of the **squared** *deviation scores*. Putting this into the language of the algebra of expectations (where "expected value" is just another name for a mean), the MD index is the *expected value of absolute deviations*, and the biased variance is the *expected value of squared deviations*:

$$MD = E(|x_i|) = \sum p_i x_i = \frac{\sum |x_i|}{N} \qquad (2.3)$$

$$S_b^2 = E(x_i^2) = \sum p_i x_i^2 = \frac{\sum x_i^2}{N} \qquad (2.4)$$

Notice that Equation 2.4 for variance as an expected value of squared deviations is, in the rightmost expression, equivalent to Equation 2.2, the *biased* formula for variance. Calculation by the expected values method does, in fact, create biased variances and covariances. Notice also that the middle expression in Equation 2.4 gives a more general way to calculate means, by the sum of the product of the values (x^2 in this case) and their probabilities, p_i. Consider the deviation scores from the data just presented $(-3, -1, 0, 0, 1, 3)$. Following Equation 2.4, we obtain the same biased variance of 3.33 that was obtained using Equation 2.2:

$$E(x_i^2) = \sum p_i x_i^2 = \left(\frac{1}{6}\right)(-3)^2 + \left(\frac{1}{6}\right)(-1)^2 + \left(\frac{2}{6}\right)(0)^2 + \left(\frac{1}{6}\right)(1)^2 + \left(\frac{1}{6}\right)(3)^2$$
$$= \frac{9}{6} + \frac{1}{6} + \frac{0}{6} + \frac{1}{6} + \frac{9}{6} = \frac{20}{6} = 3.33.$$

The biased variance can be turned into unbiased variance by multiplying it by the ratio $N/(N-1)$:

$$S_u^2 = \left(\frac{N}{(N-1)}\right)S_b^2 = \left(\frac{N}{(N-1)}\right)E(|x_i|) = \left(\frac{6}{5}\right) \cdot 3.333 = 4.$$

Although the mean deviation is a simple, readily understandable, and conceptually compelling index of spread, it is never used in practice. The variance and the standard deviation are the preferred indices of spread. One of the reasons for the preference for variance (in squared units) as a measure of spread is that, as shown in the Pythagorean theorem, squared indices have the very useful property of *additivity*. Consider a 3-4-5 triangle (i.e., a triangle with perpendicular sides of 3 and 4 units respectively, and a hypotenuse of 5 units). The sides are not additive ($3 + 4 \neq 5$, or $a + b \neq c$), but as expressed by the Pythagorean theorem the squares of the sides are additive ($3^2 + 4^2 = 5^2$, or $a^2 + b^2 = c^2$). The square of the hypotenuse is equal to the *sum of the squares* of the other two sides. Statistical theory is inherently Pythagorean, and there is a preference for indices based on the *sum of squares* and related quantities (like sum of products).

Equations 2.1 and 2.2 are *definitional* formulas for variance. They give a direct and clear expression of the definition of a variance as the average of the squared deviation scores. Although this formula is good for expressing the basic concept, it is not optimal for actual computations. There is also a computation formula, given as:

$$S^2 = \frac{\sum X^2 - \frac{\left(\sum X\right)^2}{N}}{N-1}. \tag{2.5}$$

The variance calculated by the computational formula gives the same result as the definitional formula for the simple data used above:

$$S^2 = \frac{\sum X^2 - \frac{\left(\sum X\right)^2}{N}}{N-1} = \frac{\left(8^2 + 6^2 + 5^2 + 5^2 + 4^2 + 2^2\right) - \frac{(8+6+5+5+4+2)^2}{6}}{6-1}$$

$$= \frac{170 - \frac{30^2}{6}}{5} = \frac{170 - 150}{5} = \frac{20}{5} = 4.$$

With actual empirical data, the computational formula will often be substantially more accurate. This is because the mean will seldom be a whole number, and when one rounds back the value of the mean to one or two places beyond the decimal, the definitional formula then squares deviations from that rounded value and sums them, thus compounding the rounding error. There are, in fact, three forms of the variance formula to be considered here, the simple definitional one we have been using, a more explicit version of the definitional one, and the computational formula

$$S^2 = \frac{\sum x_i^2}{N-1} = \frac{\sum \left(X_i - \overline{X}\right)^2}{N-1} = \frac{\sum X_i^2 - \frac{\left(\sum X_i\right)^2}{N}}{N-1}.$$

At this point, it will be useful to demonstrate the algebraic equivalence of the definitional formulas (the left and middle expressions) and the computational formula, the rightmost of the three expressions. This algebraic proof or demonstration will prove useful in dealing with analysis of variance and expanding it to multivariate analysis of variance in Chapter 8 and also Hotelling's T^2 in Chapter 7. Since the denominators are equivalent in these two expressions of the variance, the task can be simplified by demonstrating algebraically that the two numerators (the so-called *sums of squares*) are equivalent:

$$
\begin{aligned}
SS_{dev} &= \sum \left(X_i - \overline{X} \right)^2 \\
&= \sum \left(X_i^2 - 2X_i \overline{X} + \overline{X}^2 \right) \\
&= \sum X_i^2 - \sum 2X_i \overline{X} + \sum \overline{X}^2 \\
&= \sum X_i^2 - 2\overline{X} \sum X_i + N\overline{X}^2 \\
&= \sum X_i^2 - 2\left(\frac{\sum X_i}{N} \right) \sum X_i + N\left(\frac{\sum X_i}{N} \right)^2 \\
&= \sum X_i^2 - 2\frac{\left(\sum X_i \right)^2}{N} + \frac{\left(\sum X_i \right)^2}{N} \\
&= \sum X_i^2 - \frac{\left(\sum X_i \right)^2}{N}.
\end{aligned}
$$

The transition between line 4 and line 5 of this derivation is facilitated by remembering that the mean is equal to the sum of the scores divided by N, and that wherever $\overline{X}$ appears in the derivation, it can be replaced by

$$
\frac{\sum X_i}{N}.
$$

In addition to this algebraic demonstration of the equivalence of the two variance formulas, there is also a conceptual way to explain the components in the formulas. The fundamental "computational engine" for variance is SS, the sum of squared values. However, if one were to use the sum of squared raw values, it would not be a good index of dispersion, because it is complicated by two additional things, the mean and the sample size. The sum of squared raw values in fact reflects all three of these things: dispersion, magnitude of the mean, and magnitude of sample size. In the definitional formula, before squaring and summing, the effects of the mean are removed by subtracting the mean from each value to obtain deviation scores to be squared and summed. Sample size is adjusted for by dividing the sum of squares of deviations by N or $N - 1$.

One can, however, adjust for the mean *after* squaring and summing, and that is what is done in the computational formula. The adjustment factor is created by doing the same thing to the mean that is done to the raw scores (squaring it and then summing, but since it is a constant, summing is just multiplying it by N), and this becomes the adjustment factor ($\overline{\Sigma X}^2 = N\overline{X}^2$). In fact, there are at least three forms this adjustment factor can take, and they are all mathematically equivalent:

$$N\overline{X}^2 = 6(5)^2 = 150$$

$$\frac{\left(\sum X_i\right)^2}{N} = \frac{T^2}{N} = \frac{30^2}{6} = \frac{900}{6} = 150$$

$$T \cdot \overline{X} = T \cdot M = (30)(5) = 150.$$

Notice, in the last of the three expressions, the notation is somewhat simplified by replacing ΣX_i the sum of all raw scores with T, which stands for "total." All three of these, of course, give us the adjustment factor of 150 that was subtracted in the computational formula above to obtain the deviation sum of squares from the raw sum of squares: $170 - 150 = 20$.

There is yet another formula for calculating variance, the one that will be most useful in this book, and that is the matrix formula:

$$\text{variance} = S_x^2 = \left(\frac{1}{N-1}\right) \cdot (X'X - T \cdot M). \tag{2.6}$$

To use this method, one will have to know how to do matrix multiplication (which is explained in Chapter 3, Section 3.3.4). The matrix $\mathbf{X}$, is the 6×1 matrix of raw data:

$$\mathbf{X} = \begin{bmatrix} 8 \\ 6 \\ 5 \\ 5 \\ 4 \\ 2 \end{bmatrix}.$$

The dimensions of matrices are given as $r \times c$ (the number of rows by the number of columns). Since this matrix of data has six rows and only one column, it is referred to as a 6×1 matrix.[3] The symbol $\mathbf{X}'$ represents the transpose (see Chapter 3, Section 3.3.3) of matrix $\mathbf{X}$, which is obtained by interchanging rows and columns of the original matrix:

$$\mathbf{X}' = [8 \quad 6 \quad 5 \quad 5 \quad 4 \quad 2].$$

The matrix formula for variance (Eq. 2.4) calls for $\mathbf{X}$ to be premultiplied by its transpose, $\mathbf{X}'$. In matrix algebra, one must first check to be sure the matrices are *conformable*—if they are they can be multiplied, otherwise they cannot. Matrix algebra is not like scalar algebra where any two numbers can be multiplied. To be conformable, the number of columns in the premultiplying matrix must be equal to the number of rows in the postmultiplying matrix. If we write the dimensions of the two matrices $\mathbf{X}'$ and $\mathbf{X}$ adjacent to one another, we can see that the two "inner" dimensions, the two sixes, agree (the columns in the premultiplying matrix and the rows in the postmultiplying matrix), so they are conformable.

$$(1 \times 6)\,(6 \times 1).$$

These "inner" dimensions must agree, and the "outers" give the dimensions of the product matrix, which in this case are 1 (rows in the premultiplying matrix) and 1 (columns in the postmultiplying matrix), so the two matrices can be multiplied and the product matrix will be a 1×1, which is a single scalar number. To now do the matrix multiplication, we multiply the six elements from the premultiplying matrix $\mathbf{X}'$ by the corresponding six elements in the postmultiplying matrix $\mathbf{X}$, and sum these products to obtain the 1×1 scalar of 170:

$$\text{SSraw} = \mathbf{X}'\mathbf{X} = [8 \quad 6 \quad 5 \quad 5 \quad 4 \quad 2] \cdot \begin{bmatrix} 8 \\ 6 \\ 5 \\ 5 \\ 4 \\ 2 \end{bmatrix}$$

$$= 8^2 + 6^2 + 5^2 + 5^2 + 4^2 + 2^2 = 64 + 36 + 25 + 25 + 16 + 4 = 170.$$

[3] In the language of matrix algebra, a two-dimensional array of values is referred to as a *matrix*, a one-dimensional array of values is referred to as a *vector*, and a single value is referred to as a *scalar*. Strictly speaking, the 6×1 matrix $\mathbf{X}$ in the text above is a column vector, and the 1×6 matrix $\mathbf{X}'$ is a row vector, but since a vector is a special case of a matrix, it can also be referred to as a 1×6 matrix. Also, it is standard matrix notation to indicate matrices with bold upper case letters, and to indicate vectors with bold lower case letters. This conflicts with statistical usage in which the upper case X_i refers to raw data and the lowercase x_2 refers to deviation score data. In this chapter, we have conformed to statistical usage (with $\mathbf{X}$ used to represent a vector or a matrix of raw data and $\mathbf{x}$ used to represent a vector or a matrix of deviation score data), with whether it is a matrix or a vector being indicated by context.

It is now obvious, why the "inners" have to agree, so that elements in the rows of the premultiplying matrix will match up with the corresponding elements in the columns of the postmultiplying matrix for multiplying and summing. Notice what we have just done with this matrix multiplication—we have created the raw sum of squares of 170, the first step in obtaining a variance. Actually, matrix multiplication is an engine for efficiently obtaining sums of products (which become in fact sums of squares when the two vectors are the same, i.e., the premultiplying matrix is the transpose of the postmultiplying matrix.)

Equation 2.6 next calls for the SSraw to be converted into SSdeviation by subtracting the product of $\mathbf{T}$ and $\mathbf{M}$ from the $\mathbf{X'X}$:

$$\text{SSdev} = \mathbf{X'X} - \mathbf{T} \cdot \mathbf{M} = 170 - (30)(5) = 170 - 150 = 20.$$

Finally, this SSdeviation is turned into the variance by dividing it by the scalar quantity $N - 1$, the degrees of freedom:

$$S_x^2 = \left(\frac{1}{N-1}\right) \cdot (\mathbf{X'X} - \mathbf{T} \cdot \mathbf{M}) = \left(\frac{1}{6-1}\right) \cdot 20 = \left(\frac{1}{4}\right) \cdot 20 = 4.$$

Equation 2.6 is a matrix algebra *computational* formula for variance. One can also define a simpler *definitional* formula for the matrix approach using deviation scores rather than raw scores, analogous to the definitional formula (Eq. 2.1) that was presented in scalar notational form.

$$\text{variance} = S_x^2 = \left(\frac{1}{N-1}\right) \cdot \mathbf{x'x}. \tag{2.7}$$

The calculations from this formula are simpler, but like Equation 2.1, it has the downside of compounding rounding error when the means are not whole numbers.

$$S^2 = \left(\frac{1}{N-1}\right) \cdot \mathbf{x'x} = \left(\frac{1}{5}\right) \cdot [3 \ \ 1 \ \ 0 \ \ 0 \ \ -1 \ \ -3] \cdot \begin{bmatrix} 3 \\ 1 \\ 0 \\ 0 \\ -1 \\ -3 \end{bmatrix} = \left(\frac{1}{5}\right) \cdot 20 = 4.$$

There is also a "hybrid" matrix formula, in which only one of the two multiplied matrices has an issue with rounding error (the premultiplying deviation score matrix) while the postmultiplying matrix consists of raw scores.

$$\text{variance} = S_x^2 = \left(\frac{1}{N-1}\right) \cdot \mathbf{x'X}. \tag{2.8}$$

It might seem surprising that this formula works, but it does.

$$S_x^2 = \left(\frac{1}{N-1}\right) \cdot \mathbf{x'X} = \left(\frac{1}{5}\right) \cdot [3 \quad 1 \quad 0 \quad 0 \quad -1 \quad -3] \cdot \begin{bmatrix} 8 \\ 6 \\ 5 \\ 5 \\ 4 \\ 2 \end{bmatrix}$$

$$= \left(\frac{1}{5}\right) \cdot (24 + 6 + 0 + 0 - 4 - 6) = \left(\frac{1}{5}\right) \cdot (30 - 10) = 4.$$

This equation works because of the principles of "idempotent centering matrices," which for the curious student are explained in Chapter 3, Section 3.3.7.1.

2.5   COVARIANCE MATRICES AND CORRELATION MATRICES

Covariance is derived from the concept (and formula) for variance. It is expanded to deal with *bivariate data*, that is, a data matrix that has measurements on two variables for each of the observations, entities, or persons. Instead of the "mean of the squared deviation scores" for one variable, which defines variance, covariance is defined as the "mean of the product of the deviation scores for two variables, X and Y." In the language of the algebra of expectations, whereas the variance is *the expected value of squared deviations*, the covariance is *the expected value of the product of deviations*.

$$S_b^2 = E(x_i^2) = \sum p_i x_i^2 \tag{2.9}$$

$$\text{cov} = E(x_i y_i) = \sum p_i x_i y_i \tag{2.10}$$

Covariance is similar to variance, but instead of focusing on the spread of just one variable, it is expressive of the relationship between two variables. It must be emphasized that X and Y cannot be two separate sets of data, or else correlation/covariance cannot be determined. It must be two *variables* measured on the same set of persons, entities, or observations, so that it can be determined whether those who are high on one of the variables also tend to be high on the other. In other words, the observations must be paired, connected in a nonarbitrary way, such as two variables measured on each person in a group of persons.

Consider, for example, the following set of bivariate data, math performance scores and verbal performance scores for each of six persons, with a vector of the totals (T′) and a vector of the means (M′) shown beneath the scores.

	X Math	Y Verbal
Al	8	10
Bill	6	12
Charlie	5	11
Dan	5	7
Ed	4	8
Frank	2	6
T′=	30	54
M′=	5	9

For this bivariate data set, two variances and one covariance can be calculated. The variances can be calculated with Equation 2.5, and the covariance with Equation 2.11, which is the analogous *computational* formula for calculating covariance from raw scores.

$$\text{cov}_{xy} = \frac{\sum XY - \dfrac{(\sum X)(\sum Y)}{N}}{N-1}. \tag{2.11}$$

First, five columns are calculated: a column of X-values, a column of squared X-values, a column of Y-values, a column of squared Y-values, and a column for values of the product of X and Y. The sum for each of these five columns is calculated to provide the five input quantities called for in the formulas.

X	X^2	Y	Y^2	XY
8	64	10	100	80
6	36	12	144	72
5	25	11	121	55
5	25	7	49	35
4	16	8	64	32
2	4	6	36	12
30	170	54	514	286
$=\acute{O}X$	$=\acute{O}X^2$	$=\acute{O}Y$	$=\acute{O}Y^2$	$=\acute{O}XY$

From the raw sums of squares and the sums, variance is calculated first for X, the math score, and then for Y, the verbal score by Equation 2.5.

$$S_x^2 = \frac{\sum X^2 - \dfrac{(\sum X)^2}{N}}{N-1} = \frac{170 - \dfrac{30^2}{6}}{5} = \frac{170 - 5\cdot30}{5} = \frac{170-150}{5} = \frac{20}{5} = 4.0.$$

$$S_y^2 = \frac{\sum Y^2 - \frac{\left(\sum Y\right)^2}{N}}{N-1} = \frac{514 - \frac{54^2}{6}}{5} = \frac{514 - 9 \cdot 54}{5} = \frac{514 - 486}{5} = \frac{28}{5} = 5.6.$$

Using Equation 2.7, the covariance between the two variables is calculated from the raw sum of products and the sums of X and of Y.

$$\text{cov}_{xy} = \frac{\sum XY - \frac{\left(\sum X\right)\left(\sum Y\right)}{N}}{N-1} = \frac{340 - \frac{(54)(36)}{6}}{5}$$

$$= \frac{340 - 9 \cdot 36}{5} = \frac{340 - 324}{5} = \frac{16}{5} = 3.2.$$

These three quantities, the variance of X, the variance of Y, and the covariance between X and Y, may be collected into a variance/covariance matrix. This matrix has the two variances in the diagonal positions (cell 1,1 and cell 2,2), and the covariance in the two "mirror image" off-diagonal positions (cells 1,2 and 2,1).[4]

$$\mathbf{S} = \begin{bmatrix} s_x & \text{cov}_{xy} \\ \text{cov}_{yx} & s_y \end{bmatrix} = \begin{bmatrix} 4.0 & 3.2 \\ 3.2 & 5.6 \end{bmatrix}.$$

Now suppose that instead of just two variables, the data matrix were to have three variables:

	X Math	Y Verbal	W Logic
Al	8	10	18
Bill	6	12	18
Charlie	5	11	16
Dan	5	7	12
Ed	4	8	12
Frank	2	6	8
T'=	30	54	84
M'=	5	9	14

---

[4] Cell 1,1 is the cell at the intersection of row 1 and column 1, cell 1,2 is the cell at the intersection of row 1 and column 2, cell 2,1 is at the intersection of row 2 and column 1, and cell 2,2 is the intersection of row 2 and column 2.

To get the full 3 by 3 covariance matrix by the traditional scalar method, the process just used to get the two variances and the covariance would have to be repeated six times.

The matrix formula used above to calculate variance (Eq. 2.4) can be used to obtain all three variances and all three covariances simultaneously. The label of the equation is changed from S^2 (variance) to **S**, which as a bold capital letter signifies the entire variance/covariance matrix:

$$\mathbf{S} = \left(\frac{1}{N-1}\right) \cdot (\mathbf{X'X} - \mathbf{T} \cdot \mathbf{M'}). \tag{2.12}$$

Both **T** and also **M** are vectors rather than matrices. **T** is the column matrix (3×1) of totals, and **M'** is the row matrix of means. Notice that is no longer necessary to label the variables as X, Y, and so on, in that matrix algebra keeps track of them according to which column they are in. The entire matrix of variables is labeled **X**, but the first variable is X_1 (located in column 1 of the **X** matrix), the second is X_2 (located in column 2), and so on. The 3×3 covariance matrix for the example data is calculated according to Equation 2.12:

$$\mathbf{S} = \left(\frac{1}{N-1}\right) \cdot (\mathbf{X'X} - \mathbf{T} \cdot \mathbf{M'})$$

$$= \left(\frac{1}{5}\right) \cdot \left(\begin{bmatrix} 8 & 6 & 5 & 5 & 4 & 2 \\ 10 & 12 & 11 & 7 & 8 & 6 \\ 18 & 18 & 16 & 12 & 12 & 8 \end{bmatrix} \cdot \begin{bmatrix} 8 & 10 & 18 \\ 6 & 12 & 18 \\ 5 & 11 & 16 \\ 4 & 7 & 12 \\ 4 & 8 & 12 \\ 2 & 6 & 8 \end{bmatrix} - \begin{bmatrix} 30 \\ 54 \\ 84 \end{bmatrix} \cdot \begin{bmatrix} 5 & 9 & 14 \end{bmatrix} \right)$$

$$= \left(\frac{1}{5}\right) \cdot \left(\begin{bmatrix} 170 & 286 & 456 \\ 286 & 514 & 800 \\ 456 & 800 & 1256 \end{bmatrix} - \begin{bmatrix} 150 & 270 & 420 \\ 270 & 486 & 756 \\ 420 & 756 & 1176 \end{bmatrix} \right)$$

$$= \left(\frac{1}{5}\right) \cdot \left(\begin{bmatrix} 20 & 16 & 36 \\ 16 & 28 & 44 \\ 36 & 44 & 80 \end{bmatrix} \right) = \begin{bmatrix} 4 & 3.2 & 7.2 \\ 3.2 & 5.6 & 8.8 \\ 7.2 & 8.8 & 16 \end{bmatrix}.$$

When the data are simple (i.e., have a whole number mean), the covariance matrix can also be calculated more simply directly from deviation scores (Eq. 2.13), or by the combination of deviation scores and raw scores (Eq. 2.14), which are of course just the two corresponding variance Equations 2.7 and 2.8

expanded to accommodate multivariate data matrices rather than just a vector for a single variable.

$$S = \left(\frac{1}{N-1} \right) \cdot \mathbf{x'x}. \tag{2.13}$$

$$S = \left(\frac{1}{N-1} \right) \cdot \mathbf{x'X}. \tag{2.14}$$

Equation 2.14 is simple and efficient, and has somewhat less error compounding for noninteger means than Equation 2.13.

$$S = \left(\frac{1}{N-1} \right) \cdot \mathbf{x'X} = \left(\frac{1}{5} \right) \cdot \begin{bmatrix} 3 & 1 & 0 & 0 & -1 & -3 \\ 1 & 3 & 2 & -2 & -1 & -3 \\ 4 & 4 & 2 & -2 & -2 & -6 \end{bmatrix} \cdot \begin{bmatrix} 8 & 10 & 18 \\ 6 & 12 & 18 \\ 5 & 11 & 16 \\ 5 & 7 & 12 \\ 4 & 8 & 12 \\ 2 & 6 & 8 \end{bmatrix}$$

$$= \left(\frac{1}{5} \right) \cdot \begin{bmatrix} 20 & 16 & 36 \\ 16 & 28 & 44 \\ 36 & 44 & 80 \end{bmatrix} = \begin{bmatrix} 4 & 3.2 & 7.2 \\ 3.2 & 5.6 & 8.8 \\ 7.2 & 8.8 & 16 \end{bmatrix}.$$

Once the covariance matrix is created, the next step is to create a correlation matrix. By definition, the Pearson product moment correlation coefficient is standardized covariance. This can be done in two ways. First, one can apply the covariance formula to Z scores:

$$r = \frac{\sum Z_x Z_y}{N-1}. \tag{2.15}$$

This is conceptually meaningful—the Pearson product moment correlation coefficient (PPMCC) is defined as the ratio of the cross product of Z-scores divided by the largest value the cross product could take, which is the degrees of freedom, $N - 1$. This is conceptually simple, but has the same problem of deviation score formulas for variance and covariance—rounding can introduce inaccuracy.

The second way to calculate PPMCC is to standardize the covariance *after* it has been calculated by dividing the covariance by its two standard deviations.

$$r = \frac{\text{cov}}{s_x s_y}. \tag{2.16}$$

This formula can be converted into a definition formula that uses sums of squares and cross products, by canceling the "$N - 1$" degrees of freedom terms across the numerator and the denominator.

$$r = \frac{\text{cov}}{s_x s_y} = \frac{\left(\dfrac{\sum xy}{N-1}\right)}{\sqrt{\dfrac{\sum x^2}{N-1}}\sqrt{\dfrac{\sum y^2}{N-1}}} = \frac{\sum xy}{\sqrt{\sum x^2}\sqrt{\sum y^2}}.$$

This is the most commonly used definitional formula for the correlation coefficient, Equation 2.17.

$$r = \frac{\text{cov}}{s_x s_y} = \frac{\sum xy}{\sqrt{\sum x^2}\sqrt{\sum y^2}} \qquad r = \frac{\text{cov}}{s_x s_y}. \tag{2.17}$$

This definitional formula can also be converted into a computational formula by substituting into Equation 2.17, first the numerator term (the cross product) from Equation 2.11 to replace the definitional cross product, and then the denominator term (the sum of squares) from Equation 2.5 to replace the definitional sums of squares for both X and Y.

$$r = \frac{\left(\sum XY - \dfrac{(\sum X)(\sum Y)}{N}\right)}{\sqrt{\sum X^2 - \dfrac{(\sum X)^2}{N}}\sqrt{\sum Y^2 - \dfrac{(\sum Y)^2}{N}}}. \tag{2.18}$$

Perhaps the simplest way to think of a correlation coefficient is as the expected value of the product of standard scores.

$$r = E(Z_x Z_y) = \sum p_i Z_x Z_y. \tag{2.19}$$

In fact, this "expected values" Equation 2.19 will only work if the Z-scores are biased ones (that is created by dividing the deviation scores by N), whereas Equation 2.15 will only work if the Z-scores are unbiased ones (that is, created by dividing the deviation scores by $N - 1$). The maximum sum of products of Z-scores is N for biased Z-scores, and $N - 1$ for unbiased Z-scores. Variances, covariance, and Z-scores all come in both biased and unbiased versions, but the correlation coefficient does not. As can be seen in the derivation of Equation 2.17 above, in the formula for calculating the correlation coefficient from covariance and the two standard deviations, the degrees of freedom terms $(N - 1)$ cancel across the numerator and denominator, and are therefore not part of the formula.

Equation 2.16 is the basis of an efficient matrix way to transform a covariance matrix into a correlation matrix (a matrix of PPMCC values) through matrix algebra. Correlation matrices can often be fairly large, with 15 or 20 variables or more, so rather referring to the variables as X, Y, W, and so on, they are referred to as X_1, X_2, X_3, and so on. Consider, for example the notation of the following covariance matrix for the three variables X_1–X_3.

$$\mathbf{S} = \begin{bmatrix} \text{var}_1 & \text{cov}_{12} & \text{cov}_{13} \\ \text{cov}_{21} & \text{var}_2 & \text{cov}_{23} \\ \text{cov}_{31} & \text{cov}_{32} & \text{var}_3 \end{bmatrix}.$$

Notice that each covariance subscript indicates the row and the column in which each is located (in the order of row first, then column). For example, cov_{12}, the covariance between variable X_1, and variable X_2 is located in row 1 and column 2. The row subscript signifies the first variable in the covariance, and the column subscript signifies the second variable in the covariance. Of course, $\text{cov}_{12} = \text{cov}_{21}$, so covariance matrices are always symmetrical, with the values in the upper triangular part of the matrix (above the diagonal entries of variances) being the mirror image of the values in the lower triangular part of the matrix, as shown in the following representation of the 3×3 covariance matrix just calculated.

$$\mathbf{S} = \begin{bmatrix} \text{var}_1 & \text{cov}_{12} & \text{cov}_{13} \\ \text{cov}_{21} & \text{var}_2 & \text{cov}_{23} \\ \text{cov}_{31} & \text{cov}_{32} & \text{var}_3 \end{bmatrix} = \begin{bmatrix} 4 & 3.2 & 7.2 \\ 3.2 & 5.6 & 8.8 \\ 7.2 & 8.8 & 16 \end{bmatrix}.$$

The correlation matrix consists of each of these covariances divided by its respective standard deviations. Since the diagonal elements of the covariance matrix are variances, the standard deviations are obtained as the square root of each of these diagonal elements:

$$s_1 = \sqrt{4} = 2$$
$$s_2 = \sqrt{5.6} = 2.37$$
$$s_3 = \sqrt{16} = 4.$$

The correlation matrix for these three variables is found by dividing each covariance by the standard deviation corresponding to its row subscript and the standard deviation corresponding to its column subscript as called for in Equation 2.16.

$$R = \begin{bmatrix} \dfrac{s_1^2}{(s_1)(s_1)} & \dfrac{\text{cov}_{12}}{(s_1)(s_2)} & \dfrac{\text{cov}_{13}}{(s_1)(s_3)} \\[2ex] \dfrac{\text{cov}_{12}}{(s_1)(s_2)} & \dfrac{s_1^2}{(s_1)(s_1)} & \dfrac{\text{cov}_{12}}{(s_1)(s_2)} \\[2ex] \dfrac{\text{cov}_{12}}{(s_1)(s_2)} & \dfrac{\text{cov}_{12}}{(s_1)(s_2)} & \dfrac{s_1^2}{(s_1)(s_1)} \end{bmatrix} = \begin{bmatrix} \left(\dfrac{4}{2\cdot 2}\right) & \left(\dfrac{3.2}{2\cdot 2.37}\right) & \left(\dfrac{7.2}{2\cdot 4}\right) \\[2ex] \left(\dfrac{3.2}{2.37\cdot 2}\right) & \left(\dfrac{5.6}{2.37\cdot 2.37}\right) & \left(\dfrac{8.8}{2.37\cdot 4}\right) \\[2ex] \left(\dfrac{7.2}{4\cdot 2}\right) & \left(\dfrac{8.8}{4\cdot 2.37}\right) & \left(\dfrac{16}{4\cdot 4}\right) \end{bmatrix}$$

$$= \begin{bmatrix} 1.000 & 0.676 & 0.900 \\ 0.676 & 1.000 & 0.930 \\ 0.900 & 0.930 & 1.000 \end{bmatrix}.$$

This process can be expressed in matrix algebra by using a diagonalized matrix. A matrix is diagonalized by setting all of its off-diagonal elements to zero. The diagonalized matrix for the example covariance matrix is:

$$\mathbf{D} = \text{diag}(\mathbf{C}) = \begin{bmatrix} 4 & 0 & 0 \\ 0 & 5.6 & 0 \\ 0 & 0 & 16 \end{bmatrix}.$$

From this D matrix, one can create a standardizing matrix **E** by both taking the square root of D and also inverting it. Actually, the processes of taking the square root of a matrix or finding the inverse of a matrix are quite complex for most matrices (see Chapter 3, Sections 3.5.7 and 3.7), but fortunately, they are simple for diagonalized matrices. The square root of a diagonal matrix is found by taking the square root of each element in the diagonal, and the inverse of a diagonal matrix is found by inverting each element in its diagonal (2 becomes 1/2, etc.) as shown below.

$$\mathbf{E} = \text{inv}(\text{sqrt}(\mathbf{D})) = \text{inv}\left(\text{sqrt}\left(\begin{bmatrix} 4 & 0 & 0 \\ 0 & 5.6 & 0 \\ 0 & 0 & 16 \end{bmatrix} \right) \right)$$

$$= \text{inv}\left(\begin{bmatrix} 2 & 0 & 0 \\ 0 & 2.37 & 0 \\ 0 & 0 & 4 \end{bmatrix} \right) = \begin{bmatrix} 1/2 & 0 & 0 \\ 0 & 1/2.37 & 0 \\ 0 & 0 & 1/4 \end{bmatrix} = \begin{bmatrix} 0.5 & 0 & 0 \\ 0 & 0.423 & 0 \\ 0 & 0 & 0.25 \end{bmatrix}.$$

Matrix E is referred to as a standardizing matrix because when the covariance matrix is premultiplied by it, it standardizes the covariance matrix by rows, that is, divides each element in the row by its standard deviation.

$$\mathbf{ES} = \begin{bmatrix} 0.5 & 0 & 0 \\ 0 & 0.423 & 0 \\ 0 & 0 & 0.25 \end{bmatrix} \cdot \begin{bmatrix} 4 & 3.2 & 7.2 \\ 3.2 & 5.6 & 8.8 \\ 7.2 & 8.8 & 16 \end{bmatrix} = \begin{bmatrix} 2 & 1.6 & 3.6 \\ 1.352 & 2.366 & 3.719 \\ 1.8 & 2.2 & 4 \end{bmatrix}.$$

Notice that the product matrix on the right is the covariance matrix, with each of its rows divided by its respective standard deviation. For example, the first row of the product matrix is equal to the first row of the covariance matrix divided by two, and the last row of the product matrix is the last row of the covariance matrix divided by four.

When the covariance matrix is postmultiplied by this **E** matrix, it has the effect of standardizing the covariance matrix by columns.

$$\mathbf{SE} = \begin{bmatrix} 4 & 3.2 & 7.2 \\ 3.2 & 5.6 & 8.8 \\ 7.2 & 8.8 & 16 \end{bmatrix} \cdot \begin{bmatrix} 0.5 & 0 & 0 \\ 0 & 0.423 & 0 \\ 0 & 0 & 0.25 \end{bmatrix} = \begin{bmatrix} 2 & 1.352 & 1.8 \\ 1.6 & 2.366 & 2.2 \\ 3.6 & 3.719 & 4 \end{bmatrix}.$$

The matrix formula for a correlation matrix consists therefore of the covariance matrix both being premultiplied and also postmultiplied by the standardizing matrix E, which has the effect of standardizing the covariance matrix both by rows and also by columns, that is, dividing the covariance in each cell by its two respective standard deviations.

$$\mathbf{R} = \mathbf{ESE} \tag{2.20}$$

$$\mathbf{R} = \mathbf{ESE} = \begin{bmatrix} 0.5 & 0 & 0 \\ 0 & 0.423 & 0 \\ 0 & 0 & 0.25 \end{bmatrix} \cdot \begin{bmatrix} 4 & 3.2 & 7.2 \\ 3.2 & 5.6 & 8.8 \\ 7.2 & 8.8 & 16 \end{bmatrix} \cdot \begin{bmatrix} 0.5 & 0 & 0 \\ 0 & 0.423 & 0 \\ 0 & 0 & 0.25 \end{bmatrix}$$

$$= \begin{bmatrix} 2 & 1.6 & 3.6 \\ 1.352 & 2.366 & 3.719 \\ 1.8 & 2.2 & 4 \end{bmatrix} \cdot \begin{bmatrix} 0.5 & 0 & 0 \\ 0 & 0.423 & 0 \\ 0 & 0 & 0.25 \end{bmatrix} = \begin{bmatrix} 1.000 & 0.676 & 0.900 \\ 0.676 & 1.000 & 0.930 \\ 0.900 & 0.930 & 1.000 \end{bmatrix}.$$

The covariance matrix and the correlation matrix have geometric properties. Each of the three variables can be represented as a vector with the standard deviation as its length. For example, variable X_1 has a standard deviation of 2, therefore its vector has a length of 2. Similarly, the vector for variable X_2 has a length of 2.366, and the vector for variable X_3 has a length of 4.

The correlation coefficient between any two variables is the cosine of the angle between the vectors for those two variables. By taking the arccosine of the correlation coefficient, therefore, one can obtain the angle between the two variables. For a correlation coefficient of $r = 0.00$, the arccosine is 90°, which makes sense. When two variables are perpendicular, they are uncorrelated. Similarly, a perfect correlation of $r = 1.00$ has an arccosine of zero degrees, indicating that the two variables coincide. A correlation coefficient of 0.707 (which has an r-squared value of 0.50, 50% of the variance accounted for), has an arccosine of 45°, indicating that the relationship between the two variables is halfway between being perfectly correlated (zero degrees) and being totally uncorrelated (90°).

The angle between variables X_1 and X_2 in the example data is $\alpha_{12} = \arccos(0.676) = 47.5°$; the angle between variables X_1 and X_3 is $\alpha_{13} = \arccos(0.900) = 25.9°$, and the angle between variables X_2 and X_3 is $\alpha_{12} = \arccos(0.930) = 21.6°$. The latter two angles in fact sum to the first: $25.9° + 21.6° = 47.5°$, indicating that the vector for variable X_3 is located within the same two-dimensional plane as variable X_1 and variable X_2 as shown in the figure below. This indicates that even though the correlation matrix is of order 3 (there are three variables), its *rank* is only 2, meaning that the three variables are completely accounted for within a two-dimensional space. This is an indication of linear dependency (also called *multicollinearity*)—an indication that one of the variables is a linear combination of the other two, which is indeed the case. Notice in the original **X** matrix that for each of the six trivariate observations, the value of variable X_3 is equal to the sum of the values for the other two variables. For more information, see Chapter 3, Section 3.8, where the rank of a matrix is explained more fully.

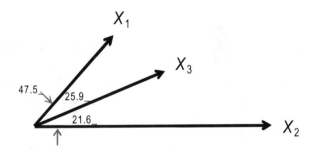

2.6 CLASSICAL PROBABILITY THEORY AND THE BINOMIAL: THE BASIS FOR STATISTICAL INFERENCE

Classical probability theory is usually illustrated and explained with such things as calculating the probability of coin tossing and throwing dice and other games of chance, but the reason it is of interest in statistical theory is because it forms the conceptual/mathematical foundation of statistical inference—inferring information about populations from sample data. The binomial distribution is the simplest case of statistical inference, and therefore can be used to demonstrate in an understandable way how the principles of probability apply to common statistical questions.

Consider the following statistical inference question. Suppose that Dr. Mason conducts a study to test the hypothesis that psychometrically measured introversion/extraversion (a relatively robust personality dimension) is recognizable from viewing videotapes of persons interacting. From 120 judgments of this kind, suppose that 103 are found to be correct. What is the probability of obtaining these results if the null hypothesis of random guessing were true? Obviously, obtaining such a high proportion correct ($P = 103/120 = 0.858$) by chance guessing is highly unlikely. However, to answer this question with precision, we will have to first review the principles of classical probability theory.

Every statistical inference question involves three kinds of data distribution: (1) one or more samples, (2) the population from which the sample(s) may have been drawn, and (3) a sampling distribution of some sample summary statistic (such as a mean or a proportion) for all possible such samples. The sample summary statistic for the example just given is a proportion, but since a proportion is a special case of a mean (the mean of a binary variable[5]), the process used for answering this proportion statistical inference question can also be generalized to inference concerning means.

The sample for the example above consists of 120 binary judgments of which 103 were correct and 17 were incorrect. This can be illustrated with a simple bar graph:

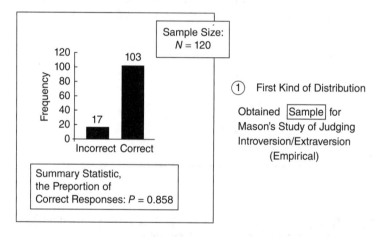

This distribution is "real." That is, it is an empirically obtained result rather than being hypothetical as are the other two distribution types.

The next distribution, the inferred population is hypothetical, and this is symbolized by enclosing it in a cloud. This population is, in fact, a null-hypothesized population, the population one would expect if indeed the judgments were made from only random guessing. The probability of a correct response in this population is symbolized by p_0, and is equal to 0.50. The probability of an incorrect response is symbolized as q_0, and is found as $q_0 = 1 - p_0 = 1 - 0.50 = 0.50$, as shown in the bar graph below representing the population distribution.

---

[5] It can be easily demonstrated that a proportion is a special case of a mean. Consider the case of five answers to questions, in which three are correct (recorded as a 1) and two are incorrect (recorded as a zero). The six data points are therefore 1, 1, 1, 0, and 0. Obviously, the proportion correct is 0.60, but that is also what one gets by the usual formula for a mean applied to these binary data values:

$$\bar{X} = \frac{\sum X}{n} = \frac{3}{5} = 0.60.$$

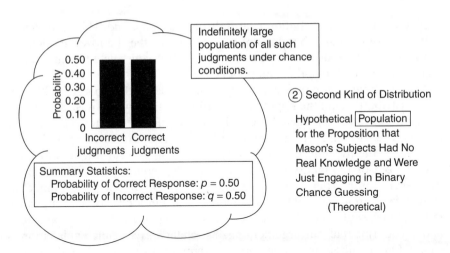

There is a third distribution implicit in this inference situation: the sampling distribution of all possible sample proportions for a sample size of 120. Implicit in this is an incredibly large sample space of all of the possible sample outcomes for a sample size of 120, of which our obtained sample with 103 correct judgments out of 120 is only one. The summary statistic used in our obtained sample is the proportion of correct judgments. If we let the proportion of correct judgments be the summary statistic for each of these many potential samples in the complete sample space, the sampling distribution of proportions given below in bar graph form expresses the relative probability of each of these possible sample proportions. This is therefore the sampling distribution of proportions for a sample size of $N = 120$.

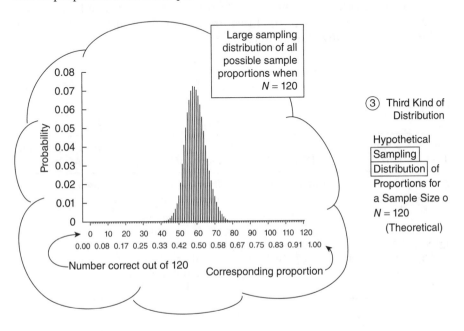

Notice that this distribution is also displayed in a cloud. This is to signify that the sampling distribution is also a product of the theorized model, the "straw man" of hypothesized chance guessing. It is constructed by inferential logic from the null hypothesized population.

Now, it should be emphasized that no one believes that this $N = 120$ sampling distribution and the population distribution from which it is constructed are true. They are constructed precisely for the purpose of showing that they are not true. They are "straw man" distributions. They are based upon the null hypothesis, the hypothesis that we intend to disprove. Our intent is to show that 103 out of 120 binary judgments ($P = 0.858$) could not possibly happen by chance, but in order to do so, we must construct a mathematical model to show how extreme our obtained outcome is within a distribution of chance (50/50) binary outcomes. A perusal of the sampling distribution of proportions above confirms that our argument against chance as a viable explanation is sound. One hundred and three correct judgments or more out of 120 is way out to the right side of this distribution, and would almost never occur in a 50/50 chance distribution like this.

Of the three kinds of distribution that make up statistical inference, only one, the sample, is real or empirical. The other two, the population and the sampling distribution, are theoretical—logical constructions based upon the null hypothesis. Two of the distributions, the sample and the population, have raw data (such as heads/tails, correct decisions/incorrect decisions, or direct measurements, such as eight inches, or an IQ score of 113) along the abscissa. The sampling distribution, on the other hand, has summary statistics along the abscissa (proportion of heads, proportion of correct decisions, mean number of inches, mean IQ, etc.). It is the theoretical distribution of the relative occurrence of all possible sample summary statistic outcomes for samples of a given size. These characteristics of the three primary kinds of distribution in inference can be summarized in a table:

The Three Kinds of Distribution	Type of Distribution	Source of Numbers	Values Along the Abscissa
① Sample	Empirical (real)	Observed Data	Raw Data
② Population	Theoretical (hypothetical)	Logical Basis, an Expression of the Null Hypothesis	Raw Data
③ Sampling Distribution of a Summary Statistic	Theoretical (hypothetical)	Logical basis, Mathematically Inferred from the Null Hypothesis	Summary Statistics

In fact, the number of possible sample outcomes for the sampling distribution of proportions shown above for samples of size $N = 120$ is two to the 120th power, which is equal to:

$$2^{120} = 1,329,227,995,784,920,000,000,000,000,000,000,000$$

which in scientific notation is 1.329E + 36. The probability of getting 103 out of 120 judgments correct by chance guessing is equal to:

$$p = \frac{226,460,004,264,655,000,000}{1,329,227,995,784,920,000,000,000,000,000,000,000}$$
$$= .0000000000000001704$$

In scientific notation, this unwieldy number becomes 1.704E-16.

While the answer to Dr. Mason's question might seem complicated and even a little mysterious, this is only because of the large numbers involved. Structurally, the probability results of the introversion/extraversion study are equivalent to the logic of coin tossing, and can be understood by the principles of classical probability theory, specifically the binomial expansion as illustrated by coin tossing. The probability of making 103 or more out of 120 introversion/extraversion judgments correct by chance is logically equivalent to the probability of getting 103 or more heads when tossing a coin 120 times. And even though this still seems complicated because of the large quantities involved, it is conceptually equivalent to the simpler task of finding the probability of getting three or more heads in four tosses of a coin, or three heads in three tosses of a coin. A review of classical probability theory and the binomial expansion can answer first the simple coin tossing questions, and then the more complex statistical inference question from the Mason introversion/extraversion study.

How likely is it to get three heads in a row in three tosses of an unbiased coin? There are two possible outcomes for the first toss: either heads, **H** (with a probability of $p = 1/2$), or tails, **T** (with a probability of $q = 1/2$). On the second toss, there are exactly the same two possible outcomes with the same probabilities, and again the same for the third toss, as shown in the branching diagram at the left below. Two outcomes times two outcomes times two outcomes, gives eight three-toss joint outcomes, as shown in the branching diagram. As shown at the right of the branching diagram, the probability of any one of these eight joint outcomes is 1/2 to the third power, or 1/8, by the multiplication rule of probabilities, which holds that the probability of two or more things happening jointly is equal to the product of the individual probabilities. Only one of these eight joint outcomes fulfils the specification of all three tosses coming up heads in three tosses of an unbiased coin, so the probability of that happening by chance (that is, when the coin is indeed unbiased) is 1/8.

Branching diagram | Pattern listing with probabilities

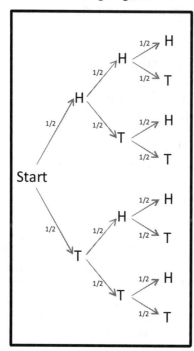

$P(H,H,H) = (\_ \times \_ \times \_) = 1/8$

$P(H,H,T) = (\_ \times \_ \times \_) = 1/8$

$P(H,T,H) = (\_ \times \_ \times \_) = 1/8$

$P(H,T,T) = (\_ \times \_ \times \_) = 1/8$

$P(T,H,H) = (\_ \times \_ \times \_) = 1/8$

$P(T,H,T) = (\_ \times \_ \times \_) = 1/8$

$P(T,T,H) = (\_ \times \_ \times \_) = 1/8$

$P(T,T,T) = (\_ \times \_ \times \_) = 1/8$

This simple example illustrates how a sampling distribution is used in hypothesis testing. The null hypothesized binary population consists of heads and tails, with the probability of heads equal to $p = 1/2$ and the probability of tails equal to $q = 1/2$. For this null hypothesized population, there are a number of possible sampling distributions of "number of heads" that could be constructed, one for each sample size. We then use logic and a branching diagram to specify the complete sample space for one of these possible sampling distributions, the one for samples of size $N = 3$, identifying the eight possible outcomes and specifying the probability that each outcome would occur given the null hypothesized population. The eight specific outcomes can be grouped into four groups in terms of the summary statistic, number of heads, to form a "sampling distribution of number of heads," which can be graphed as shown at the right. If we convert number of heads into proportion of heads, we have a sampling distribution of proportions.

Number of heads	Proportion of heads	Outcomes	Probability	
3	1.00	HHH	1/8	
2	0.67	HHT,HTH,THH	3/8	
1	0.33	HTT,THT,TTH	3/8	
0	0.00	TTT	1/8	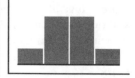

This sampling distribution of proportions is a probability model. By definition, a probability model has three characteristics: (1) the possible outcomes are mutually exclusive, (2) the outcomes are exhaustive, and (3) the probabilities sum to one. Mutually exclusive means that if one of the outcomes occurs, such as we have exactly two heads in three tosses, none of the others could have occurred, which is certainly true. Exhaustive means that the specified outcomes exhaust the possibilities; there are no other possible outcomes, which is also certainly true. The sum of the four probabilities is equal to one. All three are true, and this sampling distribution is indeed a probability model.

In tossing a fair coin, the probabilities of the binary outcomes are equal, $p = q = 1/2$. Consider an example where they are not equal. Suppose that a student takes a multiple choice test of only three items, with each item having four alternatives. What is the probability of getting all three items correct under the null hypothesis that the student has no knowledge of the subject matter but is only randomly guessing? The sample size for this question is also $N = 3$, like the coin tossing example, but now the probability of being correct on any one item is $p = 1/4$, and the probability of being incorrect is $q = 3/4$. There are still eight possible joint outcomes for this example as for the coin-tossing example, but this time the probabilities of the eight outcomes are not equal to one another, as shown in the calculations on the right below.

Branching diagram   Pattern listing with probabilities

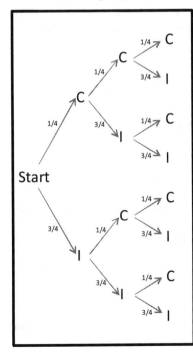

$P(C, C, C) = (\_ \times \_ \times \_) = 1/64$

$P(C, C, I) = (\_ \times \_ \times \_) = 3/64$

$P(C, I, C) = (\_ \times \_ \times \_) = 3/64$

$P(C, I, I) = (\_ \times \_ \times \_) = 9/64$

$P(I, C, C) = (\_ \times \_ \times \_) = 3/64$

$P(I, C, I) = (\_ \times \_ \times \_) = 9/64$

$P(I, I, C) = (\_ \times \_ \times \_) = 9/64$

$P(I, I, I) = (\_ \times \_ \times \_) = 27/64$

Again, the eight specific joint outcomes can be combined into four group-ings in terms of number of correct answers, to form a sampling distribution of proportions which can be graphed as shown at the right. There is only one way of getting all three correct (with its probability equal to 1/64), but there are three ways of getting two correct (CCI, CIC, and ICC), with each of them having a probability of 3/64, so by the addition rule of probability, the com-bined probability of the three is equal to 9/64. There are also three ways of getting one item correct (CII,ICI,IIC), and all three of them have a probability of 9/64, so the combined probability of the three is equal to 27/64. There is only one way to get none correct, with a probability of 27/64.

Number correct	Proportion correct	Outcomes	Probability
3	1.00	CCC	1/64
2	0.67	CCI,CIC,ICC	9/64
1	0.33	CII,ICI,IIC	27/64
0	0.00	III	27/64

This sampling distribution of proportions is also a probability model with outcomes that are mutually exclusive and exhaustive, and probabilities that sum to one.

Sampling distributions such as these can also be created algebraically. The algebraic expression is obviously where this process gets the name of "the expansion of the binomial" in that one stars with the binomial expression of $p + q$, raised to some power, and then algebra is used to carry out the expan-sion of the single binary term to each of the individual terms for that particular power:

$$(p+q)^N = (p+q)^3 = (p+q)(p^2 + 2pq + q^2) = p^3 + 3p^2q + 3pq^2 + q^3.$$

When $p = 1/2$ and $q = 1/2$, this becomes:

$$\left(\frac{1}{2}+\frac{1}{2}\right)^3 = \left(\frac{1}{2}\right)^3 + 3\left(\frac{1}{2}\right)^2\left(\frac{1}{2}\right) + 3\left(\frac{1}{2}\right)\left(\frac{1}{2}\right)^2 + \left(\frac{1}{2}\right)^3 = \frac{1}{8}+\frac{3}{8}+\frac{3}{8}+\frac{1}{8}.$$

The four terms on the right (from expanding the binomial) are indeed the probabilities of 3, 2, 1, and 0 heads, respectively.

When $p = 1/4$ and $q = 3/4$, this becomes:

$$\left(\frac{1}{4}+\frac{3}{4}\right)^3 = \left(\frac{1}{4}\right)^3 + 3\left(\frac{1}{4}\right)^2\left(\frac{3}{4}\right) + 3\left(\frac{1}{4}\right)\left(\frac{3}{4}\right)^2 + \left(\frac{3}{4}\right)^3 = \frac{1}{64}+\frac{9}{64}+\frac{27}{64}+\frac{27}{64}.$$

If the sample size were raised to four, a slightly more complex binomial expansion is created:

$$(p+q)^N = (p+q)^4 = (p+q)(p^3 + 3p^2q + 3pq^2 + q^3)$$
$$= p^4 + 4p^3q + 6pq^2 + 4p^2q + q^4.$$

The expansion obviously becomes continually more complex as the binomial is raised to successively higher powers. Notice the pattern in the powers of p and q in the expansion on the right hand side of the equation: the powers of p in the five terms decrease from 4 down to 0 from left to right, and the powers of q increase from 0 up to 4. Since zero is a possible outcome, there will always be one more term in the expansion than N, the sample size. The powers of p and q always go from N to zero and from zero to N.

Notice also the pattern of the coefficients for the terms as sample size (N) increases. The pattern is "1, 1" when $N = 1$, then becomes "1,2,1" for $N = 2$, and then "1,3,3,1" for $N = 3$, and "1,4,6,4,1" for $N = 4$. Blaise Pascal also noticed this pattern and created a triangle of values that yields the coefficients for each successively higher N value:

N=0											1										
N=1										1		1									
N=2									1		2		1								
N=3								1		3		3		1							
N=4							1		4		6		4		1						
N=5						1		5		10		10		5		1					
N=6					1		6		15		20		15		6		1				
N=7				1		7		21		35		35		21		7		1			
N=8			1		8		28		56		70		56		28		8		1		
N=9		1		9		36		84		126		126		84		36		9		1	
N=10	1		10		45		120		210		252		210		120		45		10		1

Each entry in a particular row is created by summing two values: the one to the right above the entry, and the one to the left above the entry. After all of the entries have been created by this summing process for a particular row, a "1" is placed on the left of this vector of entries and a "1" is placed on the right. This particular Pascal's triangle has 10 rows, providing coefficients for an N of 0 up to an N of 10, but obviously one could continue adding rows.

Pascal's triangle provides a more efficient strategy for the expansion of the binomial than either the branching diagram method or the algebraic expansion process just demonstrated. Suppose that we wished to create a binomial expansion for all possible outcomes and their probabilities in tossing a coin ten times. The coefficients for each term of the binomial expansion are given in the rows of this Pascal's triangle. Each probability is found by taking p to the rth power and q to the $(N-r)^{th}$ power, where r is the number of successes (the leftmost column), and N is the number of trials, 10 in this case—and then multiplying that product by its respective coefficient taken from the Pascal's triangle, as shown in the table below.

Number of Successes (r)	Probability Formula	Probability for $p=1/2$	Probability for $p=1/4$
10	$1p^{10}$	$\dfrac{1}{1024}$	$1\left(\dfrac{1}{4}\right)^{10}=\dfrac{1}{1{,}048{,}576}$
9	$10p^9q$	$\dfrac{1}{1024}$	$10\left(\dfrac{1}{4}\right)^{9}\left(\dfrac{3}{4}\right)=\dfrac{30}{1{,}048{,}576}$
8	$45p^8q^2$	$\dfrac{10}{1024}$	$45\left(\dfrac{1}{4}\right)^{8}\left(\dfrac{3}{4}\right)^{2}=\dfrac{405}{1{,}048{,}576}$
7	$120p^7q^3$	$\dfrac{120}{1024}$	$120\left(\dfrac{1}{4}\right)^{7}\left(\dfrac{3}{4}\right)^{3}=\dfrac{3240}{1{,}048{,}576}$
6	$210p^6q^4$	$\dfrac{210}{1024}$	$210\left(\dfrac{1}{4}\right)^{6}\left(\dfrac{3}{4}\right)^{4}=\dfrac{17{,}010}{1{,}048{,}576}$
5	$252p^5q^5$	$\dfrac{252}{1024}$	$252\left(\dfrac{1}{4}\right)^{5}\left(\dfrac{3}{4}\right)^{5}=\dfrac{61{,}236}{1{,}048{,}576}$
4	$210p^4q^6$	$\dfrac{210}{1024}$	$210\left(\dfrac{1}{4}\right)^{4}\left(\dfrac{3}{4}\right)^{6}=\dfrac{153{,}090}{1{,}048{,}576}$
3	$120p^3q^7$	$\dfrac{120}{1024}$	$120\left(\dfrac{1}{4}\right)^{3}\left(\dfrac{3}{4}\right)^{7}=\dfrac{262{,}440}{1{,}048{,}576}$
2	$45p^2q^8$	$\dfrac{45}{1024}$	$45\left(\dfrac{1}{4}\right)^{2}\left(\dfrac{3}{4}\right)^{8}=\dfrac{295{,}245}{1{,}048{,}576}$
1	$10pq^9$	$\dfrac{10}{1024}$	$10\left(\dfrac{1}{4}\right)\left(\dfrac{3}{4}\right)^{9}=\dfrac{196{,}830}{1{,}048{,}576}$
0	$1q^{10}$	$\dfrac{1}{1024}$	$1\left(\dfrac{3}{4}\right)^{10}=\dfrac{59{,}049}{1{,}048{,}576}$

For the first example (probabilities in column 3), with $p = 1/2$ and $q = 1/2$, the product of p and q to their respective powers are always equal to 1/2 to the 10th power, which is 1/1024. In this case, the 11 probabilities for this binomial expansion are equal to each of the 11 coefficients in their proper order divided by 1024, as shown in the third column. These 11 probability values sum to 1024/1024, or 1, and each of the 11 outcomes is mutually exclusive, and the outcomes are exhaustive of the possibilities, thus filling the qualifications for a probability model.

In the fourth column, probabilities are given for a binomial that is *nonsymmetric* with $p = 1/4$ and $q = 3/4$. The calculations are therefore a little more complex. A story problem that fits this binomial sampling distribution would be "in a test consisting of 10 multiple choice questions, with four alternatives on each ($p = 1/4$), what is the probability of each of the eleven possible numbers of items correct (from zero correct to ten correct)?" From the entries in the table, one can see that the probability of getting all 10 multiple choice answers correct is very small, 1/1,048,576. The probability of getting nine *or more* is also quite small, 31/1,048,576. This column of 11 probabilities constitutes a probability model, since the outcomes are mutually exclusive and exhaustive, and the probabilities sum to one.

As demonstrated, the calculation of any one term of the expansion of the binomial consists of two parts, the coefficient, as given by Pascal's triangle, and the product of p and q raised to their respective powers. This is shown in the formula for each term of the binomial.

$$C_r^N p^r q^{N-r} = \frac{N!}{(N-r)!r!} p^r q^{N-r}. \tag{2.21}$$

This equation can be demonstrated to give the same probabilities on any of the terms from the table above. From the fourth column ($p = 1/4$ and $q = 3/4$), the probability when $r = 8$ and $N = 10$ (that is, 8 correct multiple choice items out of 10 total items) is:

$$C_r^N p^r q^{N-r} = C_8^{10} p^8 q^2 = \frac{10!}{(10-8)!8!} \cdot \left(\frac{1}{4}\right)^8 \left(\frac{3}{4}\right)^2$$

$$= \frac{10 \cdot 9}{2 \cdot 1}\left(\frac{1}{65536}\right)\left(\frac{9}{16}\right) = 45\left(\frac{9}{1048576}\right) = \frac{405}{1048576} = 0.00039.$$

The Pascal's triangle method and the formula method each have their place. With a binomial expansion that involves a relatively small number of outcomes, when the entire probability model is to be specified, the Pascal's triangle method works well. However, a large binomial expansion (like 120 introversion/extraversion judgments) where only a small portion of the model is needed (103 or greater correct), the formula is particularly useful.

The symmetric binomial ($p = q = 1/2$) for a sample size of $N = 10$ actually fits surprisingly well to a Gaussian distribution (a normal curve) for such a small sample size, as shown in the histogram below. A symmetric binomial for $N = 30$ fits even better, but the fit for a symmetric binomial with the small sample of $N = 5$ is quite crude, as shown in the histogram below.

Even when the binomial is not symmetrical (when p is not equal to q), the fit for a Gaussian curve is not too bad. It shows a little left skew for $N = 10$, as shown below, but is close to symmetrical and reasonably Gaussian when $N = 30$. However, in the histogram for $N = 5$, the distribution is quite skewed and not a good fit to Gaussian.

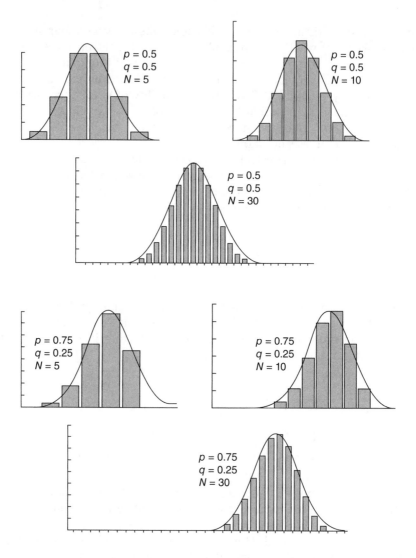

What does all of this have to do with sampling theory? It is, in fact, the mechanism by which one creates with great precision the sampling distribution of proportions or means for samples of any given size, as shown in the figure below. As shown in that figure, when the sample size is small, such as $N = 3$, the sample space is also small (eight specific outcomes), the sampling distribution of proportions is not very Gaussian in shape, and the process of creating the sampling distribution of proportions is relatively easy to specify. With an N of 4, the process of constructing the sampling distribution becomes somewhat more tedious, to specify a sample space of 16 specific possible outcomes. Also, the sampling distribution of proportions becomes more Gaussian with a smaller spread (i.e., a smaller standard error of the proportion).

When the sample size becomes large, such as 120, things change substantially. For one thing, calculations are complex and tedious, but the same logical/mathematical processes are used to obtain the exact mathematical probabilities for each of the possible outcomes. The sample space is huge. The sampling distribution of proportions is almost perfectly approximated by a normal curve. Although tedious by hand, the process of calculating the exact probability of each outcome can be done quite easily on a spreadsheet using Equation 2.21. This is, in fact, how we were able to report at the beginning of this section the number of possible sample outcomes to be 1.329E+36, and the probability of getting 103 or more correct judgments out of 120 to be 1.704E-16.

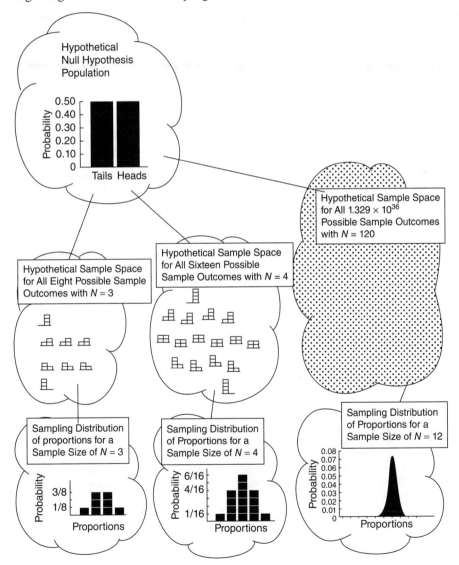

Four successively higher and more abstract ways have been demonstrated for obtaining the expansion of the binomial:

(1) the branching diagram and logic;
(2) the algebraic expansion;
(3) Pascal's triangle; and
(4) Equation 2.21, $C_r^N p^r q^{N-r}$.

The branching diagram is easiest to understand and the closest to the fundamental logic of coin tossing and the specifying of probabilities, but it is unwieldy for any but the smallest number of outcomes. Pascal's triangle and Equation 2.21 are the most abstract, but the most efficient methods and work well for larger sample sizes. There is a fifth method for obtaining binomial probabilities that is even more efficient—the normal curve approximation to the binomial, which is given by the formula:

$$z = \frac{P_{cc} - p_0}{\sqrt{\dfrac{p_0 q_0}{N}}}. \tag{2.22}$$

With this formula, we have now crossed over from classical probability theory to the practical and applied domain of statistical tests of significance. Equation 2.22, the normal curve approximation to the binomial, is also referred to as the z-*test for a single proportion*,[6] and it can be thought of as the simplest

---

[6] This is a particular application of the standard score, the Z-score. However, whereas a Z-score is a raw score minus its mean with that result divided by its standard deviation

$$Z = \frac{X_i - \bar{X}}{s}.$$

a z-test moves everything up one level and tests a sample mean minus the null hypothesized population mean, with the result divided by the standard error of the mean

$$z = \frac{\bar{X} - \mu_0}{s_{\bar{X}}} = \frac{\bar{X} - \mu_0}{\sqrt{\dfrac{s^2}{N}}}$$

The Z-score is symbolized by an upper case Z, and the z-test is symbolized by a lower case z, indicating that the z-test is at this higher level of abstraction, a sample mean within a sampling distribution of means rather than a raw score within a sample. Since a proportion is a special case of a mean, this z-test formula can also be used for proportions with an adjustment in notation that in fact yields Equation 2.13. The symbol $s_{\bar{b}}$ is used to indicate standard error of the proportion, in other words, standard error of the mean of a binary variable.

$$z = \frac{P - p_0}{s_{\bar{b}}} = \frac{P - p_0}{\sqrt{\dfrac{p_0 q_0}{N}}}$$

example of a statistical test of significance. It will, in fact, give a reasonably close approximation to the exact probabilities that are obtained from the binomial when the sample size is larger than 30. Consider, for example the exact probabilities in the table above for the 10-item multiple choice test. Suppose that one obtained 7 out of 10 items correct on that test and wishes to determine whether that could have happened by chance. From the probability entries in the table, it can be seen that the probability of getting 7 *or more* items correct by chance (i.e., under the null hypothesis of $p = 1/4$ chance guessing), is the sum of the probabilities for 7 correct, for 8 correct, for 9 correct, and for 10 correct:

$$P(7\_or\_more) = P(8\_correct) + P(8\_correct) + P(9\_correct) + P(10\_correct)$$

$$= \frac{3240}{1048576} + \frac{405}{1048576} + \frac{30}{1048576} + \frac{1}{1048576} = \frac{3676}{1048576} = 0.0035.$$

In other words, from the probability calculations, we find that this 7 or more items correct out of 10 would happen by chance only about three and a half times in a thousand.

Equation 2.13 gives an approximation to this exact probability. Four quantities enter into the formula, P_{cc}, p_0, q_0, and N. The parameter N is just sample size, which is 10; p_0 is the null hypothesized probability of being correct by chance, which is 1/4; q_0 is one minus p_0, which is ¾; and P_{cc} is the obtained sample proportion corrected for continuity (the "cc" subscript). The proportion must be corrected for continuity because the exact sampling distribution of these 11 possible proportions is discrete—there are exactly 11 possible values and nothing in between—but the Gaussian curve is a continuous distribution. If we were to use the formula with 0.70 as the entered sample proportion, we would only be including half of the probability for that bar in the bar graph, since the bar extends from 6.5 on the scale to 7.5. Instead, we must use the lower limit of the bar to include the entire probability area for $P = 0.70$, so P_{cc} is equal to 0.65, the lower limit of the proportion value. From the formula, we obtain the z result.

$$z = \frac{P_{cc} - p_0}{\sqrt{\dfrac{p_0 q_0}{N}}} = \frac{0.65 - 0.25}{\sqrt{\dfrac{(1/4)(3/4)}{10}}} = \frac{0.40}{\sqrt{\dfrac{3/16}{10}}} = \frac{0.40}{\sqrt{\dfrac{0.1875}{10}}} = \frac{0.40}{\sqrt{0.01875}} = \frac{0.40}{0.1369} = 2.921$$

Like a Z-score, the value from a z-test has meaning in terms of probability. A standard normal curve table (Table A in the statistical tables at the end of the book) can be used to identify the area above or below any z value, where the area is given in terms of proportion of total area under the normal curve, and therefore corresponds to a probability. In the table, the area beyond a z-value of 2.92 is given as 0.0018. In other words, the probability of getting a z-value this great or greater is about 1.8 in a thousand. This value approximates the exact probability of three and half in a thousand found above from the binomial calculations. The fit is not perfect, even to three places, but not too bad for an N of 10, since an N of 30 is usually the lowest sample size at which one expects the central limit theorem to have its effects.

The central limit theorem states that regardless of the shape of the population, as the sample size becomes large (at least 30 or larger), the sampling distribution of means (or proportions) will be Gaussian. Our example only has a sample size of 10, but still demonstrates the principle in that even though the population is quite skewed in shape (p = 1/4 and q = 3/4), the Gaussian curve probability from the z-test is still reasonably close to the exact probability from the binomial distribution.

Obviously the exact test is precisely correct, but the normal curve approximation is substantially more convenient to calculate, and when N is greater than 30 usually gives a good approximation to the probabilities from exact computations. The binomial exact test is like the "gold standard," and the approximate test is like paper currency. Under a gold standard, paper notes are convertible into preset, fixed quantities of gold (in theory at least). Paper currency, which is more convenient for everyday use, is to be backed up by gold as that which gives paper money its value. The binomial (for proportions) and the multinomial (for means) are the gold standard—the underlying logic on which the more convenient statistical significance tests are based. Exact probability calculations are seldom if ever used in actual practice, since the approximate tests are accurate enough for most purposes, particularly when sample size is large.

2.7   SIGNIFICANCE TESTS: FROM BINOMIAL TO Z-TESTS TO T-TESTS TO ANALYSIS OF VARIANCE

The z-test for a single proportion demonstrated at the end of the last section is one of the nine z and t significance tests to be considered here. The formula for it and the formulas for the other eight are given in Table 2.1.

The six tests in the first two columns are all large sample tests, z-tests of proportions in the first column and z-tests of means in the second column. The tests in the third column are all t-tests—tests made to deal with data from smaller samples. The three tests in the first row are all single sample tests, important pedagogically in understanding the other tests, but seldom used. The six tests in rows two and three are all two sample tests, two sample tests for independent groups in the second row, and two-sample tests for correlated groups in the third row. Test 1 (first row and first column) is the z-test of a single proportion, the normal curve approximation to the binomial just demonstrated.

TABLE 2.1. Nine _t_- and _z_-Tests: Description and Formulas

		_z_-Tests for Categorical Data (Proportions)	Quantitative Data (Tests of Means)	
			_z_-Tests: Large Sample or σ Known	_t_-Tests: Small Sample and σ Unknown
Single sample		1. The _z_-test for single proportion. $$z = \frac{P - p_0}{\sqrt{\dfrac{p_0 q_0}{N}}}$$	2. The _z_-test for a single mean. $$z = \frac{\bar{X} - \mu_0}{\sqrt{\dfrac{\sigma^2}{N}}} \text{ or } z = \frac{\bar{X} - \mu_0}{\sqrt{\dfrac{\hat{S}^2}{N}}}$$	3. The _t_-test for a single mean. $$t = \frac{\bar{X} - \mu_0}{\sqrt{\dfrac{\hat{S}^2}{N}}}$$
Two samples	Independent sample	4. The _z_-test of two proportions for independent samples. $$z = \frac{P_1 - P_2}{\sqrt{\dfrac{p_0 q_0}{N_1} + \dfrac{p_0 q_0}{N_2}}}$$	5. The _z_-test of two means for independent samples. $$z = \frac{\bar{X}_1 - \bar{X}_2}{\sqrt{\dfrac{\hat{S}_1}{N_1} + \dfrac{\hat{S}_2}{N_2}}}$$	6. The _t_ of two means for independent samples. $$t = \frac{\bar{X}_1 - \bar{X}_2}{\sqrt{\dfrac{\hat{S}_{\text{pooled}}}{N_1} + \dfrac{\hat{S}_{\text{pooled}}}{N_2}}}$$
	Correlated samples	7. The _z_-test of two proportions for correlated samples. [More simply calculated with Chi-square.]	8. The _z_-test of two means for correlated samples. $$z = \frac{\bar{D}}{\sqrt{\dfrac{\hat{S}_d^2}{N}}} \quad z = \frac{\bar{X}_1 - \bar{X}_2}{\sqrt{\dfrac{\sigma_1^2}{N} + \dfrac{\sigma_2^2}{N} - 2r\sqrt{\dfrac{\sigma_1^2}{N}}\sqrt{\dfrac{\sigma_2^2}{N}}}}$$	9. The _t_-test of two means for correlated samples. $$t = \frac{\bar{D}}{\sqrt{\dfrac{\hat{S}_d^2}{N}}}$$

There are three important principles from which all nine of these significance tests are derived:

$$(1) \quad \sigma_{\bar{X}}^2 = \frac{\sigma^2}{N}.$$

(2.23)

$$(2) \quad \sigma_{binary}^2 = pq.$$

(2.24)

$$(3) \quad \sigma_{x_1 - x_2}^2 = \sigma_{x_1}^2 + \sigma_{x_2}^2 - 2r\sigma_{x_1}\sigma_{x_2}.$$

(2.25)

DEMONSTRATION OF EQUATION 2.23—PRINCIPLE ONE: THE VARIANCE OF MEANS

Equation 2.23 holds that the variance of the sampling distribution of means is one Nth of the variance of the raw scores in the population. This is the foundational principle of all of sampling theory, and the most important thing to be learned in this section of the chapter.

$$\sigma_{\bar{X}}^2 = \frac{\sigma^2}{N}.$$

(2.23)

It makes sense that means would have less variance than raw scores, since extreme scores get combined with (on the average) less extreme scores. What is striking about this formula is that the relationship between variance of raw scores and variance of means is so systematic and simple—the *variance of means being simply one Nth of raw score variance*. If the variance of the raw scores in the population is 100, then means from a sample of size $N = 2$ will have 1/2 that much variance ($\sigma_{\bar{X}}^2 = 50$); means from a sample of $N = 4$ will have 1/4 as much variance as the population ($\sigma_{\bar{X}}^2 = 25$); means from a sample of $N = 30$ will have 1/30 as much variance as the population ($\sigma_{\bar{X}}^2 = 3.33$).

This principle can be demonstrated with a multinomial distribution. The multinomial is like the binomial except that it deals with multinary data (three or more possible values) rather than binary data. The multinomial follows similar probability principles to the binomial (the multiplication rule, the addition rule, permutations and combinations formulas for the coefficients, etc.), but in a slightly different way—variations on the same theme. Specifying the entire mechanism for the expansion of the multinomial would take us too far afield, but enough will be given to demonstrate Equation 2.23.

Imagine a population of pipes of lengths 6, 8, and 10 in, with equal numbers of each length of pipe, so that the respective probabilities of drawing each of the three are equal $p_1 = p_2 = p_3 = 1/3$.

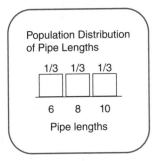

The population measurements are 6, 8, and 10, all equally likely, so the population mean can be calculated as the expected value of these measurements, and the population variance can be calculated as the expected value of the squared deviation of these measurements from their population mean.

The population mean is 8:

$$\mu = E(X) = \sum p_i X_i = \left(\frac{1}{3}\right)(6) + \left(\frac{1}{3}\right)(8) + \left(\frac{1}{3}\right)(10) = \frac{24}{3} = 8.$$

The population variance is 2.67:

$$\sigma^2 = E\left[(X - \mu)^2\right] = \sum p_i (X_i - \mu)^2$$
$$= \left(\frac{1}{3}\right)(6-8)^2 + \left(\frac{1}{3}\right)(8-8)^2 + \left(\frac{1}{3}\right)(9-8)^2 = \frac{4}{3} + \frac{0}{3} + \frac{4}{3} = \frac{8}{3} = 2.67.$$

To maintain simplicity, the sample size will be set to $N = 2$. In drawing samples of size 2 at random from these pipes, there are nine possible sample outcomes in the sample space, all equally likely, and the mean is calculated for each of the nine as shown below.

The nine samples in the sample space	The means for each of the nine samples
6, 6	6
6, 8	7
6, 10	8
8, 6	7
8, 8	8
8, 10	9
10, 6	8
10, 8	9
10, 10	10

Three of these nine equally likely samples of size $N = 2$ have a mean of 8 inches, two have a mean of 9 in, two have a mean of 7 in, one has a mean of 10 in, and one has a mean of 6 in, creating the sampling distribution of means shown in the figure below.

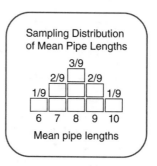

The mean of sample means ($\mu_{\overline{X}}$), that is, the mean of this sampling distribution of means, should be equal to the population mean: $\mu_{\overline{X}} = \mu$. It is calculated as the expected value of these five sample mean outcomes, and it is indeed found to be 8, the same as the population mean:

$$\mu_{\overline{X}} = E(\overline{X}) = \sum p\overline{X} = \left(\frac{1}{9}\right)(6) + \left(\frac{2}{9}\right)(7) + \left(\frac{3}{9}\right)(8) + \left(\frac{2}{9}\right)(9) + \left(\frac{1}{9}\right)(10) = \frac{72}{9} = 8.$$

We are now ready to demonstrate the foundational principle, Equation 2.23, that the variance of means is one N^{th} of the variance of raw scores. Since the population variance was found to be 2.67, the variance of the sampling distribution of means (the so-called squared "standard error of the mean") for a sample size of $N = 2$ should be half of this, or 1.33, as expressed by Equation 2.23:

$$\sigma_{\overline{X}}^2 = \frac{\sigma^2}{N} = \frac{2.67}{2} = 1.33.$$

The variance of the sampling distribution of means is calculated as the expected value of the squared deviation score of each possible sample mean from its mean-of-sample-means (the population mean), and it is indeed found to be 1.33:

$$\sigma_{\overline{X}}^2 = E\left[\left(\overline{X} - \mu_{\overline{X}}\right)^2\right] = \sum p_i \left(\overline{X} - \mu_{\overline{X}}\right)^2$$

$$= \left(\frac{1}{9}\right)(6-8)^2 + \left(\frac{2}{9}\right)(7-8)^2 + \left(\frac{3}{9}\right)(8-8)^2 + \left(\frac{2}{9}\right)(9-8)^2 + \left(\frac{1}{9}\right)(10-8)^2$$

$$= \frac{4}{9} + \frac{2}{9} + \frac{0}{9} + \frac{2}{9} + \frac{4}{9} = \frac{12}{9} = 1.33$$

DEMONSTRATION OF EQUATION 2.24—PRINCIPLE TWO: THE VARIANCE OF BINARY DATA

The second of the three major principles that form the foundation for the nine t- and z-tests, Equation 2.24, is a simplifying equation:

$$\sigma^2_{binary} = pq. \tag{2.24}$$

When data are binary rather than quantitative the usual calculations for a population variance are not necessary. The variance of a binary population is the product of p and q. Consider the example of a multiple choice exam with four choices on each item. The null hypothesized population has a probability of being correct from chance guessing of $p = 1/4$ and a probability of being incorrect of $q = 3/4$.

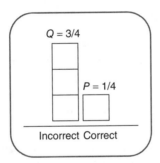

For the longhand calculation, let correct be represented as a 1 and incorrect as a 0. The relative distribution of these values in the null hypothesis population is 0,0,0,1. The population mean is of course 1/4, which is equal to the null hypothesized proportion correct, p.

$$\mu_{binary} = \frac{\sum X}{N} = \frac{0+0+0+1}{4} = \frac{1}{4}.$$

The population variance (as the expected value of the squared deviations from the population mean) is found to be 3/16:

$$\sigma^2_{binary} = E\left[(X - \mu_{binary})^2\right] = \sum p_i \left(X - \mu_{binary}\right)^2$$

$$= \left(\frac{3}{4}\right)\left(0 - \frac{1}{4}\right)^2 + \left(\frac{1}{4}\right)\left(1 - \frac{1}{4}\right)^2 = \left(\frac{3}{4}\right)\left(\frac{1}{16}\right) + \left(\frac{1}{4}\right)\left(\frac{9}{16}\right) = \frac{12}{64} = \frac{3}{16}.$$

Equation 2.24 gives this same result much more simply:

$$\sigma^2_{binary} = pq = \left(\frac{1}{4}\right)\left(\frac{3}{4}\right) = \frac{3}{16}.$$

DEMONSTRATION OF EQUATION 2.25—PRINCIPLE THREE, VARIANCE OF DIFFERENCES

The third principle, Equation 2.25, states that the variance of the difference between two variables is equal to the sum of their individual variances minus two times their covariance:

$$\sigma^2_{x_1-x_2} = \sigma^2_{x_1} + \sigma^2_{x_2} - 2r\sigma_{x_1}\sigma_{x_2} \qquad (2.25)$$

By definition, the correlation coefficient is the covariance divided by the product of standard deviations.

$$r = \frac{\text{cov}_{12}}{\sigma_{x_1}\sigma_{x_2}}.$$

Solving this equation for cov_{12}, it follows that the covariance is equal to r times the two standard deviations, and clearly the third term above is two times the covariance.

$$\text{cov}_{12} = r\sigma_{x_1}\sigma_{x_2}.$$

Notice that if the correlation coefficient r is zero, the third term disappears. It could be stated, then, that the variance of the difference between two independent variables is equal to the sum of their two individual variances.

This will be demonstrated for the general case of two correlated variables. Suppose that six persons take a dynamometer strength test at camp on a Monday (X_1), and then again on a Tuesday (X_2), and that their scores are as follows:

	X_1	X_2
Al	10	8
Bill	14	6
Chad	6	5
Dan	8	5
Ed	8	4
Frank	2	2
Totals	48	30
Means	8	5

A day at camp has taken its toll, and all six have lower strength scores on Tuesday. The Monday minus Tuesday difference scores are:

	X₁-X₂
Al	2
Bill	8
Chad	1
Dan	3
Ed	4
Frank	0
Total	18
Mean	3

For generality, Equation 2.25 is given in population/theoretical notation (with variance signified with sigma rather than s). Since this numerical demonstration is using sample data rather than population values, the formula will be used in sample form:

$$s^2_{x_1-x_2} = s^2_{x_1} + s^2_{x_2} - 2rs_{x_1}s_{x_2}$$

The variance is calculated directly on these scores (as if they were a new X variable) by matrix Equation 2.6.

$$\text{variance\_of\_differences} = S^2 = \left(\frac{1}{N-1}\right) \cdot (\mathbf{X'X - T \cdot M})$$

$$= \left(\frac{1}{6-1}\right) \cdot \left([2 \ 8 \ 1 \ 3 \ 4 \ 0] \cdot \begin{bmatrix} 2 \\ 8 \\ 1 \\ 3 \\ 4 \\ 0 \end{bmatrix} - 18 \cdot 3 \right)$$

$$= (1/5) \cdot (94 - 54) = (1/5) \cdot (404) = 8$$

A covariance matrix is calculated on X_1 and X_2 by matrix Equation 2.12.

$$S = \left(\frac{1}{N-1}\right) \cdot (\mathbf{X'X - T \cdot M'})$$

$$= \left(\frac{1}{6-1}\right) \cdot \left(\begin{bmatrix} 10 & 14 & 6 & 8 & 8 & 2 \\ 8 & 6 & 5 & 5 & 4 & 2 \end{bmatrix} \cdot \begin{bmatrix} 10 & 8 \\ 14 & 6 \\ 6 & 5 \\ 8 & 5 \\ 8 & 4 \\ 2 & 2 \end{bmatrix} - \begin{bmatrix} 48 \\ 30 \end{bmatrix} \cdot [8 \ 5] \right)$$

$$= (1/5) \cdot \left(\begin{bmatrix} 464 & 270 \\ 270 & 170 \end{bmatrix} - \begin{bmatrix} 384 & 240 \\ 240 & 150 \end{bmatrix} \right) = (1/5) \cdot \begin{bmatrix} 80 & 30 \\ 30 & 20 \end{bmatrix} = \begin{bmatrix} 16 & 6 \\ 6 & 4 \end{bmatrix}.$$

From this variance/covariance matrix, we can take the variance for X_1, the variance for X_2, and the covariance between them.

$$s_{x_1}^2 = 16$$

$$s_{x_2}^2 = 4$$

$$\text{cov} = rs_{x_1}s_{x_2} = 6$$

From the sample version of Equation 2.25 and these three quantities, we calculate the variance of the difference score of variable X_1 minus variable X_2 is 8, which agrees with the direct variance calculation on the difference scores.

$$s_{x_1-x_2}^2 = s_{x_1}^2 + s_{x_2}^2 - 2rs_{x_1}s_{x_2} = 16 + 4 - (2)(6) = 8$$

This is just one particular application of a general principle, **quadratic forms** (Chapter 3, Section 3.10.1), which will be used with a number of multivariate methods. Using matrix algebra, the difference scores are obtained by postmultiplying the matrix of X_1 and X_2 data by the vector $w = [1\ -1]$, which has the effect of subtracting X_2 from X_1.

$$X_{\text{diff}} = X \cdot w = \begin{bmatrix} 10 & 8 \\ 14 & 6 \\ 6 & 5 \\ 8 & 5 \\ 8 & 4 \\ 2 & 2 \end{bmatrix} \cdot \begin{bmatrix} 1 \\ -1 \end{bmatrix} = \begin{bmatrix} 2 \\ 8 \\ 1 \\ 3 \\ 4 \\ 0 \end{bmatrix}.$$

This X_{diff} vector is referred to as a linear combination of the two variables X_1 and X_2, and the particular linear combination is defined by w, the vector of weights. If the covariance matrix for variables X_1 and X_2 is both premultiplied and also postmultiplied by the same vector w, it yields the variance of X_{diff}, the transformed data.

$$\sigma_{x_1-x_2}^2 = w' \cdot S \cdot w = [1\ -1] \cdot \begin{bmatrix} 16 & 6 \\ 6 & 4 \end{bmatrix} \cdot \begin{bmatrix} 1 \\ -1 \end{bmatrix} = [10\ 2] \cdot \begin{bmatrix} 1 \\ -1 \end{bmatrix} = 8.$$

A quadratic form is a mathematical process that is often used for obtaining the variances and covariances of transformed variables. If algebraic symbols (rather than numerical data) are used in this particular quadratic form, it yields Equation 2.25 in sample notational form.

$$
\begin{aligned}
s^2_{x_1-x_2} = w' \cdot S \cdot w &= [1 \quad -1] \cdot \begin{bmatrix} s^2_{x_1} & rs_{x_1}s_{x_2} \\ rs_{x_1}s_{x_2} & s^2_{x_2} \end{bmatrix} \cdot \begin{bmatrix} 1 \\ -1 \end{bmatrix} \\
&= \left[(s^2_{x_1} - rs_{x_1}s_{x_2}) \quad (rs_{x_1}s_{x_2} - s^2_{x_2}) \right] \cdot \begin{bmatrix} 1 \\ -1 \end{bmatrix} \\
&= (s^2_{x_1} - rs_{x_1}s_{x_2}) - (rs_{x_1}s_{x_2} - s^2_{x_2}) = s^2_{x_1} + s^2_{x_2} - 2rs_{x_1}s_{x_2}.
\end{aligned}
$$

2.7.1 The z Test of a Single Mean

As mentioned above, the foregoing three principles are the foundation of the nine t- and z-tests in Table 2.1. Specifically, they are the principles from which the standard error term for each of the nine tests are derived. Not all tests use all three principles. Some use one of them, some use two, and some do use all three. Consider test 2, the z-test for a single mean, which only uses one.

$$
z = \frac{\overline{X} - \mu}{\sqrt{\dfrac{\hat{s}^2}{N}}} \tag{2.26}
$$

The numerator consists of a sample mean minus the population mean, and this difference comparison is divided by the standard error of the mean. The only one of the three principles used for this standard error is the first one, that the variance of means is one Nth of the variance of raw scores which can be clearly seen in the denominator of Equation 2.26 (in square root form as a standard error).

This test will be illustrated with an example. Suppose that Waterford private school wishes to test their students to see if they significantly exceed the national norms for a particular achievement test. The national score is 82. From a random sample of 134 of their students, the average score was 85.3, with a standard deviation or 5.2 (which is a variance of 27.04). The z-test is calculated from Equation 2.26.

$$
z = \frac{\overline{X} - \mu}{\sqrt{\dfrac{\hat{s}^2}{N}}} = \frac{85.3 - 82}{\sqrt{\dfrac{27.04}{134}}} = \frac{3.3}{\sqrt{0.2018}} = \frac{3.3d}{0.449} = 7.346
$$

This is well beyond the $z = 1.645$ needed for one-tailed significance at the 0.05 level, so they reject the null hypothesis. All six of the z-tests are large sample tests that are justified by the central limit theorem and require an N of at least 30 (some texts say 120, some say 200). They require the assumption of independence of observations. This means that, like the binomial, the probabilities remain the same on successive trials. Since they are justified by the central limit theorem, there are no other assumptions for the z-tests.

There is one case where this test can be used with a small sample size, and that is indicated by the alternate Equation 2.26 that is given in Table 2.1.

$$z = \frac{\overline{X} - \mu}{\sqrt{\frac{\sigma^2}{N}}}. \tag{2.26a}$$

The only difference between these two formulas is that the main Equation 2.26 uses a sample variance (as indicated by an S hat squared) to estimate the population variance in the denominator calculation of the standard error, whereas Equation 2.26a uses the actual population variance (as indicated by a sigma squared). In other words, this formula requires that the population standard deviation or variance is known. As an example, suppose that Waterford wanted to know if their incoming students have a higher IQ than the national average, but they only have a sample size of $N = 12$. However, both the population mean ($\mu = 100$) and the population standard deviation ($\sigma = 15$) are known, so they can test their obtained sample mean IQ of 104.8 for significance.

$$z = \frac{\overline{X} - \mu}{\sqrt{\frac{\sigma^2}{N}}} = \frac{104.8 - 100}{\sqrt{\frac{15^2}{12}}} = \frac{4.8}{\sqrt{\frac{225}{12}}} = \frac{4.8}{\sqrt{18.75}} = \frac{4.8}{4.33} = 1.109.$$

This does not exceed the critical ratio of 1.645 for the 0.05 level one-tailed, so the null hypothesis cannot be rejected.

The structure of Equation 2.26 is the general format for all nine of the t- and z-tests, with the numerator consisting of a sample statistic minus a population comparison value, and this difference comparison is divided by the standard error of the sampling distribution for the sample statistic. This ratio is roughly interpreted like a Z-score would be, with values around 2 (positive or negative) needed for significance (indications of rare outcomes).

$$t\text{- or } z\text{-test} = \frac{\text{sample statistic} - \text{population value}}{\text{standard error}}$$

2.7.2   The z Test of a Single Proportion

Test 1, the z-test for a single proportion was illustrated with an example at the end of the previous section, but it was applied there to an inappropriate sample (an N less than 30) simply to show the correspondence between the z-test and binomial exact probability. It will be shown here with a more realistic example. First, let's consider the form of the formula for this test.

$$z = \frac{P - p_0}{\sqrt{\dfrac{p_0 q_0}{N}}} \qquad (2.22)$$

The only difference between Equation 2.22 and the one that appeared at the end of the last section is that the sample P-value in that one was P_{cc}, corrected for continuity. This will not usually be necessary. With a large enough sample size, the effect usually would be negligible.

This formula is in fact a special case of the formula for a z-test for a single mean (Eq. 2.26). Since a proportion is a special case of a mean (the mean of a binary variable), the numerators correspond (with P being a special case of $\overline{X}$ and p_0 being a special case of μ). The denominator also corresponds, except that since this formula deals with proportions, the simpler formula for variance as $p_0 q_0$ (Eq. 2.24) applies. In other words, the standard error in this z-test uses two of the three principles, the variance of means principle (Eq. 2.23) and the variance of a proportion principle (Eq. 2.24).

Suppose that the city wishes to take a bond for supporting school development, but does not wish to propose it unless it has a reasonable chance of passing in the bond election (with a 50 percent vote required for passage), so they take a random sample of 350 voting citizens and find that 192 indicate that they would vote in favor of it. They wish to test whether the obtained proportion is enough to be significantly above the 50% $(p - 0.50)$ required for passage. The sample proportion is $P = 192/350 = 0.549$.

$$z = \frac{P - p_0}{\sqrt{\dfrac{p_0 q_0}{N}}} = \frac{0.549 - 0.500}{\sqrt{\dfrac{(0.5)(0.5)}{350}}} = \frac{0.049}{\sqrt{\dfrac{0.25}{350}}} = \frac{0.049}{\sqrt{0.000714}} = \frac{0.049}{0.0267} = 1.833.$$

This does exceed the critical ratio of 1.645 needed for one-tailed significance at the 0.01 level, so the null hypothesis can be rejected.

2.7.3   The z Test of Two Means for Independent Samples

The formula for this z-test, test 5 in Table 2.1 is given as:

$$z = \frac{\overline{X}_1 - \overline{X}_2}{\sqrt{\dfrac{\hat{s}_1^2}{N_1} + \dfrac{\hat{s}_2^2}{N_2}}}. \qquad (2.27)$$

This formula uses principle 1 (Eq. 2.23), as do all of the t- and z-tests, and since it is a two-sample test, it also uses principle 3 (Eq. 2.25). Notice in particular the covariance term at the end of that formula:

$$\sigma^2_{x_1-x_2} = \sigma^2_{x_1} + \sigma^2_{x_2} - 2r\sigma_{x_1}\sigma_{x_2}.$$

Since this z-test is for independent samples, the correlation coefficient is zero, and the covariance term at the end drops out, leaving this version of the formula as it applies in Equation 2.27:

$$\sigma^2_{x_1-x_2} = \sigma^2_{x_1} + \sigma^2_{x_2}.$$

Actually, the denominator of Equation 2.27 combines principle 3 (Eq. 2.25) and principle 1 (Eq. 2.23), the two variances have to be divided by their N values to turn them into variances of means (squared standard errors).

The denominator of this independent groups version of the formula could be summarized by saying that the standard error of the difference between two means for independent groups is equal to the square root of the sum of the squared standard errors (the one for mean 1 and the one for mean 2).

A possible example of data that would be appropriate for this test would be a laboratory study in which 200 white rats are tested for maze running times, with half being given a chance to observe from above other rats running the maze. The alternative hypothesis is that vicarious learning from observation has a significant effect on running times. Suppose that only 93 rats in the control group complete the experiment, with the mean of their running times being 22.4 seconds with a standard deviation of 2.7, and that 98 rats in the experimental group complete the experiment with the mean of their running times being 18.3 seconds with a standard deviation of 2.4. Entering these values into Equation 2.27, we obtain a z value of 2.37.

$$z = \frac{\overline{X}_1 - \overline{X}_2}{\sqrt{\dfrac{\hat{s}^2_1}{N_1} + \dfrac{\hat{s}^2_2}{N_2}}} = \frac{22.4 - 18.3}{\sqrt{\dfrac{2.7^2}{93} + \dfrac{2.4^2}{98}}} = \frac{4.1}{\sqrt{\dfrac{7.29}{93} + \dfrac{5.76}{98}}}$$

$$= \frac{4.1}{\sqrt{.0784 + 0.0588}} = \frac{4.1}{\sqrt{0.1372}} = \frac{4.1}{0.3704} = 11.07.$$

This obtained z-value far exceeds the critical ratio of 1.645 for the one-tailed 0.05 level test and also the 2.33 critical ratio for one-tailed 0.01 level test. The null hypothesis is rejected.

2.7.4 The z Test of Two Proportions for Independent Samples

In the same way that test 1, the z-test of a single proportion, is a special case of test 2, the z-test of a single mean, this test, test 4, is a special case of test 5, the z-test of two means for independent samples. It is essentially test 5 applied to binary data, and as such, it uses all three principles (Eqs. 2.23–2.25).

$$z = \frac{P_1 - P_2}{\sqrt{\dfrac{p_0 q_0}{N_1} + \dfrac{p_0 q_0}{N_2}}}. \tag{2.28}$$

The difference between these two formulas is that the proportions Equation 2.28 has proportions rather than means in the numerator and substitutes $p_0 q_0$ for the calculated variances. However, rather than having separate variances for the two samples as does Equation 2.27, this formula uses the same pooled $p_0 q_0$ values for both groups. That is because this is a test of the null hypothesis that the two proportions are equal, and since the variance of a binary variable is equal to pq, if the proportions are the same, so are the variances. The formula for combining the two proportions is:

$$p_0 = \frac{y_1 + y_2}{N_1 + N_2},$$

where y_1 is the number of positive cases out of N_1 total cases in the first group, and y_2 is the number of positive cases out of N_2 total cases in the second group.

As an example of the use of this formula, suppose that in the same vicarious learning study just described the investigators also test the proportion of rats who run the maze errorlessly in the experimental group and the control group. Of $N_1 = 93$ rats who completed the maze-running experiment in the control group, $y_1 = 23$ of them ran the maze without errors. Of the $N_2 = 98$ rats in the experimental group who completed the maze, $y_2 = 29$ of them did so without errors.

$$p_0 = \frac{y_1 + y_2}{N_1 + N_2} = \frac{23 + 29}{93 + 98} = \frac{52}{191} = 0.272$$

$$q_0 = 1 - p_0 = 1 - 0.272 = 0.728$$

$$P_1 = \frac{y_1}{N_1} = \frac{23}{93} = 0.247$$

$$P_2 = \frac{y_2}{N_2} = \frac{29}{98} = .296$$

$$z = \frac{P_1 - P_2}{\sqrt{\dfrac{p_0 q_0}{N_1} + \dfrac{p_0 q_0}{N_2}}} = \frac{0.247 - 0.296}{\sqrt{\dfrac{(0.272)(0.728)}{93} + \dfrac{(0.272)(0.728)}{98}}}$$

$$= \frac{-0.049}{\sqrt{\dfrac{0.1981}{93} + \dfrac{0.1981}{98}}} = \frac{-0.049}{\sqrt{.00213 + 0.00202}} = \frac{-0.049}{\sqrt{0.00415}} = \frac{-0.049}{0.0644} = 0.760$$

This z value did not even reach 1.00, and the null hypothesis can obviously not be rejected.

2.7.5   The z Test of Two Means for Correlated Samples

The formula for the z-test of two means for correlated samples, test 8, is the same as the corresponding formula for two means for independent samples (Eq. 2.27), except that the covariance term does not vanish.

$$ z = \frac{\overline{X}_1 - \overline{X}_2}{\sqrt{\dfrac{\hat{s}_1^2}{N_1} + \dfrac{\hat{s}_2^2}{N_2} - 2r\sqrt{\dfrac{\hat{s}_1^2}{N_1}}\sqrt{\dfrac{\hat{s}_2^2}{N_2}}}}. $$

This would be an unusually tedious formula to calculate, given the need to first obtain the correlation coefficient r between the two groups in addition to the means and variances. Fortunately, there is a much simpler formula based on difference scores that is mathematically equivalent to this complex formula. This formula uses two of the principles: the variances of means (principle 1, Eq. 2.23), and the variances of differences (principle 3, Eq. 2.25).

$$ z = \frac{\overline{D}}{\sqrt{\dfrac{\hat{s}_d^2}{N}}}. \tag{2.29} $$

This formula is computationally identical to the formula for test 9, the t-test of two means for correlated samples. However the t-test is for small samples, cannot be justified by the central limit theorem, and therefore makes additional assumptions that the corresponding z-test does not make. The use of the formula will be illustrated using test 9, the t-test.

2.7.6   The z Test of Two Proportions for Correlated Samples

This test also uses all three principles. This same test can be performed with a chi-square test, for which the formulas are substantially simpler than the proportions formula. It will not be presented here.

2.7.7   The t Test of a Single Mean

This test, test 3 in Table 2.1, is mathematically equivalent to test 2, the z-test for a single mean, but the logic of it differs substantially from the z-test. Whereas the z-test is a large sample test justified by the central limit theorem, the t-test is appropriate for small samples. In fact, there are three t-tests discussed here: the single sample t-test (also referred to as the one-sample location test), the t-test of two means for independent samples, and the t-test of two means for correlated samples.

Before 1908, significance testing of this kind could only be done with large samples. As demonstrated in this chapter, as sample size becomes large, the sampling distribution of means becomes Gaussian regardless of the shape of the population distribution. That makes it possible to test the significance of various values of the z-tests using tables of the normal curve. But when samples are small, the sampling distributions are not normally distributed and such tables cannot be used. William Sealy Gosset, a chemist working for the Guinness brewery in Ireland, who published under the pen name of Student, came up with a way to apply such tests to small samples. He discovered that when the population has a known shape, Gaussian, the sampling distribution of the means will have a specifiable shape. For very small samples, the shape is highly leptokurtic. He called these leptokurtic distributions t distributions, and he created distributions and tabled values for every sample size (every degrees of freedom) up to the point where the sample size is large enough to begin approximating a Gaussian (normal) distribution. The first t-test was published by "Student" in Biometrika in 1908.

Whereas all six of the z-tests only make one assumption, independence of observations, the single sample t-test makes two assumptions, (1) independence of observations and (2) a Gaussian population. Because the formula is the same as the z-test of a single mean, test 2, and because it was illustrated with example data there, it will not be illustrated with data here.

$$t = \frac{\overline{X} - \mu}{\sqrt{\dfrac{\hat{s}^2}{N}}}. \tag{2.30}$$

2.7.8 The t Test of Two Means for Independent Samples

The formula for this test, test 6 in Table 2.1, is similar to the formula for test 5, the z-test of two means for independent samples, but the way in which it differs is instructive.

$$t = \frac{\overline{X}_1 - \overline{X}_2}{\sqrt{\dfrac{\hat{s}^2_{pooled}}{N_1} + \dfrac{\hat{s}^2_{pooled}}{N_2}}}. \tag{2.31}$$

The difference in formulas is that whereas the corresponding z-test uses the two group variances in the denominator, this test uses a pooled estimate of variance in both places. That is because this test makes the assumption of homoscedasticity, equality of variances of the two groups at the population level. This test, therefore, makes three assumptions: (1) independence of observations (which all of the t- and z-tests make), (2) Gaussian population (which all three t-tests make), and (3) homoscedasticity. The pooled estimate of the variance is found by summing the SS (sums of squares) values for the two groups, and then dividing that by the sum of the degrees of freedom for the two groups.

$$S^2_{pooled} = \frac{SS_1 + SS_2}{df_1 + df_2}.$$

Suppose that in a covert test of prejudice you have two randomly assigned groups (experimental and control) instructed to read and evaluate a common essay on the question of strengthening U.S. immigration laws. Accompanying the essay is a description of the author and his credentials. The description is changed somewhat for the experimental group to indicate that the author belongs to an unpopular political group. It is hypothesized that the experimental group will rate the content of the essay lower because of the unpopular group association created in their minds. Given the following data for seven subjects in each group, calculate the independent groups t-test of the difference between the two group means.

control group	experimental group
11	11
12	7
8	7
9	6
9	5
8	5
6	1

The two means are calculated:

$$\overline{X}_1 = \frac{T_1}{N_1} = \frac{11+12+8+9+9+8+6}{7} = \frac{63}{7} = 9.$$

$$\overline{X}_2 = \frac{T_2}{N_2} = \frac{11+7+7+6+5+5+1}{7} = \frac{42}{7} = 6.$$

The two sum of squares values are obtained:

$$SS_1 = X_1'X_1 - T_1 M_1 = \begin{bmatrix} 11 & 12 & 8 & 9 & 9 & 8 & 6 \end{bmatrix} \cdot \begin{bmatrix} 11 \\ 12 \\ 8 \\ 9 \\ 9 \\ 8 \\ 6 \end{bmatrix} - 63 \cdot 7 = 591 - 567 = 24.$$

$$SS2_2 = X_2' X_2 - T_2 M_2 = \begin{bmatrix} 11 & 7 & 7 & 6 & 5 & 5 & 1 \end{bmatrix} \cdot \begin{bmatrix} 11 \\ 7 \\ 7 \\ 6 \\ 5 \\ 5 \\ 1 \end{bmatrix} - 42 \cdot 7 = 306 - 254 = 54.$$

The pooled estimate of the variance is calculated:

$$S_{pooled}^2 = \frac{SS_1 + SS_2}{df_1 + df_2} = \frac{24 + 54}{6 + 6} = \frac{78}{12} = 6.5.$$

The t-ratio is calculated from Equation 2.31.

$$t = \frac{\overline{X}_1 - \overline{X}_2}{\sqrt{\dfrac{\hat{S}_{pooled}^2}{N_1} + \dfrac{\hat{S}_{pooled}^2}{N_2}}} = \frac{9 - 6}{\sqrt{\dfrac{6.5}{7} + \dfrac{6.5}{7}}} = \frac{3}{\sqrt{\dfrac{13}{7}}} = \frac{3}{\sqrt{1.857}} = \frac{3}{1.363} = 2.20$$

The critical t ratio for 0.05 level one-tailed test with 12 degrees of freedom is found in Table B of the statistical tables at the end of the book.

$$t_{0.05, 12\ df,\ one\text{-}tailed} = 1.782.$$

The null hypothesis can be rejected at the 0.05 level.

2.7.9 The *t* Test of Two Means for Correlated Samples

The formula for this test, test 9 in Table 2.1, is the same as the formula for test 6, the z-test of two means for correlated samples. The only difference between the two tests that the sample size for which they are appropriate, the assumptions for the two tests, and the kind of table in which one evaluates the outcomes.

Whereas the correlated z-test requires a large sample size, the correlated t-test is appropriate for small sample sizes. The z-test only makes one assumption, independence of observations, but the t-test makes that assumption also plus the additional assumption of a Gaussian population, and of course the t-test result is looked up in a t table, whereas the z-test result is looked up in a table of the normal curve.

The formula is very simple indeed. Once one has created difference scores, the formula for processing them is essentially the same as that used for the single group t- or z-tests.

$$t = \frac{\overline{D}}{\sqrt{\frac{\hat{s}_d^2}{N}}}.$$

(2.32)

Difference scores are calculated either by subtracting the second set of values from the first or vice versa. One's choice may be affected by the fact that it is easier computationally to work with data when most of the values are positive. The value of t will not be affected by which group of scores is subtracted from which, only the sign, and all that is necessary is to ensure that one keeps track of whether the direction of difference is in accord with what was hypothesized.

There are two-ways data sets can be correlated. They can be repeated measures data sets, where each person is his own control and is measured twice, perhaps once under control conditions and once under experimental conditions. Of course, counterbalancing is important in this kind of design. The second way is to have randomized blocks data. The experimental and control groups are two separate groups of persons, but the subjects in the two groups are paired in some nonarbitrary way. It is only necessary to have a robust way of pairing them so that a correlation between the groups can be established in order to reduce the error term. The blocks could be husband and wife pairs, or they could be paired up by some kind of pretest that puts the two highest scores in a pair, the next two highest in another pair, and so on.

Suppose that we have a randomized blocks study where a pretest of language acquisition skill is used to create pairs of subjects, and then we test the effectiveness of two different methods of teaching German. We obtain the following data from six pairs of subjects. The larger the test score value, the stronger the performance on the test of German fluency. We predict that the second method of instruction will be superior.

	method one	method two
pair 1	8	10
pair 2	6	11
pair 3	5	9
pair 4	5	6
pair 5	4	7
pair 6	2	5

From preliminary examination of the paired data, we see that the second method does have higher scores than the first in accord with our hypothesis. We calculate the difference scores by subtracting the scores for method 1 from the scores for method 2.

$D_i = X_{i2} - X_{i1}$
10-8=2
11-6=5
9-5=4
6-5=1
7-4=3
5-2=3

The mean and then the variance of the difference scores are calculated.

$$\overline{D} = \frac{T_d}{N} = \frac{2+5+4+1+3+3}{6} = \frac{18}{6} = 3$$

$$\hat{S}_d^2 = \left(\frac{1}{n-1}\right)SS_{\text{diffscores}} = \left(\frac{1}{6-1}\right)\left(D'D - T_d\overline{D}\right)$$

$$= \left(\frac{1}{5}\right)\left([2 \quad 5 \quad 4 \quad 1 \quad 3 \quad 3] \cdot \begin{bmatrix} 2 \\ 5 \\ 4 \\ 1 \\ 3 \\ 3 \end{bmatrix} - 18 \cdot 3\right) = \left(\frac{1}{5}\right)(64-54) = \frac{10}{5} = 2.00.$$

The t-ratio is calculated from Equation 2.32.

$$t = \frac{\overline{D}}{\sqrt{\frac{\hat{S}_d^2}{N}}} = \frac{3}{\sqrt{\frac{2}{6}}} = \frac{3}{\sqrt{0.333}} = \frac{3}{0.577} = 5.20.$$

The critical t-ratio for a 0.05 level one-tailed test with 5 degrees of freedom is

$$t_{0.05, 5\,df, \text{one-tailed}} = 2.015.$$

We exceed this critical ratio, so we reject H_o at the 0.05 level.

2.7.10 Assumptions and Sampling Distributions of the Nine Tests

As shown in Table 2.2, the six z-tests have only one assumption, the independence of observation assumption. All six of them are large sample tests with a Gaussian sampling distribution, justified by the central limit theorem.

The central limit theorem makes three assertions, all three of which have been demonstrated earlier in this chapter:

(1) Regardless of the shape of the population, as sample size becomes large (at least 30 or larger), the sampling distribution of means (or proportions) will assume a Gaussian shape.

TABLE 2.2. Nine *t*- and *z*-Tests: Assumptions and Sampling Distributions

		z-Tests for Categorical Data (proportions)	Quantitative Data (Tests of Means)	
			z-Tests: Large Sample or σ Known	t-Tests: Small Sample and σ Unknown
Single sample		1. The test for a single proportion. **Assumptions:** 1. Independence of observations [No other assumptions, CLT] **Sampling distribution:** *Gaussian*	2. The z-test for a single mean. **Assumptions:** 1. Independence of observations [No other assumptions, CLT] **Sampling distribution:** *Gaussian*	3. The t-test for a single mean. **Assumptions:** 1. Independence of observations 2. Gaussian population **Sampling distribution:** *t-shaped*
Two samples	Independent samples	4. The z-test of two proportions for independent samples. **Assumptions:** 1. Independence of observations [No other assumptions, CLT] **Sampling distribution:** *Gaussian*	5. The z-test of two means for independent samples. **Assumptions:** 1. Independence of observations **Sampling distribution:** *Gaussian*	6. The t-test of two means for independent samples. **Assumptions:** 1. Independence of observations 2. Gaussian population 3. Homoscedasticity **Sampling distribution:** *t-shaped*
	Correlated samples	7. The z-test of two proportions for correlated samples. **Assumptions:** 1. Independence of observations **Sampling distribution:** *Gaussian*	8. The z-test of two means for correlated samples. **Assumptions:** 1. Independence of observations [No other assumptions, CLT] **Sampling distribution:** *Gaussian*	6. The t-test of two means for correlated samples. **Assumptions:** 1. Independence of observations 2. Gaussian population **Sampling distribution:** *t-shaped*

(2) The mean of the sampling distributions of means will be equal to the population mean: $\mu_{\bar{X}} = \mu$.
(3) The variance of the sampling distribution of means is one Nth of the variance of the raw scores in the population.

$$\sigma_{\bar{X}}^2 = \frac{\sigma^2}{N}.$$

The three t-tests are small sample tests and do not depend upon the central limit theorem to establish the shape of their sampling distributions. They make the assumption of a Gaussion population (in addition to the independence of observations assumption), and their sampling distributions are a family of leptokurtic distributions, one for each level of degrees of freedom, with lower degrees of freedom corresponding to more leptokurtic distributions, and higher levels of degrees of freedom corresponding to sampling distributions that are closer to Gaussian.

Tables 2.1 and 2.2 illustrate the connection between classical probability theory and applied statistical testing methods, such as the t-test and analysis of variance (ANOVA). In the upper left corner of these two tables are the methods that grow directly out of the binomial and the multinomial and the principles of classical probability theory—the z-test of a single proportion (which is also the normal curve approximation to the binomial) and the z-test of a single mean (which is also the normal curve approximation to the multinomial). Three principles were identified that account for the differences in the formulas of sampling distributions for the nine z and t-tests in the table. On the lower right of the table are the two-sample t-tests that make connection with ANOVA. For two-sample data, the F-test value in the ANOVA model is the square of the t-value in the corresponding t-test. ANOVA is based upon the same theoretical rationale as the t-tests, and both of them find their logical roots in classical probability theory.

2.8   MATRIX APPROACH TO ANALYSIS OF VARIANCE

In the matrix algebra approach, analysis of variance is a straightforward extension of the simple logic of variance. Consider the three-group ANOVA data below with four observations in each of the three cells, and means and totals below the cells.

group 1	group 2	group3
2	3	7
2	4	9
3	6	12
5	7	12

	group 1	group 2	group3		
T_j =	12	20	40	T =	72
M_j =	3	5	10	M =	6

These three groups can be placed in a single vector, and a total variance can be calculated on it using Equation 2.6, the matrix formula for variance.

2
2
3
5
3
4
6
7
7
9
12
12

T =	72
M =	6

$$\text{variance} = S_x^2 = \left(\frac{1}{N-1}\right) \cdot (\mathbf{X'X} - \mathbf{T} \cdot \mathbf{M}). \tag{2.6}$$

Actually, all we need is the SStotal from this stacked vector, so only part of the formula will be needed:

$$\text{SST} = (\mathbf{X'X} - \mathbf{T} \cdot \mathbf{M})$$

$$= [2 \ \ 2 \ \ 3 \ \ 5 \ \ 3 \ \ 4 \ \ 6 \ \ 7 \ \ 7 \ \ 9 \ \ 12 \ \ 12] \cdot \begin{bmatrix} 2 \\ 2 \\ 3 \\ 5 \\ 3 \\ 4 \\ 6 \\ 7 \\ 7 \\ 9 \\ 12 \\ 12 \end{bmatrix} - (72) \cdot (6)$$

$$= 570 - 432 = 138.$$

This SST (sum of squares total) ANOVA formula can be thought of as consisting of two components, the $\mathbf{X'X}$ component, which is the raw data component (which we will label Q_{data}), and the $\mathbf{TM}$ component, the product of the total **T** and the mean **M**, which we will label Q_{total}. The symbol "Q" just stands for "quantity," or individual terms in the ANOVA sums of squares equations. The SST equation just given can thus be broken down into its two terms, its two Q components.

$$Q_{tot} = \mathbf{T} \cdot \mathbf{M} = (72) \cdot (6) = 432. \tag{2.33}$$

$$Q_{data} = \mathbf{X'X} = \begin{bmatrix} 2 & 2 & 3 & 5 & 3 & 4 & 6 & 7 & 7 & 9 & 12 & 12 \end{bmatrix} \cdot \begin{bmatrix} 2 \\ 2 \\ 3 \\ 5 \\ 3 \\ 4 \\ 6 \\ 7 \\ 7 \\ 9 \\ 12 \\ 12 \end{bmatrix} = 570. \tag{2.34}$$

With one more component, the Qgroup component, we have all that we need to calculate the one-way ANOVA sum of squares formulas. It is calculated in a manner similar to the other two, but with cell group means and cell group totals as its vectors.

$$Q_{group} = T_j' M_j = \begin{bmatrix} 12 & 20 & 40 \end{bmatrix} \cdot \begin{bmatrix} 3 \\ 5 \\ 10 \end{bmatrix} = 5365. \tag{2.35}$$

The key to how to combine these Q components together to get all of the sums of squares needed for a given ANOVA model is found in the degrees of freedom. It is fairly easy, even with a complex ANOVA design, to work out the degrees of freedom for each SS component. From the various additive degrees of freedom in the model one can see what the Q components are for constructing each SS value for the ANOVA summary table. Consider this small table of only degrees of freedom and the Q components needed for each term. In the degrees of freedom expressions for this simple one-way ANOVA, the symbol N stands for all of the datapoints, so it is associated with Qraw. The number 1 corresponds to the grand mean and the grand total of which there is only one, so this is associated with Q_{total}. The subscript being used here for group means is j, and the counter for group means is k, so k is associated with the Qgroup component, as shown in the tabled comparisons of degrees of freedom and Q formulas.

Degrees of freedom and Q formulas

Source	SS	df
Between	Qgroup-Qtotal	k-1
Within	Qdata-Qgroup	N-k
Total	Qdata-Qtotal	N-1

From these SS formulas (written in terms of Q components), one can calculate each of the SS terms for the analysis of variance:

$$\text{SSB} = Q_{\text{group}} - Q_{\text{tot}} = 536 - 432 = 104. \tag{2.36}$$

$$\text{SSW} = Q_{\text{data}} - Q_{\text{group}} = 570 - 536 = 34. \tag{2.37}$$

$$\text{SST} = Q_{\text{data}} - Q_{\text{tot}} = 570 - 432 = 138. \tag{2.38}$$

The sums of squares and the degrees of freedom are collected into an ANOVA summary table, and then the mean square (MS terms) are calculated by dividing the SS values by their corresponding df values. Finally, the MS for the hypothesis being tested (between groups in this case) is divided by the MS for error (within groups in this case) to produce an F-ratio. The critical F-ratio in the tabled values corresponding to the numerator and denominator degrees of freedom is selected, and the obtained F-ratio is tested against it to determine whether there are significant intergroup differences.

ANOVA Summary Table

Source	SS	df	MS	F
Between	104	2	52	13.76
Within	34	9	3.78	
Total	138	11		

The final step in analysis of variance is to determine whether the obtained F-ratio is statistically significant. We first determine the significance level at which we wish to test the null hypothesis. Suppose that we wish to test it with an alpha level of $\alpha = 0.01$. We turn to the table of F-ratios in Table C in the statistical tables at the end of book and look up the table's value for the 0.01 level. We have 2 degrees of freedom in the numerator of the F-ratio and 9 degrees of freedom in the denominator, so we examine the second column of Table C and go down until we get to a denominator df value of 9, which is on the second page of the table. The tabled value of the 0.01 level is seen to be 8.02. The way we write this critical ratio is $F_{0.01, 2, 9} = 8.02$. We compare our obtained F-ratio of 13.76 with this, and see that we have easily exceeded the critical ratio, so we reject the null hypothesis at the 0.01 level.

2.9 SUMMARY

We have reviewed seven areas of elementary statistics that are important to have as well-practiced habits as you begin to approach the next level of statistical methods, multivariate methods, and a variety of modern regression methods. We have discussed first the four kinds of measurement scales. We will see that some of the methods in the book will deal with data that are fully quantitative, that involve interval/ratio data. Others will use categorical variables, either in the independent variables or in the dependent variables. It is important to be clear about what are the properties of each of these kinds of data as we begin the quest.

The second habit is reflected in the three types of central tendency, the mean, median, and the mode, but has to do with shapes of distributions, whether Gaussian or skewed, or with positive or negative kurtosis. Many of the methods in this book assume that the data are indeed fully quantitative and that they correspond fairly well to a Gaussian distribution. We will encounter ways to inspect data for serious departures from the expected distributional properties.

The third statistical concept that should be habitual is variance, where the variance is defined as the mean of the squared deviation scores. This is probably the most central and most crucial of the concepts, for upon it nearly all of the others are built. In connection with variance, we began to introduce matrix algebra as a simplifying computational strategy that makes the higher multivariate methods computationally possible.

Similar to variance is the fourth habit, covariance. Covariance is a generalization of the concept of variance. Whereas variance is the average, or more precisely the expected value, of the *squared* deviation scores, covariance is the expected value of the *product* of deviation scores, the product between two variables. Both variance and also the Pearson product moment correlation coefficient are special cases of covariance. The variance is the covariance of a variable with itself, and the correlation coefficient is "standardized covariance," that is, the covariance of standardized variables.

The fifth habit is the principles of classical probability theory, and more particularly the computational methods and the concepts of the binomial distribution. We learn the binomial so that we can actually understand the fundamental logic of what we are doing when we calculate a statistical significance test, and the nine t and z statistical significance tests are the sixth habit. We get the connection between classical probability theory and the significance tests when we see that each of the nine tests of Section 2.7, is in one way or another actually based upon the logic of the binomial (and its generalization the multinomial). The simplest of the z-tests is a normal curve approximation to the binomial. The binomial is an exact test. That is, it gives us exact probabilities of defined events. But with large data sets, it can become quite tedious, so we use an approximation to binomial probabilities based on the normal curve. We can use this normal curve approximation for many sampling distributions because of the *central limit theorem*. The central limit theorem holds that regardless of the shape of the population, as sample size becomes large, the sampling distributon of means will approach Gaussian shape.

The last of the seven habits is to use the analysis of variance method and to understand its concepts. It is in fact a generalization of the logic of the *t*-test, and for that reason it also makes the same assumptions as the *t*-test: independence of observations, a normally distributed population, and in the case of the independent groups ANOVA (or the independent groups *t*-test), equality of variance in the subgroups. It has been demonstrated that both the *t*-test and analysis of variance are robust with respect to violation of the normality assumption (Box, 1953), and under some circumstances robust to violation of the equality of variance assumption, such that these tests are often successfully used even when the assumptions are not strictly met. The situation is well summarized in the following abstract from Boneau (1960):

> There has been an increase lately in the use of the frequently less powerful nonparametric statistical techniques particularly when the data do not meet the assumptions for t or analysis of variance. By means of empirical analysis and a mathematical frame of reference it is demonstrated that the probability values for both t, and by generalization for F are scarcely influenced when the data do not meet the required assumptions. One exception to this conclusion is the situation where there exists unqueal variances and unequal sample sizes. In this case the probability values will be quite different from the nominal values.

We introduced the use of matrix algebra to accomplish the calculations of analysis variance more efficiently. At the beginning of Chapter 8, the chapter on multivariate extensions of analysis of variance, we will pick up where this chapter leaves off, with the matrix equations for analysis of variance. We will see that those matrix equations are easy to expand to accommodate the multivariate approach to data. We are now ready to move on to a more in-depth treatment of matrix algebra, and then to each of the multivariate and regression methods in the remaining chapters.

STUDY QUESTIONS

A. Essay Questions

1. What is variance? Make your definition so precise that one could calculate a variance from your verbal definition.

2. Many statistics are based on the concept of expected values. Explain the concepts of variance, covariance, and Pearson Product Moment Correlation Coefficient in terms of expected values.

3. Explain how the Pearson Product Moment Correlation Coefficient, the covariance, and variance are related to one another.

4. Write the formulas (definitional, computational, and matrix) for obtaining the SSCP matrix, the covariance matrix, and the correlation matrix from raw data.

5. Write the formulas (definitional, computational, and matrix) for the sums of squares for each term in an ANOVA model.

6. Explain the "centering matrix" $(I-(1/n)J)$. What is its place in the computation of a covariance matrix?

7. Explain the rationale of Student's t-test, and explain how it grows out of the logic of the z-test for means.

8. Explain how the one-way analysis of variance partitions or "decomposes" the variation, that is, the sums of squares. Also show how these partitions of SS, the partitions of the df values, and also the MS values and the F-ratio are collected into an ANOVA Summary Table.

B. Calculation Questions

1. For the simplest case data given below, calculate the variance by both of the methods given in the text, the definitional formula and the computational formula, and show that they give the same result:

X
8
6
5
5
4
2

2. Now double each of the scores in the X-score distribution of question 1 above and calculate the variance and also the standard deviation. How does each of these quantities change? Explain

3. For the simplest case data given below, calculate the Pearson correlation coefficient by four methods (the computational equation; the definitional equation; the Z-score equation; and the matrix equation):

X	Y
8	16
6	12
5	10
5	10
4	8
2	4

4. For the data matrix $\mathbf{X}$ given below, use matrix algebra to accomplish each of the successive transformations of the data from raw data to SSCP_{raw} matrix $(\mathbf{X'X})$, to $\text{SSCP}_{\text{deviation}}$ matrix, to covariance matrix $(\mathbf{S})$, to correlation matrix $(\mathbf{R})$.

$$\mathbf{X} = \begin{bmatrix} 8 & 12 & 20 \\ 6 & 8 & 14 \\ 5 & 10 & 15 \\ 5 & 16 & 21 \\ 4 & 4 & 8 \\ 2 & 10 & 12 \end{bmatrix}.$$

5. For the data below, calculate the two sample independent groups t-test. Suppose that you expect the experimental group mean to be greater than the control group mean. You therefore use a one-tailed test. Is the difference between means statistically significant (with an alpha level of 0.05)?

Control group	Experimental group
11	10
7	11
7	7
6	8
5	8
5	7
1	5

6. Calculate a one-way ANOVA (AV1) on the data in question 5 above, computing the SS values by the matrix method. Compare the results to the two-sample independent groups t-test that you calculated on question 6 above. Is the square of the t-ratio equal to the F-ratio?

7. For the data of question 5 above, write out the decomposed matrices for the linear model shown below, and show how the sum of squares of the entries in each of the component matrices is related to the SS values in the ANOVA summary table.

$$X_{ij} = \mu + \alpha_j + \varepsilon_{ij}.$$

C. Data Analysis Questions

1. For the data matrix $\mathbf{X}$ given below, use the matrix functions of Excel to accomplish each of the successive transformations of the data from raw data

to SSCP$_{raw}$ matrix (**X′X**), to SSCP$_{deviation}$ matrix, to covariance matrix (**S**), to correlation matrix (**R**). Compare your results with those from the "by hand" calculator computations in question 4 of section B above.

$$\mathbf{X} = \begin{bmatrix} 8 & 12 & 20 \\ 6 & 8 & 14 \\ 5 & 10 & 15 \\ 5 & 16 & 21 \\ 4 & 5 & 9 \\ 2 & 10 & 12 \end{bmatrix}.$$

2. Using SAS, SPSS, or Stata calculate the variances and the correlation matrix for the data in question 1 and compare your results.

3. Using SAS, SPSS, or Stata calculate the one-way ANOVA for the data in question 5 of section B. Compare your results.

REFERENCES

Boneau, C. A. 1960. The effects of violations of assumptions underlying the t test. *Psychological Bulletin*, 57(1), 49–64.

Box, G. E. P. 1953. Non-normality and tests on variance. *Biometrika*, *40*, 318–335.

Covey, S. R. 1989. *The Seven Habits of Highly Effective People*. New York: Free Press.

CHAPTER THREE

FUNDAMENTALS OF MATRIX ALGEBRA

3.1 INTRODUCTION

Matrix algebra is a branch of mathematics, usually taught in a linear algebra course. However, in this chapter, the interest in matrix algebra is not for its own sake, but only as a tool for accomplishing multivariate data analysis. Matrix algebra is a highly efficient medium of expression, and is thus uniquely well suited to the task. Many statistics (such as variance, covariance, and correlation coefficients) are direct outcomes of simple matrix operations. The strategy in this chapter is to demonstrate only as much matrix algebra as will be needed in the other chapters, and to illustrate the matrix algebra operations as much as possible using applications of them to statistical computations.

In Chapter 1, it was mentioned that multivariate statistics can be viewed simply as the same univariate methods that you already know (t-test, ANOVA, correlation, etc.), but re-written in *matrix algebra*, and then expanded to accommodate multiple dependent variables. This has a twofold advantage. First, matrix expression of most of these methods is more elegant and efficient than ordinary algebraic expression. The second reason for expressing them in matrix algebra is the most important one for our purposes: in matrix form, they can readily be converted into multivariate data analysis methods by the addition of more columns to the matrices to accommodate multiple dependent variables.

Multivariate Analysis for the Biobehavioral and Social Sciences: A Graphical Approach,
First Edition. Bruce L. Brown, Suzanne B. Hendrix, Dawson W. Hedges, Timothy B. Smith.
© 2012 John Wiley & Sons, Inc. Published 2012 by John Wiley & Sons, Inc.

3.2 DEFINITIONS AND NOTATION

A *matrix* is a rectangular or square array of numbers arranged in rows and columns. Matrices are thus two-dimensional, rows by columns. Boldface upper-case letters are used to represent matrices. A *vector* is a matrix consisting of only a single row or a single column. Vectors are thus one-dimensional. Boldface lower case letters are used to represent vectors. Column format is the standard form for vectors, and when a vector is in row form, it is referred to as the *transpose* vector. For example, the column form of a vector is referred to as **b** as shown below, and the row form of that same vector is referred to as **b′**, where the "prime" sign after the vector name indicates *transpose*. A *scalar* is a single real number. If a matrix is of size n by p (where n is the number of rows and p is the number of columns), then a vector is an n by 1 matrix or a 1 by n matrix. A scalar could be considered to be a 1 by 1 matrix.

$$b = \begin{bmatrix} 2 \\ 1 \\ 4 \end{bmatrix}, \qquad b' = \begin{bmatrix} 2 & 1 & 4 \end{bmatrix}.$$

For two matrices to be considered equal, they must have exactly the same number of rows and the same number of columns, and all of the corresponding elements must be equal. For example, matrix **D** and matrix **E** below are equal.

$$D = \begin{bmatrix} 2 & 3 & 7 \\ 4 & 1 & 3 \end{bmatrix}, \qquad E = \begin{bmatrix} 2 & 3 & 7 \\ 4 & 1 & 3 \end{bmatrix}.$$

3.3 MATRIX OPERATIONS AND STATISTICAL QUANTITIES

3.3.1 Addition and Subtraction

In order for matrices to be *conformable* for addition or subtraction, they must be of exactly the same size. That is, they must have exactly the same number of rows and columns as one another. The sum of the two matrices is then just the sum of the corresponding elements in the two matrices as shown with matrix **C** below created by summing matrix **A** and matrix **B**.

$$C = A + B = \begin{bmatrix} 5 & 3 \\ 1 & 6 \\ 3 & 7 \end{bmatrix} + \begin{bmatrix} 3 & 2 \\ 2 & 1 \\ 2 & 2 \end{bmatrix} = \begin{bmatrix} 8 & 5 \\ 3 & 7 \\ 5 & 9 \end{bmatrix}.$$

Similarly, to subtract one matrix from another, one merely subtracts each of the corresponding elements, as shown below, where matrix **D** is created by subtracting matrix **B** from matrix **A**:

$$\mathbf{D} = \mathbf{A} - \mathbf{B} = \begin{bmatrix} 5 & 3 \\ 1 & 6 \\ 3 & 7 \end{bmatrix} - \begin{bmatrix} 3 & 2 \\ 2 & 1 \\ 2 & 2 \end{bmatrix} = \begin{bmatrix} 2 & 1 \\ -1 & 5 \\ 1 & 5 \end{bmatrix}.$$

This can be readily applied to data analysis. Suppose one has a matrix of empirical data with six observations units (such as six persons) and two variables. This can be represented in a 6×2 matrix (a matrix with six rows and two columns), where the six rows represent the observation units and the two columns represent the two variables.

$$\mathbf{X} = \begin{bmatrix} 8 & 12 \\ 6 & 16 \\ 5 & 10 \\ 5 & 8 \\ 4 & 10 \\ 2 & 4 \end{bmatrix}.$$

To obtain deviation scores from these raw scores, we first obtain the mean for each variable (which is 5 for the first variable and 10 for the second), and then place these two means in a 6×2 matrix (in order to be conformable with the raw data) and subtract it from the raw data matrix.

$$\mathbf{X} - \bar{\mathbf{X}} = \begin{bmatrix} 8 & 12 \\ 6 & 16 \\ 5 & 10 \\ 5 & 8 \\ 4 & 10 \\ 2 & 4 \end{bmatrix} - \begin{bmatrix} 5 & 10 \\ 5 & 10 \\ 5 & 10 \\ 5 & 10 \\ 5 & 10 \\ 5 & 10 \end{bmatrix} = \begin{bmatrix} 3 & 2 \\ 1 & 6 \\ 0 & 0 \\ 0 & -2 \\ -1 & 0 \\ -3 & -6 \end{bmatrix} = \mathbf{x}.$$

Notice that the symbol **x** (a bold, lower case letter x) has been used to represent this 6×2 deviation score matrix. This is a minor violation of the usual convention of using upper case bold letters to represent matrices and lower case bold letters to represent vectors, but it conforms to the convention in statistical notation of using a lower case x to represent deviation scores. The statistical convention takes precedence here without too much confusion, since it is clear from the context that **x** refers to a matrix not a vector.

3.3.2 Scalar Multiplication

To multiply a matrix by a scalar, each element in the matrix is multiplied by that scalar. For example, if a 3×2 matrix **A** were multiplied by the scalar p, and p were equal to 4, the resultant matrix would contain each of the elements of **A** multiplied by 4:

$$p \cdot \mathbf{A} = 4 \cdot \begin{bmatrix} 2 & 6 \\ -1 & 4 \\ 3 & 3 \end{bmatrix} = \begin{bmatrix} 8 & 24 \\ -4 & 16 \\ 12 & 12 \end{bmatrix}.$$

3.3.3 Transpose of a Matrix

The *transpose* of a matrix is obtained by interchanging rows and columns. Thus the columns of matrix **B** are the rows of matrix **B′**, and the rows of matrix **B** are the columns of matrix **B′**. For example, consider the following matrix:

$$\mathbf{B} = \begin{bmatrix} 2 & 1 & 3 \\ 4 & 6 & 5 \\ 7 & 2 & 3 \\ 1 & 3 & 2 \end{bmatrix}.$$

This matrix is a 4×3 matrix, meaning that it has four rows and three columns. To obtain its transpose, column 1 (with four elements) becomes row 1, column 2 (also with four elements) becomes row 2, and column 3 (also with four elements) becomes row 3. The **B′** matrix thus is a 3×4 matrix, since it has three rows of four column entries as shown below:

$$\mathbf{B'} = \begin{bmatrix} 2 & 4 & 7 & 1 \\ 1 & 6 & 2 & 3 \\ 3 & 5 & 3 & 2 \end{bmatrix}.$$

3.3.4 Matrix Multiplication

The first principle of matrix multiplication is that the matrices must be *conformable* in order to multiply them. That is, matrix algebra is not like scalar algebra where any two numbers can be multiplied. Not every possible pair of matrices can be multiplied. In matrix algebra, one must first check to be sure the matrices are conformable—if they are, they can be multiplied, otherwise they cannot. To be conformable, the number of *columns* in the premultiplying matrix must be equal to the number of *rows* in the postmultiplying matrix. Matrices are specified as $r \times c$, where r is the number of rows and c is the number of columns. For example, a 2×3 matrix has two rows and three columns. Matrix **A** below is a 2×3 matrix. By contrast, a 3×4 matrix has three rows and four columns. Matrix **B** below is a 3×4 matrix.

$$\mathbf{A} = \begin{bmatrix} 5 & 3 & 7 \\ 6 & 2 & 4 \end{bmatrix}$$

$$\mathbf{B} = \begin{bmatrix} 2 & 3 & 4 & 6 \\ 1 & 3 & 2 & 1 \\ 5 & 2 & 4 & 3 \end{bmatrix}.$$

These two matrices can be multiplied as **AB**, since the number of columns in the premultiplying matrix (**A**) is 3, and the number of rows in the postmultiplying matrix (**B**) is also 3, so the two matrices are conformable to multiplying as **AB**. However, they cannot be multiplied in the reverse order **BA**, since in that order, the number of columns in the premultiplying matrix is 4, and the number of rows in the postmultiplying matrix is only 2, so the two matrices are not conformable. The resultant matrix **C**, obtained by multiplying **A** and **B** (**C** = **AB**), will have the number of rows of the premultiplying matrix and the number of columns of the postmultiplying matrix, so in this case, **C** would be a 2×4 matrix. If we write the dimensions of the two matrices **A** and **B** adjacent to one another:

$$(2 \times 3)(3 \times 4)$$

we see that the two "inner" dimensions (the two threes) agree, so the two are conformable in this order and can be multiplied. The two "outer" dimensions (the 2 and the 4) give the dimensions of the matrix that results from multiplying, that is, a 2×4 matrix will result.

Vectors are a simpler kind of matrix, so we will first use vectors to demonstrate how matrix multiplication is done. Suppose that we have a 1x3 row vector $\mathbf{r}'$:

$$\mathbf{r}' = \begin{bmatrix} 5 & 3 & 7 \end{bmatrix}.$$

And a 3×1 column vector **s**:

$$s = \begin{bmatrix} 2 \\ 1 \\ 5 \end{bmatrix}.$$

The process of premultiplying **s** by $\mathbf{r}'$ is to obtain the sum of products of corresponding values in the two vectors as shown below:

$$\mathbf{r}'s = \begin{bmatrix} 5 & 3 & 7 \end{bmatrix} \cdot \begin{bmatrix} 2 \\ 1 \\ 5 \end{bmatrix} = 5 \cdot 2 + 3 \cdot 1 + 7 \cdot 5 = 48.$$

It is now obvious why the number of columns in the premultiplying matrix must be equal to the number of rows in the postmultiplying matrix. The matrix multiplying process is to take a row from the premultiplying matrix, lay it alongside a column from the postmultiplying matrix, multiply corresponding elements, and sum them up. This can only be done if there are the same number of elements in the row of the premultiplying matrix and the column of the post-multiplying matrix. Now we will premultiply matrix **B** by matrix **A** to obtain matrix **C**, and illustrate the process when we have multiple rows in the premultiplying matrix and multiple columns in the postmultiplying matrix:

$$
\mathbf{C} = \mathbf{A} \cdot \mathbf{B} = \begin{bmatrix} 5 & 3 & 7 \\ 6 & 2 & 4 \end{bmatrix} \cdot \begin{bmatrix} 2 & 3 & 4 & 6 \\ 1 & 3 & 2 & 1 \\ 5 & 2 & 4 & 3 \end{bmatrix}
$$

$$
= \begin{bmatrix} 5 \cdot 2 + 3 \cdot 1 + 7 \cdot 5 & 5 \cdot 3 + 3 \cdot 3 + 7 \cdot 2 & 5 \cdot 4 + 3 \cdot 2 + 7 \cdot 4 & 5 \cdot 6 + 3 \cdot 1 + 7 \cdot 3 \\ 6 \cdot 2 + 2 \cdot 1 + 4 \cdot 5 & 6 \cdot 3 + 2 \cdot 3 + 4 \cdot 2 & 6 \cdot 4 + 2 \cdot 2 + 4 \cdot 4 & 6 \cdot 6 + 2 \cdot 1 + 4 \cdot 3 \end{bmatrix}
$$

$$
= \begin{bmatrix} 48 & 38 & 54 & 54 \\ 34 & 32 & 44 & 50 \end{bmatrix}.
$$

One can see why the resultant matrix has dimensions with the number of rows of the premultiplying matrix and the number of columns of the postmultiplying matrix. This premultiplying matrix **A** has two, rows, and the postmultiplying **B** has four columns, and when we obtain sums of products of the two rows of **A** with the four columns of **B**, the resulting eight *sums of products* are organized into a 2×4 matrix **C**.

The impact of the order of matrix multiplication is particularly obvious in multiplying vectors. Suppose that a 4×1 vector **f** is premultiplied by its transpose **f'**:

$$
\mathbf{f'f} = \begin{bmatrix} 3 & -1 & 2 & 4 \end{bmatrix} \cdot \begin{bmatrix} 3 \\ -1 \\ 2 \\ 4 \end{bmatrix} = 30.
$$

This is called a *dot product* because the product consists of a single scalar number. This is the product of a 1×4 vector and a 4×1 vector. The "inner" dimensions (the fours) are conformable, and the "outer" dimensions (the ones) define the size of the resultant product, a 1×1 scalar.

If the order is reversed, and vector **f** is postmultiplied by its transpose, the result is a *matrix product*, so called because the product of these two vectors is a matrix.

$$
\mathbf{f} \cdot \mathbf{f'} = \begin{bmatrix} 3 \\ -1 \\ 2 \\ 4 \end{bmatrix} \cdot \begin{bmatrix} 3 & -1 & 2 & 4 \end{bmatrix} = \begin{bmatrix} 9 & -3 & 6 & 12 \\ -3 & 1 & -2 & -4 \\ 6 & -2 & 4 & 8 \\ 12 & -4 & 8 & 16 \end{bmatrix}.
$$

In this arrangement, the multiplication is a 4×1 vector times a 1×4 vector. The "inner" dimensions (the ones) conform, and the "outer" dimensions (the fours) define the size of the resultant product, a 4×4 matrix. Order makes all the difference. In ordinary algebra, the *commutative law* holds, and $a \cdot b = b \cdot a$, but clearly this law does not hold for matrix algebra.

Variance is one of the central concepts of statistics. It is defined as the sum of squared deviation scores divided by the degrees of freedom, or algebraically as:

$$s^2 = \frac{\sum x^2}{N-1}.$$

Another central concept in statistics is covariance, which is defined as the sum of products of deviation scores (across two variables) divided by the degrees of freedom. Algebraically this is:

$$\text{cov} = \frac{\sum x \cdot y}{N-1}.$$

Covariance is the more general of these two concepts. Likewise, the covariance formula is the more general of the two formulas. Variance can be thought of as a special case of covariance—the covariance of a variable with itself. In other words, when the x and the y are the same variable, then $\sum xy = \sum xx = \sum x^2$ and the covariance formula becomes the formula for variance. The "computational engine" for both variance and covariance is the *sum of products*, but in the case of variance, since the x and the y are the same variable, it is the *sum of squares*, which is a special case of the sum of products. As just demonstrated, the essence of matrix multiplication is to fill the product matrix with *sums of products* of the rows of the premultiplying matrix and the columns of the postmultiplying matrix. Matrix multiplication is thus an elegant and efficient way of calculating variances and covariance simultaneously. This will next be illustrated.

Consider the 6×2 deviation score matrix (with six observation units and two variables) that was calculated in Section 3.3.1 above:

$$x = \begin{bmatrix} 3 & 2 \\ 1 & 6 \\ 0 & 0 \\ 0 & -2 \\ -1 & 0 \\ -3 & -6 \end{bmatrix}.$$

If it is premultiplied by its transpose, the resultant matrix **B** contains in its two diagonal elements the *sum of squares* for each of the two variables (20 and 80), and the off-diagonal elements contain the *sum of products* for the two variables (30).

$$\text{CSSCP} = \mathbf{B} = x'x = \begin{bmatrix} 3 & 1 & 0 & 0 & -1 & -3 \\ 2 & 6 & 0 & -2 & 0 & -6 \end{bmatrix} \begin{bmatrix} 3 & 2 \\ 1 & 6 \\ 0 & 0 \\ 0 & -2 \\ -1 & 0 \\ -3 & -6 \end{bmatrix} = \begin{bmatrix} 20 & 30 \\ 30 & 80 \end{bmatrix}.$$

These are the numerators from the variance formula (in the diagonal elements) and from the covariance formula (in the off-diagonal elements), respectively. This is referred to as a *CSSCP* matrix, which is an acronym for *centered sum of squares and cross-products* matrix. "Cross-products" is a statistical term meaning "sum of products." *Centered* refers to the fact that it is created from deviation scores (which are centered around zero). All that needs to be done to this matrix to convert it into a variance/covariance matrix is to divide each element by the degrees of freedom, $N - 1$, to convert them into variances and covariances. This will be shown in the next section.

3.3.5 Division by a Scalar

There is actually no operation of dividing one matrix by another. However, the equivalent of division in matrix algebra is to multiply one matrix by the inverse of another. Section 3.7 below explains how to obtain the inverse of a matrix. This is not a trivial problem. Algorithms for obtaining matrix inverses are one of the major achievements in matrix algebra.

However, the inverse of a scalar *is* easy to accomplish, and division of a matrix by a scalar is correspondingly easy. For example, the inverse of 8 is 1/8. As shown in Section 3.3.2, to multiply a matrix by a scalar, one simply multiplies each element of the matrix by that scalar. Dividing a matrix by a scalar therefore consists of dividing each element of the matrix by the scalar, or, equivalently, multiplying each element of the matrix by the inverse of that scalar. As an example, to divide matrix $\mathbf{D}$ by eight, one multiplies each element of $\mathbf{D}$ by 1/8:

$$(1/8) \cdot \mathbf{D} = (1/8) \cdot \begin{bmatrix} 12 & 10 & 9 \\ 5 & 16 & 7 \\ 8 & 6 & 24 \end{bmatrix} = \begin{bmatrix} 1.5 & 1.3 & 1.1 \\ 0.6 & 2 & 0.9 \\ 1 & 0.8 & 3 \end{bmatrix}.$$

This process can be used to convert a CSSCP matrix into a variance/covariance matrix. One simply multiplies the CSSCP matrix by the inverse of the degrees of freedom. For the CSSCP matrix given in Section 3.3.4 above, there were six units of observation in the raw data matrix **X**, so the degrees of freedom is $N - 1 = 6 - 1 = 5$. The variance/covariance matrix **C** is therefore found by taking 1/5th of the CSSCP matrix **B**:

$$\mathbf{C} = (1/5) \cdot \mathbf{B} = (1/5) \cdot \begin{bmatrix} 20 & 30 \\ 30 & 80 \end{bmatrix} = \begin{bmatrix} 4 & 6 \\ 6 & 16 \end{bmatrix}.$$

Since a variance is a special case of a covariance, a variance/covariance matrix could be referred to simply as a *covariance matrix*. That convention will be followed throughout the remainder of the book.

From the first diagonal element of this covariance matrix, it can be seen that the variance of variable one is 4. This accords with what is found from the algebraic formula:

$$s_x^2 = \frac{\sum x^2}{N-1} = \frac{\left[3^2 + 1^2 + 0^2 + 0^2 + (-1)^2 + (-3)^2 \right]}{6-1} = \frac{20}{5} = 4.$$

The second diagonal element gives the variance of variable two as 16, which also accords with the value obtained using the familiar variance formula:

$$s_y^2 = \frac{\sum y^2}{N-1} = \frac{\left[2^2 + 6^2 + 0^2 + (-2)^2 + 0^2 + (-6)^2 \right]}{6-1} = \frac{80}{5} = 16.$$

The off-diagonal elements each give the covariance between the two variables as 6, which accords with the value obtained from the formula for covariance:

$$\text{cov}_{xy} = \frac{\sum x \cdot y}{N-1} = \frac{\left[(3) \cdot (2) + (1) \cdot (6) + (0) \cdot (0) + (0) \cdot (-2) + (-1) \cdot (0) + (-3) \cdot (-6) \right]}{6-1}$$
$$= \frac{30}{5} = 6.$$

Covariance matrices are always "mirror image," with the elements above the diagonal being equal to the corresponding elements below the diagonal, because $\text{cov}_{xy} = \text{cov}_{yx}$.

In this section, it has been demonstrated that one can divide a matrix by a scalar by multiplying the matrix by the inverse of that scalar. In Section 3.7, several algorithms will be demonstrated for the much more difficult task of obtaining the inverse of a matrix. When one matrix, for example, matrix **H**, is multiplied by the inverse of another matrix, for example, matrix $\mathbf{E}^{-1}$, this is the matrix algebra analog of dividing matrix **H** by matrix **E**.

3.3.6 Symmetric Matrices and Diagonal Matrices

A symmetric matrix is a square matrix in which the upper triangular portion (the portion above the diagonal) is the mirror image of the lower triangular portion. A symmetric matrix can be produced by either premultiplying or postmultiplying a matrix by its transpose. For example, consider the 4×2 matrix $\mathbf{A}$ given below:

$$\mathbf{A} = \begin{bmatrix} 6 & 2 \\ 3 & 5 \\ 4 & 3 \\ 1 & 2 \end{bmatrix}.$$

When this matrix is premultiplied by its transpose, the resultant matrix is a 2×2 square symmetric matrix. The single value above the diagonal is equivalent to the single value below the diagonal. The matrix is therefore symmetrical.

$$\mathbf{A'A} = \begin{bmatrix} 6 & 3 & 4 & 1 \\ 2 & 5 & 3 & 2 \end{bmatrix} \cdot \begin{bmatrix} 6 & 2 \\ 3 & 5 \\ 4 & 3 \\ 1 & 2 \end{bmatrix} = \begin{bmatrix} 62 & 41 \\ 41 & 42 \end{bmatrix}.$$

Similarly, when this matrix is postmultiplied by its transpose, the resultant matrix is a 4×4 square symmetric matrix.

$$\mathbf{AA'} = \begin{bmatrix} 40 & 28 & 30 & 10 \\ 28 & 34 & 27 & 13 \\ 30 & 27 & 25 & 10 \\ 10 & 13 & 10 & 5 \end{bmatrix}.$$

The six values in the upper triangular portion of the matrix are equivalent to the six values in the lower triangular portion. That is, the upper triangular portion of the matrix is the mirror image of the lower triangular portion. This is always true. Multiplying a matrix by its transpose in either order will always produce a symmetric matrix, with a lower triangular portion that will be the mirror image of the upper triangular portion.

A diagonal matrix is a square matrix that has zeros in all of the off-diagonal entries. For example, the matrix $\mathbf{E}$ below is a 4×4 diagonal matrix:

$$\mathbf{E} = \begin{bmatrix} 3 & 0 & 0 & 0 \\ 0 & 5 & 0 & 0 \\ 0 & 0 & 6 & 0 \\ 0 & 0 & 0 & 4 \end{bmatrix}.$$

Diagonalization is an operation that can be performed on any square matrix. For example, we might begin with a matrix $\mathbf{A}$:

$$\mathbf{A} = \begin{bmatrix} 4 & 5 & 2 \\ 5 & 16 & 7 \\ 2 & 7 & 9 \end{bmatrix}.$$

that we wish to diagonalize:

$$\mathbf{D} = \text{diag}(\mathbf{A}) = \begin{bmatrix} 4 & 0 & 0 \\ 0 & 16 & 0 \\ 0 & 0 & 9 \end{bmatrix}.$$

Matrix $\mathbf{D}$ that was just created is the *diagonalized* form of matrix $\mathbf{A}$. It is produced by simply zeroing out every nondiagonal element of $\mathbf{A}$. Diagonal matrices can be used to weight rows or columns of other matrices. Premultiplying by a diagonal matrix weights the rows, and postmultiplying by a diagonal matrix weights the columns. Suppose, for example, that you wish to double the elements in the first row of matrix $\mathbf{G}$, triple the elements in the second row, and quadruple the elements in the third row. You premultiply $\mathbf{G}$ by a diagonal matrix $\mathbf{H}$ that has those weights (2, 3, and 4), and it has the desired effect:

$$\mathbf{H} \cdot \mathbf{G} = \begin{bmatrix} 2 & 0 & 0 \\ 0 & 3 & 0 \\ 0 & 0 & 4 \end{bmatrix} \cdot \begin{bmatrix} 1 & 2 & 3 \\ 4 & 5 & 6 \\ 7 & 8 & 9 \end{bmatrix} = \begin{bmatrix} 2 & 4 & 6 \\ 12 & 15 & 18 \\ 28 & 32 & 36 \end{bmatrix}.$$

If the matrix $\mathbf{G}$ is postmultiplied by the diagonal weighting matrix $\mathbf{H}$, it multiplies each of the three columns of $\mathbf{G}$ by those weights:

$$\mathbf{G} \cdot \mathbf{H} = \begin{bmatrix} 1 & 2 & 3 \\ 4 & 5 & 6 \\ 7 & 8 & 9 \end{bmatrix} \cdot \begin{bmatrix} 2 & 0 & 0 \\ 0 & 3 & 0 \\ 0 & 0 & 4 \end{bmatrix} = \begin{bmatrix} 2 & 6 & 12 \\ 8 & 15 & 24 \\ 14 & 24 & 36 \end{bmatrix}.$$

Diagonal matrices are particularly useful in converting a covariance matrix into a correlation matrix. By definition, the Pearson product moment correlation coefficient is a *standardized covariance*. That is, it is the covariance of standard scores:

$$r = \frac{\sum z_x z_y}{N-1}.$$

In this formula, standardization is accomplished before the *sum of products* operation and dividing by $N - 1$. Alternately, one can first calculate a covariance and then standardize it after the fact by just dividing it by its two standard deviations, the one for x and the one for y:

$$r_{xy} = \frac{\text{cov}_{xy}}{s_x s_y}.$$

For the data matrix given in Section 3.3.1, the covariance was found to be 6 (in Section 3.3.5), and the two variances were found to be 4 and 16 (giving standard deviations of 2 and 4 for x and y, respectively). The correlation coefficient between variable x and variable y is therefore:

$$r_{xy} = \frac{\text{cov}_{xy}}{s_x s_y} = \frac{6}{(2) \cdot (4)} = \frac{6}{8} = 0.75.$$

To accomplish this same kind of standardization for an entire covariance matrix using matrix algebra, requires a *standardizing matrix*, **E**, which is a diagonal matrix of reciprocals of standard deviations. The first step is to diagonalize the covariance matrix **C**:

$$\text{diag}(\mathbf{C}) = \text{diag} \begin{bmatrix} 4 & 6 \\ 6 & 16 \end{bmatrix} = \begin{bmatrix} 4 & 0 \\ 0 & 16 \end{bmatrix}.$$

The next step is to obtain the square root of this diagonal matrix. It is quite difficult to find the square root of most matrices, but the square root of a diagonal matrix is found simply by taking the square root of each element in the diagonal:

$$\text{sqrt}(\text{diag}(\mathbf{C})) = \text{sqrt} \cdot \begin{bmatrix} 4 & 0 \\ 0 & 16 \end{bmatrix} = \begin{bmatrix} 2 & 0 \\ 0 & 4 \end{bmatrix}.$$

Finally, the inverse of this matrix is obtained. It is quite difficult to find the inverse of most matrices, but the inverse of a diagonal matrix is found simply by taking the inverse of each element in the diagonal of the matrix:

$$\mathbf{E} = \text{inverse}(\text{sqrt}(\text{diag}(\mathbf{C}))) = \text{inverse} \cdot \begin{bmatrix} 2 & 0 \\ 0 & 4 \end{bmatrix} = \begin{bmatrix} 1/2 & 0 \\ 0 & 1/4 \end{bmatrix}.$$

When the covariance matrix is both premultiplied and also postmultiplied by this standardizing matrix **D**, the resultant matrix is a correlation matrix **R**.

$$\mathbf{R} = \mathbf{E} \cdot \mathbf{C} \cdot \mathbf{E} = \begin{bmatrix} 1/2 & 0 \\ 0 & 1/4 \end{bmatrix} \cdot \begin{bmatrix} 4 & 6 \\ 6 & 16 \end{bmatrix} \cdot \begin{bmatrix} 1/2 & 0 \\ 0 & 1/4 \end{bmatrix} = \begin{bmatrix} 2 & 3 \\ 1.5 & 4 \end{bmatrix} \cdot \begin{bmatrix} 1/2 & 0 \\ 0 & 1/4 \end{bmatrix}$$

$$= \begin{bmatrix} 1 & 0.75 \\ 0.75 & 1 \end{bmatrix}.$$

Premultiplying by D standardizes the C matrix by rows, and postmultiplying by D standardizes it by columns. The obtained correlation coefficient between the first variable and the second variable of 0.75 agrees with what was obtained from the algebraic formula. This method is particularly efficient when there are three or more variables:

$$
\mathbf{R} = \mathbf{E} \cdot \mathbf{C} \cdot \mathbf{E} =
\begin{bmatrix} 1/2 & 0 & 0 \\ 0 & 1/4 & 0 \\ 0 & 0 & 1/3 \end{bmatrix}
\cdot
\begin{bmatrix} 4 & 6 & 4 \\ 6 & 16 & 6 \\ 4 & 6 & 9 \end{bmatrix}
\cdot
\begin{bmatrix} 1/2 & 0 & 0 \\ 0 & 1/4 & 0 \\ 0 & 0 & 1/3 \end{bmatrix}
$$

$$
=
\begin{bmatrix} 2 & 3 & 2 \\ 1.5 & 4 & 1.5 \\ 1.33 & 2 & 3 \end{bmatrix}
\cdot
\begin{bmatrix} 1/2 & 0 & 0 \\ 0 & 1/4 & 0 \\ 0 & 0 & 1/3 \end{bmatrix}
=
\begin{bmatrix} 1 & 0.75 & 0.67 \\ 0.75 & 1 & 0.5 \\ 0.67 & 0.5 & 1 \end{bmatrix}.
$$

With three variables, as shown, there are three correlation coefficients above and three below the diagonal. With four variables, there would be six correlation coefficients above and six below the diagonal, and so on.

3.3.7 The Identity Matrix and the *J* Matrix

An *identity matrix* is a diagonal matrix with ones down the diagonal. It is represented by the matrix symbol **I**. Since **I** is a diagonal matrix, it is a square matrix. The reason it is called an identity matrix is because when some other matrix is multiplied by it, the other matrix remains unchanged, that is, it retains its identity.

The identity matrix is the matrix algebra analogue of the number 1. Just as multiplying any number by 1 does not change the number, that is, its identity remains unchanged:

$$8 \times 1 = 8,$$

so, analogously, multiplying any matrix by **I** leaves the matrix unchanged:

$$
\mathbf{C} \cdot \mathbf{I} =
\begin{bmatrix} 4 & 3 & 6 \\ 3 & 9 & 4 \\ 6 & 4 & 16 \end{bmatrix}
\cdot
\begin{bmatrix} 1 & 0 & 0 \\ 0 & 1 & 0 \\ 0 & 0 & 1 \end{bmatrix}
=
\begin{bmatrix} 4 & 3 & 6 \\ 3 & 9 & 4 \\ 6 & 4 & 16 \end{bmatrix}
= \mathbf{C}.
$$

The identity matrix has only one parameter, the order of the matrix. Since it is square, the order gives both the number of rows and also the number of columns. And the order must be selected to be conformable with the matrix that is to be multiplied by it. For example, if one has a 6×3 matrix **Y** and one wishes to multiply it by an identity matrix, the order of the identity matrix will depend on whether one wishes to premultiply the **Y** matrix by it or postmultiply the **Y** matrix by it. To postmultiply **Y** by **I**, **I** must be a 3×3 matrix (a matrix of order 3):

$$\mathbf{Y \cdot I} = \begin{bmatrix} 9 & 5 & 5 \\ 7 & 4 & 11 \\ 6 & 5 & 10 \\ 6 & 8 & 9 \\ 5 & 6 & 7 \\ 3 & 2 & 6 \end{bmatrix} \cdot \begin{bmatrix} 1 & 0 & 0 \\ 0 & 1 & 0 \\ 0 & 0 & 1 \end{bmatrix} = \begin{bmatrix} 9 & 5 & 5 \\ 7 & 4 & 11 \\ 6 & 5 & 10 \\ 6 & 8 & 9 \\ 5 & 6 & 7 \\ 3 & 2 & 6 \end{bmatrix} = \mathbf{Y}.$$

To premultiply **Y** by **I**, **I** must be a matrix of order 6:

$$\mathbf{I \cdot Y} = \begin{bmatrix} 1 & 0 & 0 & 0 & 0 & 0 \\ 0 & 1 & 0 & 0 & 0 & 0 \\ 0 & 0 & 1 & 0 & 0 & 0 \\ 0 & 0 & 0 & 1 & 0 & 0 \\ 0 & 0 & 0 & 0 & 1 & 0 \\ 0 & 0 & 0 & 0 & 0 & 1 \end{bmatrix} \cdot \begin{bmatrix} 9 & 5 & 5 \\ 7 & 4 & 11 \\ 6 & 5 & 10 \\ 6 & 8 & 9 \\ 5 & 6 & 7 \\ 3 & 2 & 6 \end{bmatrix} = \begin{bmatrix} 9 & 5 & 5 \\ 7 & 4 & 11 \\ 6 & 5 & 10 \\ 6 & 8 & 9 \\ 5 & 6 & 7 \\ 3 & 2 & 6 \end{bmatrix} = \mathbf{Y}.$$

The **J** matrix is a matrix of ones. When a matrix such as the 6×3 matrix **Y** just shown is pre-multiplied by a 6×6 **J** matrix, the effect is to create a 6×3 matrix of the sums of each of the columns **Y** (where each row is just a repeat of these three column sums).

$$\mathbf{J \cdot Y} = \begin{bmatrix} 1 & 1 & 1 & 1 & 1 & 1 \\ 1 & 1 & 1 & 1 & 1 & 1 \\ 1 & 1 & 1 & 1 & 1 & 1 \\ 1 & 1 & 1 & 1 & 1 & 1 \\ 1 & 1 & 1 & 1 & 1 & 1 \\ 1 & 1 & 1 & 1 & 1 & 1 \end{bmatrix} \cdot \begin{bmatrix} 9 & 5 & 5 \\ 7 & 4 & 11 \\ 6 & 5 & 10 \\ 6 & 8 & 9 \\ 5 & 6 & 7 \\ 3 & 2 & 6 \end{bmatrix} = \begin{bmatrix} 36 & 30 & 48 \\ 36 & 30 & 48 \\ 36 & 30 & 48 \\ 36 & 30 & 48 \\ 36 & 30 & 48 \\ 36 & 30 & 48 \end{bmatrix} = \mathbf{T}.$$

If one wished to simply obtain the column sums, premultiplying by a 1×6 **j′** vector would give a single vector of the column sums, which would usually suffice:

$$\mathbf{j' \cdot Y} = \begin{bmatrix} 1 & 1 & 1 & 1 & 1 & 1 \end{bmatrix} \cdot \begin{bmatrix} 9 & 5 & 5 \\ 7 & 4 & 11 \\ 6 & 5 & 10 \\ 6 & 8 & 9 \\ 5 & 6 & 7 \\ 3 & 2 & 6 \end{bmatrix} = \begin{bmatrix} 36 & 30 & 48 \end{bmatrix} = \mathbf{t'}.$$

Similarly, postmultiplying **Y** by a 3×3 **J** matrix creates a 6×3 matrix of row sums of **Y** (where each column is just a repeat of the same six row sums).

$$
\mathbf{Y} \cdot \mathbf{J} =
\begin{bmatrix}
9 & 5 & 5 \\
7 & 4 & 11 \\
6 & 5 & 10 \\
6 & 8 & 9 \\
5 & 6 & 7 \\
3 & 2 & 6
\end{bmatrix}
\cdot
\begin{bmatrix}
1 & 1 & 1 \\
1 & 1 & 1 \\
1 & 1 & 1
\end{bmatrix}
=
\begin{bmatrix}
19 & 19 & 19 \\
22 & 22 & 22 \\
21 & 21 & 21 \\
23 & 23 & 23 \\
18 & 18 & 18 \\
11 & 11 & 11
\end{bmatrix}
= \mathbf{T}.
$$

If one wished to simply obtain the row sums, postmultiplying by a 3×1 **j** vector would give a single vector of the row sums:

$$
\mathbf{Y} \cdot \mathbf{j} =
\begin{bmatrix}
9 & 5 & 5 \\
7 & 4 & 11 \\
6 & 5 & 10 \\
6 & 8 & 9 \\
5 & 6 & 7 \\
3 & 2 & 6
\end{bmatrix}
\cdot
\begin{bmatrix}
1 \\
1 \\
1
\end{bmatrix}
=
\begin{bmatrix}
19 \\
22 \\
21 \\
23 \\
18 \\
11
\end{bmatrix}
= \mathbf{t}.
$$

There are occasions where a **J** matrix is needed rather than a **j** vector, because repeats of the sums are needed. An example of this use of a **J** matrix is shown in which we create a matrix of deviation scores from a matrix of raw scores:

$$
y = \mathbf{Y} - (1/n)\mathbf{J} \cdot \mathbf{Y} =
\begin{bmatrix}
9 & 5 & 5 \\
7 & 4 & 11 \\
6 & 5 & 10 \\
6 & 8 & 9 \\
5 & 6 & 7 \\
3 & 2 & 6
\end{bmatrix}
- (1/6) \cdot
\begin{bmatrix}
1 & 1 & 1 & 1 & 1 & 1 \\
1 & 1 & 1 & 1 & 1 & 1 \\
1 & 1 & 1 & 1 & 1 & 1 \\
1 & 1 & 1 & 1 & 1 & 1 \\
1 & 1 & 1 & 1 & 1 & 1 \\
1 & 1 & 1 & 1 & 1 & 1
\end{bmatrix}
\cdot
\begin{bmatrix}
9 & 5 & 5 \\
7 & 4 & 11 \\
6 & 5 & 10 \\
6 & 8 & 9 \\
5 & 6 & 7 \\
3 & 2 & 6
\end{bmatrix}
$$

$$
=
\begin{bmatrix}
9 & 5 & 5 \\
7 & 4 & 11 \\
6 & 5 & 10 \\
6 & 8 & 9 \\
5 & 6 & 7 \\
3 & 2 & 6
\end{bmatrix}
- (1/6) \cdot
\begin{bmatrix}
36 & 30 & 48 \\
36 & 30 & 48 \\
36 & 30 & 48 \\
36 & 30 & 48 \\
36 & 30 & 48 \\
36 & 30 & 48
\end{bmatrix}
=
\begin{bmatrix}
9 & 5 & 5 \\
7 & 4 & 11 \\
6 & 5 & 10 \\
6 & 8 & 9 \\
5 & 6 & 7 \\
3 & 2 & 6
\end{bmatrix}
-
\begin{bmatrix}
6 & 5 & 8 \\
6 & 5 & 8 \\
6 & 5 & 8 \\
6 & 5 & 8 \\
6 & 5 & 8 \\
6 & 5 & 8
\end{bmatrix}
$$

$$
=
\begin{bmatrix}
3 & 0 & -3 \\
1 & -1 & 3 \\
0 & 0 & 2 \\
0 & 3 & 1 \\
-1 & 1 & -1 \\
-3 & -3 & -2
\end{bmatrix}.
$$

3.3.7.1 *Obtaining Covariance Matrices Using I and J* A method will now be derived and demonstrated for obtaining covariance matrices directly from raw data (rather than from deviation scores) using an identity matrix and a **J** vector. The formula just used for obtaining deviation scores from raw scores can be reworked to obtain an equation for a *centering matrix*, that is, a matrix that will "center" the raw scores by turning them into deviation scores (which are centered around zero):

$$y = \mathbf{Y} - (1/n)\mathbf{J} \cdot \mathbf{Y} = \mathbf{I} \cdot \mathbf{Y} - (1/n)\mathbf{J} \cdot \mathbf{Y} = [\mathbf{I} - (1/n)\mathbf{J}] \cdot \mathbf{Y} = \mathbf{Q}_{\text{centering}} \cdot \mathbf{Y}.$$

From this we see that the centering matrix (symbolized as $\mathbf{Q}_{\text{centering}}$) can be obtained by the formula:

$$\mathbf{Q}_{\text{centering}} = \mathbf{I} - (1/n)\mathbf{J}.$$

This matrix is an *n* by *n* matrix, which for the 6×3 raw data matrix just given is a 6×6 matrix:

$$\mathbf{Q}_{\text{centering}} = \mathbf{I} - (1/n)\mathbf{J} = \begin{bmatrix} 1 & 0 & 0 & 0 & 0 & 0 \\ 0 & 1 & 0 & 0 & 0 & 0 \\ 0 & 0 & 1 & 0 & 0 & 0 \\ 0 & 0 & 0 & 1 & 0 & 0 \\ 0 & 0 & 0 & 0 & 1 & 0 \\ 0 & 0 & 0 & 0 & 0 & 1 \end{bmatrix} - (1/6) \begin{bmatrix} 1 & 1 & 1 & 1 & 1 & 1 \\ 1 & 1 & 1 & 1 & 1 & 1 \\ 1 & 1 & 1 & 1 & 1 & 1 \\ 1 & 1 & 1 & 1 & 1 & 1 \\ 1 & 1 & 1 & 1 & 1 & 1 \\ 1 & 1 & 1 & 1 & 1 & 1 \end{bmatrix}.$$

$$= \begin{bmatrix} 1 & 0 & 0 & 0 & 0 & 0 \\ 0 & 1 & 0 & 0 & 0 & 0 \\ 0 & 0 & 1 & 0 & 0 & 0 \\ 0 & 0 & 0 & 1 & 0 & 0 \\ 0 & 0 & 0 & 0 & 1 & 0 \\ 0 & 0 & 0 & 0 & 0 & 1 \end{bmatrix} - \begin{bmatrix} \frac{1}{6} & \frac{1}{6} & \frac{1}{6} & \frac{1}{6} & \frac{1}{6} & \frac{1}{6} \\ \frac{1}{6} & \frac{1}{6} & \frac{1}{6} & \frac{1}{6} & \frac{1}{6} & \frac{1}{6} \\ \frac{1}{6} & \frac{1}{6} & \frac{1}{6} & \frac{1}{6} & \frac{1}{6} & \frac{1}{6} \\ \frac{1}{6} & \frac{1}{6} & \frac{1}{6} & \frac{1}{6} & \frac{1}{6} & \frac{1}{6} \\ \frac{1}{6} & \frac{1}{6} & \frac{1}{6} & \frac{1}{6} & \frac{1}{6} & \frac{1}{6} \\ \frac{1}{6} & \frac{1}{6} & \frac{1}{6} & \frac{1}{6} & \frac{1}{6} & \frac{1}{6} \end{bmatrix}$$

$$
= \begin{bmatrix}
\dfrac{5}{6} & -\dfrac{1}{6} & -\dfrac{1}{6} & -\dfrac{1}{6} & -\dfrac{1}{6} & -\dfrac{1}{6} \\
-\dfrac{1}{6} & \dfrac{5}{6} & -\dfrac{1}{6} & -\dfrac{1}{6} & -\dfrac{1}{6} & -\dfrac{1}{6} \\
-\dfrac{1}{6} & -\dfrac{1}{6} & \dfrac{5}{6} & -\dfrac{1}{6} & -\dfrac{1}{6} & -\dfrac{1}{6} \\
-\dfrac{1}{6} & -\dfrac{1}{6} & -\dfrac{1}{6} & \dfrac{5}{6} & -\dfrac{1}{6} & -\dfrac{1}{6} \\
-\dfrac{1}{6} & -\dfrac{1}{6} & -\dfrac{1}{6} & -\dfrac{1}{6} & \dfrac{5}{6} & -\dfrac{1}{6} \\
-\dfrac{1}{6} & -\dfrac{1}{6} & -\dfrac{1}{6} & -\dfrac{1}{6} & -\dfrac{1}{6} & \dfrac{5}{6}
\end{bmatrix}
$$

When n is large, this matrix can become quite unwieldy, and it is not an efficient computational strategy. It is presented here because it has important theoretical properties that lead to efficient computational strategy. Note that is does, in fact, center matrix $\mathbf{Y}$ by turning it into a matrix of deviation scores $\mathbf{y}$, when $\mathbf{Y}$ is premultiplied by it:

$$
\mathbf{y} = \mathbf{Q}_{\text{centering}} \cdot \mathbf{Y} =
\begin{bmatrix}
\dfrac{5}{6} & -\dfrac{1}{6} & -\dfrac{1}{6} & -\dfrac{1}{6} & -\dfrac{1}{6} & -\dfrac{1}{6} \\
-\dfrac{1}{6} & \dfrac{5}{6} & -\dfrac{1}{6} & -\dfrac{1}{6} & -\dfrac{1}{6} & -\dfrac{1}{6} \\
-\dfrac{1}{6} & -\dfrac{1}{6} & \dfrac{5}{6} & -\dfrac{1}{6} & -\dfrac{1}{6} & -\dfrac{1}{6} \\
-\dfrac{1}{6} & -\dfrac{1}{6} & -\dfrac{1}{6} & \dfrac{5}{6} & -\dfrac{1}{6} & -\dfrac{1}{6} \\
-\dfrac{1}{6} & -\dfrac{1}{6} & -\dfrac{1}{6} & -\dfrac{1}{6} & \dfrac{5}{6} & -\dfrac{1}{6} \\
-\dfrac{1}{6} & -\dfrac{1}{6} & -\dfrac{1}{6} & -\dfrac{1}{6} & -\dfrac{1}{6} & \dfrac{5}{6}
\end{bmatrix}
\cdot
\begin{bmatrix}
9 & 5 & 5 \\
7 & 4 & 11 \\
6 & 5 & 10 \\
6 & 8 & 9 \\
5 & 6 & 7 \\
3 & 2 & 6
\end{bmatrix}
$$

$$
= \begin{bmatrix}
3 & 0 & -3 \\
1 & -1 & 3 \\
0 & 0 & 2 \\
0 & 3 & 1 \\
-1 & 1 & -1 \\
-3 & -3 & -2
\end{bmatrix}.
$$

This centering matrix has an interesting property. When you square it by multiplying it by itself, the product is equal to the root. That is, when you multiply it by itself, it gives back itself as the product:

$$\mathbf{Q} \cdot \mathbf{Q} = \begin{bmatrix} \frac{5}{6} & -\frac{1}{6} & -\frac{1}{6} & -\frac{1}{6} & -\frac{1}{6} & -\frac{1}{6} \\ -\frac{1}{6} & \frac{5}{6} & -\frac{1}{6} & -\frac{1}{6} & -\frac{1}{6} & -\frac{1}{6} \\ -\frac{1}{6} & -\frac{1}{6} & \frac{5}{6} & -\frac{1}{6} & -\frac{1}{6} & -\frac{1}{6} \\ -\frac{1}{6} & -\frac{1}{6} & -\frac{1}{6} & \frac{5}{6} & -\frac{1}{6} & -\frac{1}{6} \\ -\frac{1}{6} & -\frac{1}{6} & -\frac{1}{6} & -\frac{1}{6} & \frac{5}{6} & -\frac{1}{6} \\ -\frac{1}{6} & -\frac{1}{6} & -\frac{1}{6} & -\frac{1}{6} & -\frac{1}{6} & \frac{5}{6} \end{bmatrix} \cdot \begin{bmatrix} \frac{5}{6} & -\frac{1}{6} & -\frac{1}{6} & -\frac{1}{6} & -\frac{1}{6} & -\frac{1}{6} \\ -\frac{1}{6} & \frac{5}{6} & -\frac{1}{6} & -\frac{1}{6} & -\frac{1}{6} & -\frac{1}{6} \\ -\frac{1}{6} & -\frac{1}{6} & \frac{5}{6} & -\frac{1}{6} & -\frac{1}{6} & -\frac{1}{6} \\ -\frac{1}{6} & -\frac{1}{6} & -\frac{1}{6} & \frac{5}{6} & -\frac{1}{6} & -\frac{1}{6} \\ -\frac{1}{6} & -\frac{1}{6} & -\frac{1}{6} & -\frac{1}{6} & \frac{5}{6} & -\frac{1}{6} \\ -\frac{1}{6} & -\frac{1}{6} & -\frac{1}{6} & -\frac{1}{6} & -\frac{1}{6} & \frac{5}{6} \end{bmatrix}.$$

$$ = \begin{bmatrix} \frac{5}{6} & -\frac{1}{6} & -\frac{1}{6} & -\frac{1}{6} & -\frac{1}{6} & -\frac{1}{6} \\ -\frac{1}{6} & \frac{5}{6} & -\frac{1}{6} & -\frac{1}{6} & -\frac{1}{6} & -\frac{1}{6} \\ -\frac{1}{6} & -\frac{1}{6} & \frac{5}{6} & -\frac{1}{6} & -\frac{1}{6} & -\frac{1}{6} \\ -\frac{1}{6} & -\frac{1}{6} & -\frac{1}{6} & \frac{5}{6} & -\frac{1}{6} & -\frac{1}{6} \\ -\frac{1}{6} & -\frac{1}{6} & -\frac{1}{6} & -\frac{1}{6} & \frac{5}{6} & -\frac{1}{6} \\ -\frac{1}{6} & -\frac{1}{6} & -\frac{1}{6} & -\frac{1}{6} & -\frac{1}{6} & \frac{5}{6} \end{bmatrix} = \mathbf{Q}$$

Matrices that have this property are referred to as *idempotent*. That is, the centering matrix $\mathbf{Q}$ is said to be *idempotent*, since squaring it leaves it unchanged. It is this property that makes it useful in statistical analysis. The simplest way to obtain a *CSSCP* matrix is to premultiply the deviation scores $\mathbf{x}$ matrix by its transpose:

$$\text{CSSCP} = \mathbf{x'x} \tag{3.1}$$

$$\mathbf{x'x} = \begin{bmatrix} 2 & 1 & 0 & -3 \\ 1 & 0 & 1 & -2 \\ -3 & 2 & 1 & 0 \end{bmatrix} \cdot \begin{bmatrix} 2 & 1 & -3 \\ 1 & 0 & 2 \\ 0 & 1 & 1 \\ -3 & -2 & 0 \end{bmatrix} = \begin{bmatrix} 14 & 8 & -4 \\ 8 & 6 & -2 \\ -4 & -2 & 14 \end{bmatrix}.$$

But with real data, this is problematic, since the deviation scores have rounding error, and multiplying deviation scores compounds the rounding error by squaring and multiplying it. Calculating the CSSCP matrix by taking the raw SSCP matrix $\mathbf{X'X}$ and subtracting a matrix from it that corrects for the mean, $\mathbf{tm'}$ gives a more accurate result:

$$\text{CSSCP} = \mathbf{X'X} - \mathbf{tm'} \tag{3.2}$$

$$\mathbf{X'X} - \mathbf{tm'} = \begin{bmatrix} 8 & 7 & 6 & 3 \\ 5 & 4 & 5 & 2 \\ 6 & 11 & 10 & 9 \end{bmatrix} \cdot \begin{bmatrix} 8 & 5 & 6 \\ 7 & 4 & 11 \\ 6 & 5 & 10 \\ 3 & 2 & 9 \end{bmatrix} - \begin{bmatrix} 24 \\ 16 \\ 36 \end{bmatrix} \cdot \begin{bmatrix} 6 & 4 & 9 \end{bmatrix}$$

$$= \begin{bmatrix} 158 & 104 & 212 \\ 104 & 70 & 142 \\ 212 & 142 & 338 \end{bmatrix} - \begin{bmatrix} 144 & 96 & 216 \\ 96 & 64 & 144 \\ 216 & 144 & 324 \end{bmatrix} = \begin{bmatrix} 14 & 8 & -4 \\ 8 & 6 & -2 \\ -4 & -2 & 14 \end{bmatrix}.$$

Using Equation 3.1 creates squares and products of the rounding error, thus compounding the rounding error. The computational Equation 3.2 that corrects for the mean after the fact is a precise and accurate way to obtain a CSSCP matrix. So, the choice seems to be between a formula that is simple but prone to rounding error problems, and one that is complex but precise. In fact, because of the idempotent nature of the centering matrix, there is a third alternative.

$$\text{CSSCP} = \mathbf{X'x} \tag{3.3}$$

$$= \begin{bmatrix} 8 & 7 & 6 & 3 \\ 5 & 4 & 5 & 2 \\ 6 & 11 & 10 & 9 \end{bmatrix} \cdot \begin{bmatrix} 2 & 1 & -3 \\ 1 & 0 & 2 \\ 0 & 1 & 1 \\ -3 & -2 & 0 \end{bmatrix} = \begin{bmatrix} 14 & 8 & -4 \\ 8 & 6 & -2 \\ -4 & -2 & 14 \end{bmatrix}$$

This formula is nearly as simple as the Equation 3.1, but more accurate and precise when the means are not whole numbers, since there is only rounding error on one of the matrices being multiplied. It is surprising that this formula works. It does because of the idempotent property of the centering matrix. This can be seen algebraically if we first note that the transpose of a product of two matrices is equal to the product of the transposes of the matrices in reversed order, that is: $(\mathbf{AB})' = \mathbf{B'A'}$:

$$\text{CSSCP} = \mathbf{x'x} = (\mathbf{Q_cX})' \cdot (\mathbf{Q_cX}) = \mathbf{X'Q_c'Q_cX} = \mathbf{X'Q_cX} = \mathbf{X'x}.$$

In this derivation, the simplification can be seen in the transition from the fourth step in the equation sequence to the fifth, where one of the $\mathbf{Q}_c$ values drops out because centering matrices are idempotent, and therefore $\mathbf{Q}_c\mathbf{Q}_c = \mathbf{Q}_c$.

3.3.7.2 Summary of Principles Relevant to Covariance and Correlation Matrices

A set of observations should be made about the relationship between matrix operations and the various matrices in the process of creating covariance matrices and correlation matrices:

(1) The SSCP matrix, $\mathbf{X'X}$, is conceptually very simple. It is basic matrix multiplication applied to an $\mathbf{X}$ matrix (that is to a data matrix with observation units as rows and variables as columns). It could be said that the SSCP matrix is a "special case" of matrix multiplication: matrix multiplication applied to a particular kind of matrix, an "observations by variables" data matrix.

(2) The CSSCP matrix, $\mathbf{x'x}$ or $\mathbf{X'x}$, is a further special case (or a restricted case) of the SSCP matrix. It is matrix multiplication of data matrices that are centered.

(3) The covariance matrix is an even more restricted special case of the CSSCP matrix. It is the CSSCP matrix divided by its degrees of freedom. Variances are the diagonal terms in this matrix, and covariances are the off-diagonal terms.

(4) The correlation matrix is the most restricted, most specific special case of this progression of matrices. It is the covariance matrix calculated on standard scores, or equivalently, the covariance matrix adjusted for its standard deviations.

Matrix multiplication is therefore clearly well fitted to the task of efficiently calculating these statistical quantities.

3.3.7.3 Relevance to Euclidean Geometry and Pythagorean Theorem

Matrix multiplication is clearly well fitted to efficiently calculating statistics, such as covariances and correlations. Both matrix algebra and also covariance structures are firmly based upon Euclidean geometry and the Pythagorean theorem. You will recall that the Pythagorean Theorem applies to a right triangle, and states that if a and b are the lengths of the legs of the triangle, and c is the length of the hypotenuse, then $c^2 = a^2 + b^2$, and the length of c can be calculated by taking the square root of the sum of the squares of the other two sides.

Since a column vector of length n can be thought of as the coordinates of a vector in n dimensional space, one could calculate the length of that vector by taking the square root of the sum of the squared elements in the vector. When a column vector (e.g., the data vector $\mathbf{x}$) is premultiplied by its own transpose, the resultant dot product is the sum of squares of the elements in that vector:

$$\mathbf{x}'\mathbf{x} = \begin{bmatrix} 8 & 6 & 5 & 5 & 4 & 2 \end{bmatrix} \cdot \begin{bmatrix} 8 \\ 6 \\ 5 \\ 5 \\ 4 \\ 2 \end{bmatrix} = 8 \cdot 8 + 6 \cdot 6 + 5 \cdot 5 + 5 \cdot 5 + 4 \cdot 4 + 2 \cdot 2 = 170.$$

So the *length* of a vector can be calculated by taking the square root of this sum of squares. This is the Pythagorean, or more precisely, *generalized Pythagorean* theorem. That is, just as $c^2 = a^2 + b^2$, one can square and sum any number of "sides" or vector values, and the square root of that sum of squares is the length of the vector in n-dimensional space. This is also closely related to variance, since variance is like the sum of squares of the vector values, and the square root of the variance—the standard deviation—is like the length of the vector. The exact nature of this relationship will become clear in later chapters.

3.4 PARTITIONED MATRICES AND ADJOINED MATRICES

3.4.1 Adjoined Matrices

At times, it is useful to adjoin or concatenate two or more matrices into one matrix, or in other words, to put them together into one matrix. In order to adjoin matrices, they must have the same row dimension for horizontal or column dimension. For instance, if we have L, a 4×4 matrix and **M**, a 4×5 matrix, we can create **L**|**M**, an adjoined matrix:

$$\mathbf{L} = \begin{bmatrix} 3 & 1 & 9 & 0 \\ 6 & 2 & 7 & 4 \\ 2 & 0 & 5 & 1 \\ 1 & 6 & 3 & 8 \end{bmatrix}$$

$$\mathbf{M} = \begin{bmatrix} 5 & 0 & 5 & 7 & 4 \\ 8 & 2 & 3 & 3 & 2 \\ 2 & 6 & 4 & 0 & 5 \\ 4 & 1 & 7 & 2 & 1 \end{bmatrix}$$

$$\mathbf{L}\,|\,\mathbf{M} = \begin{bmatrix} 3 & 1 & 9 & 0 & 5 & 0 & 5 & 7 & 4 \\ 6 & 2 & 7 & 4 & 8 & 2 & 3 & 3 & 2 \\ 2 & 0 & 5 & 1 & 2 & 6 & 4 & 0 & 5 \\ 1 & 6 & 3 & 8 & 4 & 1 & 7 & 2 & 1 \end{bmatrix}.$$

We can also adjoin matrices below other matrices, as long as the dimensions match up with the dimensions of the matrix or matrices above them. The two or more separate sections of the adjoined matrix are referred to as submatrices, and a vertical bar is positioned between the submatrices to keep the original matrices separated. The advantage to adjoining matrices is that you can perform one operation on the adjoined matrix and have resultant matrix that is also adjoined and includes the results for both or all of the original matrices.

3.4.2 Partitioned Matrices

Sometimes it is helpful to take a large matrix and partition it into smaller submatrices either vertically or horizontally or both. For instance, suppose we have a 4×5 matrix, $\mathbf{M}$:

$$\mathbf{M} = \begin{bmatrix} 5 & 0 & 5 & 7 & 4 \\ 8 & 2 & 3 & 3 & 2 \\ 2 & 6 & 4 & 0 & 5 \\ 4 & 1 & 7 & 2 & 1 \end{bmatrix}.$$

And we want to partition it into four submatrices: $\mathbf{M}_1$, a 2×2 matrix; $\mathbf{M}_2$, a 2×3 matrix; $\mathbf{M}_3$, a 2×2 matrix; and $\mathbf{M}_4$, a 2×3 matrix.

$$\mathbf{M}_1 = \begin{bmatrix} 5 & 0 \\ 8 & 2 \end{bmatrix}$$

$$\mathbf{M}_2 = \begin{bmatrix} 5 & 7 & 4 \\ 3 & 3 & 2 \end{bmatrix}$$

$$\mathbf{M}_3 = \begin{bmatrix} 2 & 6 \\ 4 & 1 \end{bmatrix}$$

$$\mathbf{M}_4 = \begin{bmatrix} 4 & 0 & 5 \\ 7 & 2 & 1 \end{bmatrix}$$

One particular time when partitioning a matrix can be helpful is when a horizontally adjoined matrix with two submatrices is premultiplied by its transpose, which is a vertically adjoined matrix with two submatrices, resulting in a large matrix that contains four natural submatrices corresponding to the product of each combination of the original two submatrices.

3.5   TRIANGULAR SQUARE ROOT MATRICES

3.5.1   Triangular Matrices

A triangular matrix is a matrix that has zeros in either the upper triangular portion of the matrix that is above the diagonal, or in the lower triangular portion of the matrix that is below the diagonal. Matrix **E** below is an example of an upper triangular matrix, and matrix **F** is an example of a lower triangular matrix:

$$\mathbf{E} = \begin{bmatrix} 3 & 2 & 5 & 2 \\ 0 & 5 & 1 & 3 \\ 0 & 0 & 6 & 2 \\ 0 & 0 & 0 & 4 \end{bmatrix}$$

$$\mathbf{F} = \begin{bmatrix} 4 & 0 & 0 & 0 \\ 2 & 9 & 0 & 0 \\ 8 & 7 & 11 & 0 \\ 3 & 4 & 1 & 8 \end{bmatrix}.$$

3.5.2   The Cholesky Method for Finding a Triangular Square Root Matrix

It is often useful to find the triangular square root of a symmetric matrix. In fact, a number of the multivariate statistical methods require this (canonical correlation, discriminant analysis, etc.). The primary algorithm for finding triangular square root matrices is the Cholesky method, named after its developer, Lieutenant Andre' Louis Cholesky of the French Navy.

Suppose that you wish to find the triangular root of the square positive definite (*positive definite* is explained in Section 3.8) matrix **A** below.

$$\mathbf{A} = \begin{bmatrix} 4 & 6 & 2 \\ 6 & 16 & 3 \\ 2 & 3 & 9 \end{bmatrix}.$$

Each element in this matrix is indicated by its row and column location. For example, element a_{13} in this matrix is the first row and third column, and has the value "2." Element a_{11} is the first diagonal element and has the value "4," and so on. To find the first diagonal element in the triangular root matrix **T**, we follow Rencher's (2002) equation from page 26:

$$t_{11} = \sqrt{a_{11}} = \sqrt{4} = 2.$$

Notice that this first diagonal element of the **T** matrix is just the square root of the corresponding element in the **A** matrix. The remaining elements in this first row of the **T** matrix are found by dividing the corresponding elements in the **A** matrix by the t_{11} element just calculated:

$$t_{12} = \frac{a_{12}}{t_{11}} = \frac{6}{2} = 3$$

$$t_{13} = \frac{a_{13}}{t_{11}} = \frac{2}{2} = 1.$$

The second diagonal element (and the third, and in fact all remaining diagonal elements) are found according to the formula given by Rencher (page 26):

$$t_{ii} = \sqrt{a_{ii} - \sum_{k=1}^{i-1} t_{ki}^2} \,.$$

This is the algebraic way of saying that the **T** matrix diagonal elements (after the first) are found by taking the square root of a quantity. That quantity is the corresponding element in the **A** matrix, adjusted by subtracting from it the sum of squares of all the elements in the **T** matrix that are above the location of the new desired element. For diagonal element t_{22}, this comes out to be:

$$t_{22} = \sqrt{a_{22} - \sum_{k=1}^{1} t_{k2}^2} = \sqrt{16 - 3^2} = \sqrt{7}.$$

The remaining entries in the second row are found from Rencher's (2002) formula:

$$t_{ij} = \frac{a_{ij} - \sum_{k=1}^{i-1} t_{ki} t_{kj}}{t_{ii}}.$$

This is the algebraic way of saying that each remaining row entry in the **T** matrix to the right of the diagonal element is found by taking the corresponding element in the **A** matrix, subtracting a quantity from it, and then dividing the result by the just-computed diagonal element for that row. The quantity that is subtracted is just the sum of the products of the **T** matrix elements above the desired new element multiplied by the corresponding **T** matrix elements above the diagonal element. For the remaining element in the second row of our example matrix, this comes out to be:

$$t_{23} = \frac{a_{23} - \sum\limits_{k=1}^{i-1} t_{k2}t_{k3}}{t_{22}} = \frac{3 - 3 \cdot 1}{\sqrt{7}} = \frac{0}{\sqrt{7}} = 0.$$

Now that this entry has been calculated, the final diagonal element t_{33} can be calculated using the formula:

$$t_{33} = \sqrt{a_{33} - \sum\limits_{k=1}^{2} t_{k3}^2} = \sqrt{9 - (1^2 + 0^2)} = \sqrt{9 - 1} = \sqrt{8}.$$

Since this is an upper triangular root matrix, all of the elements below the diagonal are zero. We now have calculated all of the nonzero elements, so the matrix can be constructed:

$$\mathbf{T} = \begin{bmatrix} 2 & 3 & 1 \\ 0 & \sqrt{7} & 0 \\ 0 & 0 & \sqrt{8} \end{bmatrix}.$$

We can now check this *triangular square root* matrix by premultiplying it by its transpose and demonstrating that it is indeed the square root of matrix $\mathbf{A}$, since this product yields matrix $\mathbf{A}$:

$$\mathbf{T'T} = \begin{bmatrix} 2 & 0 & 0 \\ 3 & \sqrt{7} & 0 \\ 1 & 0 & \sqrt{8} \end{bmatrix} \cdot \begin{bmatrix} 2 & 3 & 1 \\ 0 & \sqrt{7} & 0 \\ 0 & 0 & \sqrt{8} \end{bmatrix} = \begin{bmatrix} 4 & 6 & 2 \\ 6 & 16 & 3 \\ 2 & 3 & 9 \end{bmatrix}.$$

Notice that *order of multiplication* matters with triangular square root matrices. When this upper triangular square root matrix $\mathbf{T}$ is postmultiplied by its transpose rather than premultiplied by it, the resulting product is not equal to the original matrix $\mathbf{A}$:

$$\mathbf{TT'} = \begin{bmatrix} 2 & 3 & 1 \\ 0 & \sqrt{7} & 0 \\ 0 & 0 & \sqrt{8} \end{bmatrix} \cdot \begin{bmatrix} 2 & 0 & 0 \\ 3 & \sqrt{7} & 0 \\ 1 & 0 & \sqrt{8} \end{bmatrix} = \begin{bmatrix} 14 & 3\sqrt{7} & \sqrt{8} \\ 3\sqrt{7} & 7 & 0 \\ \sqrt{8} & 0 & 8 \end{bmatrix} \neq \mathbf{A}.$$

3.6 DETERMINANTS

Only square matrices have determinants. The determinant is a single value that summarizes descriptive information about the matrix. The properties that it summarizes are the easiest to see with a simple 2×2 matrix. The determinant is found as the product of the diagonal elements minus the product of the off-diagonal elements. For example, the determinant of a simple 2×2 matrix is found to be:

$$|\mathbf{S}| = \begin{vmatrix} 4 & 1 \\ 1 & 9 \end{vmatrix} = 4 \cdot 9 - 1 \cdot 1 = 36 - 1 = 35.$$

The calculation of the determinant of a matrix becomes much more challenging when the size of the matrix is 3×3 or larger. We will now demonstrate three methods for obtaining determinants for larger matrices. The first two methods will work only with 3×3 matrices, but the third will work for larger matrices (even though it becomes insufferably tedious for larger matrices). The best method for finding a determinant in actual practice is, of course, just using the determinant function in Excel. This is done by entering the instruction:

$$= \text{MDETERM(F5:H7)}$$

into a cell of the spreadsheet, where "B3:F7" is the location of a square matrix on the spreadsheet (in this case, a 5×5 matrix with the upper left corner at cell B3 and the lower right corner at cell F7). This instruction is completed with a "control-shift-enter" instruction. In today's world, Excel (or some other matrix language application) is the only reasonable way to create determinants for anything 4×4 or larger.

The first of the three hand calculation methods for 3×3 matrices given below is referred to as the *bent diagonals method*. Another version of this same approach is the *matrix extension method*. A third method that will be useful also in obtaining the inverse of a matrix (Section 3.6) is the *method of cofactors*.

3.6.1 The Bent Diagonals Method

Algebraically, the determinant of a 3×3 matrix is given by:

$$|\mathbf{A}| = a_{11}a_{22}a_{33} + a_{12}a_{23}a_{31} + a_{13}a_{32}a_{21} - a_{31}a_{22}a_{13} - a_{32}a_{23}a_{11} - a_{33}a_{12}a_{21}.$$

In words, this is the sum of the products of the two positive bent diagonals and the one positive diagonal, minus each of the two negative bent diagonals and the negative diagonal. This is shown schematically as:

We will now use this scheme to find the determinant of the following 3×3 matrix:

$$\mathbf{S} = \begin{bmatrix} 4 & 5 & 1 \\ 5 & 9 & 2 \\ 1 & 2 & 7 \end{bmatrix}.$$

We first create the product of the three elements of the positive diagonal:

$$a_{11}a_{22}a_{33} = 4 \cdot 9 \cdot 7 = 252.$$

and similar products for the two positive bent diagonals:

$$a_{12}a_{23}a_{31} = 5 \cdot 2 \cdot 1 = 10$$
$$a_{13}a_{32}a_{21} = 1 \cdot 2 \cdot 5 = 10.$$

Next, we create the product of the three elements of the negative diagonal:

$$a_{31}a_{22}a_{13} = 1 \cdot 9 \cdot 1 = 9,$$

and similar products for the two negative bent diagonals:

$$a_{32}a_{23}a_{11} = 2 \cdot 2 \cdot 4 = 16$$
$$a_{33}a_{12}a_{21} = 7 \cdot 5 \cdot 5 = 175.$$

The three negative diagonal products are now subtracted from the sum of the three positive diagonal products, and the determinant is found to be 72.

$$|\mathbf{A}| = a_{11}a_{22}a_{33} + a_{12}a_{23}a_{31} + a_{13}a_{32}a_{21} - a_{31}a_{22}a_{13} - a_{32}a_{23}a_{11} - a_{33}a_{12}a_{21}$$
$$= 252 + 10 + 10 - 9 - 16 - 175 = 272 - 200 = 72.$$

3.6.2 The Matrix Extension Method

There is another method for 3×3 matrices that is equivalent to Rencher's bent diagonals method, but it is conceptually easier. We will refer to it as *the "matrix extension" diagonals scheme* because it also involves positive and negative diagonals, but it requires that we create an extension to the matrix. For a 3×3 matrix, we create the extension by placing copies of columns 1 and 2 on the right of column 3, thus creating a 3×5 matrix:

$$\mathbf{S}_{\text{extension}} = \begin{bmatrix} 4 & 5 & 1 \\ 5 & 9 & 2 \\ 1 & 2 & 7 \end{bmatrix}\begin{bmatrix} 4 & 5 \\ 5 & 9 \\ 1 & 2 \end{bmatrix}$$

Now we do not need any "bent diagonals." The whole thing can be done with ordinary linear diagonals. The sums of products for the three positive linear diagonals are:

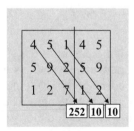

and sums of products for the three negative linear diagonals are:

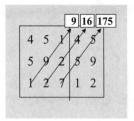

We subtract the sum of the three negative linear diagonals from the sum of the three positive diagonals, and obtain the determinant:

$$\mathbf{S} = [252 + 10 + 10] - [9 + 16 + 175] = 272 - 200 = 72.$$

3.6.3 The Method of Cofactors

Another method that can be used to calculate the determinant of a matrix is the method of cofactors. This method works for any size matrix, but can be quite tedious for anything larger than a 4×4 matrix. To do this method, we first select one row or column that we will use to do the expansion. Let us select the first row of $\mathbf{S}$, which has the elements 4, 5, and 1.

$$\mathbf{S} = \begin{bmatrix} 4 & 5 & 1 \\ 5 & 9 & 2 \\ 1 & 2 & 7 \end{bmatrix}.$$

We then take each element of this row (or column) one at a time and calculate the cofactor for each element. We do this by ignoring the rest of the elements that are on the same row or column as the element, and we are left with a matrix that has one fewer rows and one fewer columns than the original matrix that is referred to as the minor of the element. So, for the first element of **S**, 4, we ignore the rest of the elements on that row (5 and 1), and the other elements that are in the same column (5 and 1), and we are left with the matrix

$$\mathbf{S}_{11} = \begin{bmatrix} 9 & 2 \\ 2 & 7 \end{bmatrix}.$$

which is the minor for the first element of S, s_{zz}.

The determinant of this new, smaller matrix is then multiplied by either a 1 or a −1 depending on the corresponding element's position in the matrix, to produce the cofactor of the selected element. The first element of the first row gets multiplied by 1, and then the second element on the first row gets multiplied by −1, with all of the elements being assigned alternately either a 1 or −1 in a checkerboard pattern as shown below:

$$\begin{bmatrix} 1 & -1 & 1 \\ -1 & 1 & -1 \\ 1 & -1 & 1 \end{bmatrix}$$

So the cofactor for element s_{11} is just the determinant of the smaller matrix calculated with the formula shown above for a 2×2 matrix:

$$\begin{vmatrix} 9 & 2 \\ 2 & 7 \end{vmatrix} = 9*7 - 2*2 = 63 - 4 = 59.$$

The cofactor for element s_{12} is:

$$-1* \begin{vmatrix} 5 & 2 \\ 1 & 7 \end{vmatrix} = -1*(5*7 - 1*2) = -1*(35 - 2) = -33.$$

And the cofactor for element s_{13} is:

$$\begin{vmatrix} 5 & 9 \\ 1 & 2 \end{vmatrix} = 5*2 - 1*9 = 10 - 9 = 1.$$

The determinant of the original matrix is then calculated by multiplying every element of the selected row or column by its cofactor and summing across the elements of the selected row or column:

$$|\mathbf{S}| = \begin{vmatrix} 4 & 5 & 1 \\ 5 & 9 & 2 \\ 1 & 2 & 7 \end{vmatrix} = 4*(59) + 5*(-33) + 1*(1) = 236 + -165 + 1 = 72.$$

Although we selected the first row around which to perform the expansion, we would get the same answer by expanding around any other row or column.

3.6.4 Meaning of the Trace and the Determinant of a Covariance Matrix

To make the meaning of a determinant clear in a context well known to us, we will consider the determinant of a covariance matrix. If matrix **S** below were a covariance matrix, the diagonal entries indicate that the variance of the first variable is 4, and the variance of the second variable is 9. The covariance between the variables is 1:

$$|\mathbf{S}| = \begin{vmatrix} 4 & 1 \\ 1 & 9 \end{vmatrix}.$$

The Pearson product moment correlation coefficient is therefore:

$$r = \frac{\text{cov}}{\sqrt{s_x^2 s_y^2}} = \frac{1}{\sqrt{4 \cdot 9}} = \frac{1}{6} = 0.167.$$

It will be remembered that in the geometry of covariance matrices, each variable can be represented as a vector. The standard deviations (2 and 3 for this matrix) are the lengths of the vectors for the two variables, and the PPMCC is the cosine of the angle between them. A PPMCC of 0.167 is therefore indicative that the angle between these two vectors is 80.4 degrees:

$$\alpha = \arccos(.167) = 80.4°.$$

This is a large angle, almost as far apart as the vectors could possibly be (90°), so we see that the geometric figure created by these two vectors has a large *volume*. As Rencher explains in Chapter 3, the determinant of a covariance matrix gives the *generalized variance* of that covariance matrix. When the generalized variance is large, the *volume* of the geometric figure for that set of variables is large.

Let us consider now the determinant of a similar covariance matrix with exactly the same variances, but a much larger covariance. (When the covariance is large, it is an indication that those two variances are redundant with one another. They are overlapping, or redundant, so the *volume* or *generalized variance* of the matrix is low.)

$$|\mathbf{S}_b| = \begin{vmatrix} 4 & 5 \\ 5 & 9 \end{vmatrix} = 4 \cdot 9 - 5 \cdot 5 = 36 - 25 = 11.$$

We see here that the determinant for this matrix is considerably less than for the former one where there was not so much overlap. We see also that the angle between the two variables in this second covariance matrix is small, meaning that the two vectors are close together, creating a geometric figure with smaller volume.

$$r = \frac{\text{cov}}{\sqrt{s_x^2 s_y^2}} = \frac{5}{\sqrt{4 \cdot 9}} = \frac{5}{6} = 0.833$$

$$\alpha = \arccos(.833) = 33.6°.$$

The determinant, therefore, is an index of the volume of the geometric figure in many-dimensional-space that can be created from any square matrix.

The *trace* of a matrix is just the sum of the diagonal entries of that matrix. Since only square matrices have a diagonal, then by definition only square matrices can have a trace. Consider matrix H below:

$$\mathbf{H} = \begin{bmatrix} 9 & 2 & 3 \\ 2 & 16 & 8 \\ 3 & 8 & 12 \end{bmatrix}.$$

Its trace is the sum of its diagonal elements:

$$tr(\mathbf{H}) = 9 + 16 + 12 = 37.$$

The *trace*, like the *determinant* is a single value that summarizes descriptive information about a matrix. It is particularly useful when the matrix in question is a *covariance matrix*. The diagonal elements in a covariance matrix are variances, so the trace of a covariance matrix indicates the total number of variance units in that set of variables. The off-diagonal elements of the covariance matrix, the covariances, indicate how much overlap there is in the variances among pairs of the variables.

In matrix **S** above, the two diagonal elements are 4 and 9, so the trace is just the sum of these two, 4 + 9 = 13. The trace of a covariance matrix is the *total variance* of the data set represented by that matrix. Obviously, matrix **S** and matrix $\mathbf{S_b}$ above have the same trace, even though they do not have the same determinant. We could say, then, that they have the *same total variance*. However, the *generalized variance* of **S** is considerably larger than the *generalized variance* of $\mathbf{S_b}$. *Generalized variance* takes into account the redundancy implied by the covariance structure, whereas *total variance* does not. *Total variance* is just the total number of variance units in the data set without regard to covariance structure.

3.7 MATRIX INVERSION

Some square matrices have an inverse matrix, which is a matrix that can be multiplied by the square matrix and result in an identity matrix. For a real number, a, the inverse is indicated as a^{-1}, which is equal to 1/a, and the product of a number and its inverse (a*1/a) is equal to 1. Since we cannot divide by a matrix, we indicate the inverse matrix of **A** as $\mathbf{A^{-1}}$, and we can show that $\mathbf{A^{-1}}$ is the inverse of **A** by showing that the following equations are true:

$$\mathbf{AA}^{-1} = \mathbf{I}, \mathbf{A}^{-1}\mathbf{A} = \mathbf{I}.$$

Note that the matrix $\mathbf{A}^{-1}$ and also the matrix $\mathbf{I}$ are both square matrices that have the same dimensions as $\mathbf{A}$. If the inverse matrix exists, then it is unique and it can be pre- or postmultiplied by A to get the identity matrix. The inverse of a matrix will exist when the determinant of the matrix is not equal to zero. Matrices that have an inverse are called nonsingular or invertible matrices. Matrices that do not have an inverse are called singular.

Inverse matrices are particularly useful when we want to solve a matrix equation that has the product of a square matrix and the matrix we want to solve for on the same side of the equation. For instance, if we want to solve for the vector $\mathbf{x}$ in the following equation:

$$\mathbf{Ax} = \mathbf{B}.$$

Because we cannot divide by matrices, we cannot divide both sides of the equation by $\mathbf{A}$, but we can instead multiply both sides of the equation by the inverse of $\mathbf{A}$ to eliminate A from the left side of the equation as follows:

$$\mathbf{A}^{-1}\mathbf{Ax} = \mathbf{A}^{-1}\mathbf{B}$$

$$\mathbf{Ix} = \mathbf{A}^{-1}\mathbf{B}$$

$$\mathbf{x} = \mathbf{A}^{-1}\mathbf{B},$$

thus solving for $\mathbf{x}$.

Calculating an inverse matrix is not a trivial process, but various algorithms are available for inverting a square matrix.

3.7.1 Matrix Inversion by the Method of Cofactors

One method for calculating the inverse of a matrix A is to first calculate the matrix of cofactors. This is done by replacing each element of matrix A with its cofactor, which is the determinant of its minor submatrix multiplied by either a 1 or a –1 (as described in Section 3.6.3). Then we just take the transpose of this new matrix of cofactors and divide it by the determinant of the original matrix $\mathbf{A}$.

We will calculate the inverse of matrix $\mathbf{A}$. First, we will calculate the determinant to make sure that the inverse of matrix A exists. We will use the matrix extension method.

$$|\mathbf{A}| = \begin{bmatrix} 4 & 6 & 2 \\ 6 & 16 & 3 \\ 2 & 3 & 9 \end{bmatrix} \begin{matrix} 4 & 6 \\ 6 & 16 \\ 2 & 3 \end{matrix} = 4*16*9 + 6*3*2 + 2*6*3 -$$

$$2*16*2 - 3*3*4 - 9*6*6$$

$$= 576 + 36 + 36 - 64 - 36 - 324 = 224.$$

Then we will calculate the matrix of cofactors which we will call **C**:

$$
\mathbf{C} =
\begin{bmatrix}
\begin{vmatrix} 16 & 3 \\ 3 & 9 \end{vmatrix} & -\begin{vmatrix} 6 & 3 \\ 2 & 9 \end{vmatrix} & \begin{vmatrix} 6 & 16 \\ 2 & 3 \end{vmatrix} \\[2mm]
-\begin{vmatrix} 6 & 2 \\ 3 & 9 \end{vmatrix} & \begin{vmatrix} 4 & 2 \\ 2 & 9 \end{vmatrix} & -\begin{vmatrix} 4 & 6 \\ 2 & 3 \end{vmatrix} \\[2mm]
\begin{vmatrix} 6 & 2 \\ 16 & 3 \end{vmatrix} & -\begin{vmatrix} 4 & 2 \\ 6 & 3 \end{vmatrix} & \begin{vmatrix} 4 & 6 \\ 6 & 16 \end{vmatrix}
\end{bmatrix}
$$

$$
\mathbf{C} =
\begin{bmatrix}
16*9-3*3 & -(6*9-2*3) & 6*3-2*16 \\
-(6*9-3*2) & 4*9-2*2 & -(4*3-2*6) \\
6*3-16*2 & -(4*3-6*2) & 4*16-6*6
\end{bmatrix}
$$

$$
\mathbf{C} =
\begin{bmatrix}
135 & -48 & -14 \\
-48 & 32 & 0 \\
-14 & 0 & 28
\end{bmatrix}.
$$

We then take the transpose of **C**, **C′** (which happens to be equal to **C** for this example since the original matrix was symmetric) and divide by the determinant of **A**, which is 224:

$$
\mathbf{C'}/|\mathbf{A}| =
\begin{bmatrix}
135 & -48 & -14 \\
-48 & 32 & 0 \\
-14 & 0 & 28
\end{bmatrix}
\Big/ 224 =
\begin{bmatrix}
135/224 & -48/224 & -14/224 \\
-48/224 & 32/224 & 0/224 \\
-14/224 & 0/224 & 28/224
\end{bmatrix}
$$

$$
\mathbf{A}^{-1} =
\begin{bmatrix}
0.603 & -0.214 & -0.063 \\
-0.214 & 0.143 & 0 \\
-0.063 & 0 & 0.125
\end{bmatrix}.
$$

Finally, it can be verified that this is indeed the inverse by multiplying **A** by it in either order (since both are symmetric), and showing that the resultant product matrix is an identity matrix **I**:

$$
\mathbf{A} \cdot \mathbf{A}^{-1} =
\begin{bmatrix}
4 & 6 & 2 \\
6 & 16 & 3 \\
2 & 3 & 9
\end{bmatrix}
\cdot
\begin{bmatrix}
0.603 & -0.214 & -0.063 \\
-0.214 & 0.143 & 0 \\
-0.063 & 0 & 0.125
\end{bmatrix}
=
\begin{bmatrix}
1 & 0 & 0 \\
0 & 1 & 0 \\
0 & 0 & 1
\end{bmatrix}.
$$

3.7.2 Matrix Inversion by the Cholesky Method

Harry H. Harman (1967) on pages 42–44 of his classic book *Modern Factor Analysis*, demonstrates how to create the inverse of a matrix using the Cholesky algorithm. The process is a straightforward extension of the Cholesky algorithm given in Section 3.5.2 above. You adjoin an identity matrix onto the right side of the target matrix (matrix **A** in this case) and then perform the Cholesky algorithm on this larger matrix. Since our **A** matrix in Section 3.5.2 above was a 3×3 matrix, as is the adjoined identity matrix, this creates a 3x6 matrix to be *Choleskyed*:

$$\mathbf{A}|\mathbf{I} = \begin{bmatrix} 4 & 6 & 2 & | & 1 & 0 & 0 \\ 6 & 16 & 3 & | & 0 & 1 & 0 \\ 2 & 3 & 9 & | & 0 & 0 & 1 \end{bmatrix}.$$

The schematic design of the process is as follows:

A	**I**
T	**(T′)⁻¹**

The top 3×6 matrix is the target matrix **A** adjoined to an identity matrix **I**. The bottom 3×6 matrix is the upper triangular **T** matrix that is to be extracted by the algorithm, adjoined to the lower triangular **(T′)⁻¹** matrix that is to be extracted in parallel. For the **A** matrix of Section 3.5.2 above, the resultant matrices are:

$$\mathbf{A} = \begin{array}{|ccc|ccc|} \hline 4 & 6 & 2 & 1 & 0 & 0 \\ 6 & 16 & 3 & 0 & 1 & 0 \\ 2 & 3 & 9 & 0 & 0 & 1 \\ \hline \end{array} = \mathbf{I}$$

$$\mathbf{T} = \begin{array}{|ccc|ccc|} \hline 2 & 3 & 1 & 0.5 & 0 & 0 \\ 0 & \sqrt{7} & 0 & -\dfrac{1.5}{\sqrt{7}} & \dfrac{1}{\sqrt{7}} & 0 \\ 0 & 0 & \sqrt{8} & -\dfrac{0.5}{\sqrt{8}} & 0 & \dfrac{1}{\sqrt{8}} \\ \hline \end{array} = \mathbf{(T′)^{-1}}$$

The computations of each of the new elements in the **(T′)⁻¹** matrix are given in what follows. The formulas for the Cholesky process are, of course, the same as those used in Section 3.5.2 above, but we already have the diagonal element for each of the three rows, so we only use two of the formulas, the one that gives the remaining elements in the first row and the one that gives the remaining elements in the successive rows (second, third, etc.). For the first row, the additional entries t_{14}, t_{15}, and t_{16} are:

$$t_{14} = \frac{a_{14}}{t_{11}} = \frac{1}{2} = 0.5$$

$$t_{15} = \frac{a_{15}}{t_{11}} = \frac{0}{2} = 0$$

$$t_{16} = \frac{a_{16}}{t_{11}} = \frac{0}{2} = 0.$$

For the second row, the additional entries t_{24}, t_{25}, and t_{26} are:

$$t_{24} = \frac{a_{24} - \sum\limits_{k=1}^{1} t_{k2} t_{k4}}{t_{22}} = \frac{0 - (3)(0.5)}{\sqrt{7}} = -\frac{1.5}{\sqrt{7}}$$

$$t_{25} = \frac{a_{25} - \sum\limits_{k=1}^{1} t_{k2} t_{k5}}{t_{22}} = \frac{1 - (3)(0)}{\sqrt{7}} = \frac{1}{\sqrt{7}}$$

$$t_{26} = \frac{a_{26} - \sum\limits_{k=1}^{1} t_{k2} t_{k6}}{t_{22}} = \frac{0 - (3)(0)}{\sqrt{7}} = \frac{0}{\sqrt{7}} = 0.$$

For the third row, the additional entries t_{34}, t_{35}, and t_{36} are:

$$t_{34} = \frac{a_{24} - \sum\limits_{k=1}^{1} t_{k3} t_{k4}}{t_{22}} = \frac{0 - \left((1)(0.5) + (0)\left(\frac{-1.5}{\sqrt{7}}\right)\right)}{\sqrt{8}} = -\frac{0.5}{\sqrt{8}}$$

$$t_{35} = \frac{a_{25} - \sum\limits_{k=1}^{1} t_{k3} t_{k5}}{t_{22}} = \frac{0 - \left((1)(0) + (0)\left(\frac{1}{\sqrt{7}}\right)\right)}{\sqrt{8}} = \frac{0}{\sqrt{8}} = 0$$

$$t_{36} = \frac{a_{26} - \sum\limits_{k=1}^{1} t_{k3} t_{k6}}{t_{22}} = \frac{1 - ((1)(0) + (0)(0))}{\sqrt{8}} = \frac{1}{\sqrt{8}}.$$

In Section 3.5.2 above, we demonstrated that the upper triangular root matrix $\mathbf{T}$ is indeed the square root of the $\mathbf{A}$ matrix. We premultiplied $\mathbf{T}$ by its transpose, and the resultant product matrix was equal to $\mathbf{A}$:

$$\mathbf{T'T = A}.$$

Similarly, it can be demonstrated that the lower triangular inverse root matrix $(\mathbf{T'})^{-1}$ is the inverse of $\mathbf{T}$ by executing either of the following two equations and obtaining the identity matrix $\mathbf{I}$ as the product:

$$(\mathbf{T'})^{-1}\mathbf{T'} = \mathbf{I}$$
$$\mathbf{T}^{-1}\mathbf{T} = \mathbf{I}.$$

Both equations are true, but since the second is simpler, it will be used for the demonstration:

$$\mathbf{T}^{-1}\cdot\mathbf{T} = \begin{bmatrix} 0.5 & \dfrac{-1.5}{\sqrt{7}} & \dfrac{-0.5}{\sqrt{8}} \\ 0 & \dfrac{1}{\sqrt{7}} & 0 \\ 0 & 0 & \dfrac{1}{\sqrt{8}} \end{bmatrix} \cdot \begin{bmatrix} 2 & 3 & 1 \\ 0 & \sqrt{7} & 0 \\ 0 & 0 & \sqrt{8} \end{bmatrix} = \begin{bmatrix} 1 & 0 & 0 \\ 0 & 1 & 0 \\ 0 & 0 & 1 \end{bmatrix}.$$

However, the inverse of the root matrix **T** was not the purpose for this algorithm. The purpose was to create the inverse of the original matrix **A**. This is obtained by multiplying the $\mathbf{T}^{-1}$ inverse matrix by its transpose according to either of the following formulae:

$$\mathbf{A}^{-1} = (\mathbf{T}')^{-1}\mathbf{T}^{-1}$$

$$\mathbf{A}^{-1} = \mathbf{T}^{-1}(\mathbf{T}')^{-1}$$

The second of these will be used:

$$\mathbf{A}^{-1} = \mathbf{T}^{-1}\cdot(\mathbf{T}')^{-1} = \begin{bmatrix} 0.5 & \dfrac{-1.5}{\sqrt{7}} & \dfrac{-0.5}{\sqrt{8}} \\ 0 & \dfrac{1}{\sqrt{7}} & 0 \\ 0 & 0 & \dfrac{1}{\sqrt{8}} \end{bmatrix} \cdot \begin{bmatrix} 0.5 & 0 & 0 \\ \dfrac{-1.5}{\sqrt{7}} & \dfrac{1}{\sqrt{7}} & 0 \\ \dfrac{-0.5}{\sqrt{8}} & 0 & \dfrac{1}{\sqrt{8}} \end{bmatrix}$$

$$= \begin{bmatrix} 0.603 & -0.214 & -0.063 \\ -0.214 & 0.143 & 0 \\ -0.063 & 0 & 0.125 \end{bmatrix}.$$

Finally, as shown above, it can be verified that this is indeed the inverse by multiplying **A** by our calculated matrix, and showing that the resultant product matrix is an identity matrix **I**:

$$\mathbf{A}\cdot\mathbf{A}^{-1} = \begin{bmatrix} 4 & 6 & 2 \\ 6 & 16 & 3 \\ 2 & 3 & 9 \end{bmatrix} \cdot \begin{bmatrix} 0.603 & -0.214 & -0.063 \\ -0.214 & 0.143 & 0 \\ -0.063 & 0 & 0.125 \end{bmatrix} = \begin{bmatrix} 1 & 0 & 0 \\ 0 & 1 & 0 \\ 0 & 0 & 1 \end{bmatrix}.$$

3.8 RANK OF A MATRIX

The rank of a matrix is defined as the number of linearly independent rows (or columns) of the matrix. So the largest that a rank can be is the smallest of the number of rows or the number of columns. If one or more rows can be written as a linear combination of another row or rows, then the matrix is not of full rank, and it is called singular. For instance if the third row of a matrix can be written as the first row multiplied by a constant, or as the sum of the second and third rows, then the matrix is not of full rank. For a square matrix, if the determinant is zero, then the matrix is not of full rank. A matrix that is of full rank is also called a nonsingular matrix. A matrix of rank r can be represented as vectors in r dimensional space.

A matrix is positive definite if each of its upper left submatrices has a positive determinant. In order for this to hold, it is clear that a positive definite matrix must also be of full rank since the determinant of the matrix must be positive. A positive definite matrix is somewhat analogous to a positive real number. In Section 3.5.2, we were calculating the triangular square root of a matrix that was designated as a positive definite matrix. This is similar to requiring a non-negative number when taking a square root in order to get a real value. A matrix is positive semi-definite if each of its upper left submatrices has a non-negative (positive or 0) determinant.

3.9   ORTHOGONAL VECTORS AND MATRICES

For two vectors to be orthogonal means they are perpendicular to one another, or in other words, they have a zero correlation with one another. To test whether two vectors are orthogonal to one another, one obtains their sum of products. If the sum of products is zero, the two matrices are orthogonal. If the sum of products is not zero, they are not orthogonal. Consider, for example, the three column vectors $\mathbf{a_1}$, $\mathbf{a_2}$, and $\mathbf{a_3}$ below.

$$a_1 = \begin{bmatrix} 1 \\ 1 \\ 1 \\ -3 \end{bmatrix} \quad a_2 = \begin{bmatrix} 1 \\ 1 \\ -2 \\ 0 \end{bmatrix} \quad a_3 = \begin{bmatrix} 3 \\ 1 \\ -2 \\ -1 \end{bmatrix}.$$

Vector $\mathbf{a_1}$ is orthogonal to $\mathbf{a_2}$, but neither $\mathbf{a_1}$ nor $\mathbf{a_2}$ is orthogonal to $\mathbf{a_3}$. This can be shown by obtaining the *sum of products* for each of the three possible pairings of these three vectors. (The matrix algebra way of obtaining the sum of products between two vectors is to premultiply one by the transpose of the other, to obtain the *dot product*, which is equal to the sum of products.)

$$\mathbf{a_1'} \cdot \mathbf{a_2} = \begin{bmatrix} 1 & 1 & 1 & -3 \end{bmatrix} \cdot \begin{bmatrix} 1 \\ 1 \\ -2 \\ 0 \end{bmatrix} = 1 \cdot 1 + 1 \cdot 1 + 1 \cdot (-2) + (-3) \cdot 0 = 1 + 1 - 2 + 0 = 0$$

$$\mathbf{a'_1 \cdot a_3} = \begin{bmatrix} 1 & 1 & 1 & -3 \end{bmatrix} \cdot \begin{bmatrix} 3 \\ 1 \\ -2 \\ -1 \end{bmatrix} = 1 \cdot 3 + 1 \cdot 1 + 1 \cdot (-2) + (-3) \cdot (-1) = 3 + 1 - 2 + 3 = 5$$

$$\mathbf{a'_2 \cdot a_3} = \begin{bmatrix} 1 & 1 & -2 & 0 \end{bmatrix} \cdot \begin{bmatrix} 3 \\ 1 \\ -2 \\ -1 \end{bmatrix} = 1 \cdot 3 + 1 \cdot 1 + (-2) \cdot (-2) + 0 \cdot (-1) = 3 + 1 + 4 + 0 = 8.$$

These column vectors can all be collected into a matrix **A**:

$$\mathbf{A} = \begin{bmatrix} 1 & 1 & 3 \\ 1 & 1 & 1 \\ 1 & -2 & -2 \\ -3 & 0 & -1 \end{bmatrix}.$$

To find out whether the three column vectors of this matrix are orthogonal to one another, the matrix **A** is premultiplied by its transpose. Each off-diagonal element of the product matrix that is zero indicates that the vector corresponding to the row of that zero entry and the vector corresponding to its column are orthogonal to one another.

$$\mathbf{A' \cdot A} = \begin{bmatrix} 1 & 1 & 1 & -3 \\ 1 & 1 & -2 & 0 \\ 3 & 1 & -2 & -1 \end{bmatrix} \cdot \begin{bmatrix} 1 & 1 & 3 \\ 1 & 1 & 1 \\ 1 & -2 & -2 \\ -3 & 0 & -1 \end{bmatrix} = \begin{bmatrix} 12 & 0 & 5 \\ 0 & 6 & 8 \\ 5 & 8 & 15 \end{bmatrix}.$$

For example, we see from this product matrix that the only zero entry is in row 1, column 2, indicating that vector a1 and vector a2 are the only ones that are orthogonal to one another as demonstrated above.

3.10 QUADRATIC FORMS AND BILINEAR FORMS

Quadratic forms is the name given to a matrix algebra process by which a square matrix, such as a covariance matrix **C**, is premultiplied by a conformable vector, and also postmultiplied by that same vector to obtain a scalar. Bilinear forms are similar, except that the matrix is premultiplied by one set of "weights" and postmultiplied by a different set of weights.

3.10.1   Quadratic Forms

Suppose that we have a 3×3 covariance matrix $\mathbf{C}$ that was obtained from a 30×3 data matrix $\mathbf{A}$, with six persons each being measured on three variables.

$$\mathbf{C} = \begin{bmatrix} 4 & 6 & 4 \\ 6 & 16 & 6 \\ 4 & 6 & 9 \end{bmatrix}.$$

Suppose that we postmultiply matrix $\mathbf{A}$ by a 3×1 vector of ones, $\mathbf{w} = [1\ 1\ 1]$. This has the effect of creating a new vector $\mathbf{t_w}$ that is a 6×1 vector of the sums of the three variables for each of the six persons. We calculate the variance of this new totaled variable and find it to be $S_w^2 = 61$. We can also obtain the variance of that simple linear combination vector $\mathbf{t}$ by premultiplying and postmultiplying the covariance matrix $\mathbf{C}$ by that same 3×1 vector of ones, and we see that it does indeed come to 61, the same value obtained by directly calculating the variance of vector $\mathbf{t}$.

$$S_w^2 = \mathbf{w}'\mathbf{Cw} = \begin{bmatrix} 1 & 1 & 1 \end{bmatrix} \begin{bmatrix} 4 & 6 & 4 \\ 6 & 16 & 6 \\ 4 & 6 & 9 \end{bmatrix} \begin{bmatrix} 1 \\ 1 \\ 1 \end{bmatrix} = 61.$$

Now suppose that we wish to obtain a different linear combination of the matrix $\mathbf{A}$. We wish to use weights of $\mathbf{v} = [1\ 3\ -2]$, and we find that this linear combination vector $\mathbf{t_v}$ has a variance of $S_v^2 = 132$. Using the quadratic forms method to calculate the variance of the linear combination, we confirm that it matches the one directly calculated.

$$S_v^2 = \mathbf{v}'\mathbf{Cv} = \begin{bmatrix} 1 & 3 & -2 \end{bmatrix} \begin{bmatrix} 4 & 6 & 4 \\ 6 & 16 & 6 \\ 4 & 6 & 9 \end{bmatrix} \begin{bmatrix} 1 \\ 3 \\ -2 \end{bmatrix} = 132.$$

3.10.2   Bilinear Forms

In the two quadratic forms examples just given, suppose that you calculate the covariance between vector $\mathbf{t_w}$ and vector $\mathbf{t_v}$, and it comes out to 60. To obtain the covariance between two weighted sums, the covariance matrix of the original variables is premultiplied by one of the vectors of weights and postmultiplied by the other, and this does indeed produce the correct covariance.

$$S_v^2 = \mathbf{w}'\mathbf{Cv} = \begin{bmatrix} 1 & 1 & 1 \end{bmatrix} \begin{bmatrix} 4 & 6 & 4 \\ 6 & 16 & 6 \\ 4 & 6 & 9 \end{bmatrix} \begin{bmatrix} 1 \\ 3 \\ -2 \end{bmatrix} = 60.$$

3.10.3 Covariance Matrix Transformation

Finally, it is possible to combine quadratic forms and bilinear forms into one single process by which covariance matrices of observed variables can be transformed into covariance matrices of transformed variables. Suppose that we adjoin the two vectors of weights to create a 3×2 matrix of weights, $\mathbf{W}$.

$$\mathbf{W} = \mathbf{w} \mid \mathbf{v} = \begin{pmatrix} 1 & 1 \\ 1 & 3 \\ 1 & -2 \end{pmatrix} = \begin{bmatrix} 1 & 1 \\ 1 & 3 \\ 1 & -2 \end{bmatrix}.$$

We now premultiply and postmultiply the covariance matrix $\mathbf{C}$ by the matrix of transforming weights, and create a new covariance matrix $\mathbf{C_w}$, that has the variance of the first linear combination variable in the first diagonal entry, the variance of the second linear combination variable in the second diagonal entry, and the covariance between the two in the off-diagonal entries.

$$\mathbf{C_w} = \mathbf{W'CW} = \begin{bmatrix} 1 & 1 & 1 \\ 1 & 3 & -2 \end{bmatrix} \begin{bmatrix} 4 & 6 & 4 \\ 6 & 16 & 6 \\ 4 & 6 & 9 \end{bmatrix} \begin{bmatrix} 1 & 1 \\ 1 & 3 \\ 1 & -2 \end{bmatrix} = \begin{bmatrix} 61 & 60 \\ 60 & 132 \end{bmatrix}.$$

This process is, of course, reversible, and the new covariance matrix $\mathbf{C_w}$ of transformed scores can be transformed back into the covariance matrix $\mathbf{C}$ of observed variables.

$$\mathbf{C} = \mathbf{W'C_wW} = \begin{bmatrix} 1 & 1 \\ 1 & 3 \\ 1 & -2 \end{bmatrix} \begin{bmatrix} 61 & 60 \\ 60 & 132 \end{bmatrix} \begin{bmatrix} 1 & 1 & 1 \\ 1 & 3 & -2 \end{bmatrix} = \begin{bmatrix} 4 & 6 & 4 \\ 6 & 16 & 6 \\ 4 & 6 & 9 \end{bmatrix}.$$

Quadratic forms have many important applications in statistical methods and theory. One of the great historical papers in statistics, for example, is Box's (1954) application of quadratic forms to the problem of alpha inflation in significance testing. Most of chapter 13 of Rencher and Schaalje's (2008) linear models book is devoted to a detailed explanation of quadratic forms and bilinear forms, and a discussion of the many ways they are useful in statistical methods. We used quadratic forms in the beginning of Section 2.7 in conjunction with an explanation of Equation 2.25, the formula for obtaining the variance of the difference between two means. We also used bilinear and quadratic forms in Chapter 4, Section 4.5.4, where "factor analysis as data transformation" is explained, and the discussion of Equation 4.12 explains how that formula is an application of quadratic and bilinear forms. In fact, anytime we are transforming a group of observed variables to a group of latent variables (or back again), quadratic and bilinear forms are the process by which the covariance matrices are transformed.

3.11   EIGENVECTORS AND EIGENVALUES

For a certain class[1] of square matrices $\mathbf{A}$, there exist certain vectors $\mathbf{x}$ such that when the matrix is multiplied by them, the resultant vector is equal to vector $\mathbf{x}$ times a constant λ:

$$\mathbf{A} \cdot \mathbf{x} = \mathbf{x} \cdot \lambda.$$

The vector $\mathbf{x}$ is an *eigenvector* of matrix $\mathbf{A}$, and λ is the *eigenvalue* corresponding to that eigenvector. In German, "eigen" means "same" or "self." It is thus a "self" vector, or one that gives itself back (multiplied by a constant) when the target matrix $\mathbf{A}$ is multiplied by it. This will be illustrated with the matrix $\mathbf{A}$ used in the previous section (3.10).

$$\mathbf{A} = \begin{bmatrix} 4 & 6 & 2 \\ 6 & 16 & 3 \\ 2 & 3 & 9 \end{bmatrix}.$$

Note that the $\mathbf{x_1}$ vector given below has the property that was just described. When $\mathbf{A}$ is postmultiplied by $\mathbf{x_1}$, the resultant vector is equal to the $\mathbf{x_1}$ vector multiplied by a constant.

$$\mathbf{A} \cdot \mathbf{x_1} = \begin{bmatrix} 4 & 6 & 2 \\ 6 & 16 & 3 \\ 2 & 3 & 9 \end{bmatrix} \cdot \begin{bmatrix} 0.374 \\ 0.872 \\ 0.316 \end{bmatrix} = \begin{bmatrix} 7.360 \\ 17.144 \\ 6.205 \end{bmatrix} = \begin{bmatrix} 0.374 \\ 0.872 \\ 0.316 \end{bmatrix} \cdot (19.662) = x_1 \cdot \lambda_1.$$

This indicates that $\mathbf{x_1}$ is an eigenvector of $\mathbf{A}$, and λ_1 is the associated eigenvalue. Actually, the matrix $\mathbf{A}$ has additional eigenvectors and eigenvalues. Since the rank of the matrix is 3, there are three non-zero eigenvalues with associated eigenvectors:

---

[1] This class of matrices is referred to as "positive definite or positive semi-definite." Covariance matrices, correlation matrices, and SSCP matrices are of this type. A positive definite square matrix is one in which all of the eigenvalues are positive. A positive semi-definite matrix is one in which all of the eigenvalues are positive or zero. The case that is excluded from our discussion of eigenvalues and eigenvectors is the case where one or more of the eigenvalues is negative. Such matrices are mathematically tractable, but will have eigenvectors with imaginary numbers. Although such matrices have applications in physics, they are not used in the multivariate statistical methods presented here. The number of positive eigenvalues indicates the rank of a matrix. A positive semidefinite matrix, therefore, is a singular matrix, one in which its rank is less than its order. A singular matrix has a zero determinant and its inverse is undefined. It can however, be factored by the method of eigenvalues and eigenvectors, and is meaningful and acceptable in many multivariate methods, such as some of the factor analysis models in Chapter 4.

$$\mathbf{A} \cdot x_1 = x_1 \cdot \lambda_1$$
$$\mathbf{A} \cdot x_2 = x_2 \cdot \lambda_2$$
$$\mathbf{A} \cdot x_3 = x_3 \cdot \lambda_3.$$

The second eigenvector and its associated eigenvalue is:

$$\mathbf{A} \cdot x_2 = \begin{bmatrix} 4 & 6 & 2 \\ 6 & 16 & 3 \\ 2 & 3 & 9 \end{bmatrix} \cdot \begin{bmatrix} -0.025 \\ -0.331 \\ 0.943 \end{bmatrix} = \begin{bmatrix} -0.197 \\ -2.611 \\ 7.449 \end{bmatrix} = \begin{bmatrix} -0.025 \\ -0.331 \\ 0.943 \end{bmatrix} \cdot (7.895) = x_2 \cdot \lambda_2.$$

And the third eigenvector and its associated eigenvalue is:

$$\mathbf{A} \cdot x_3 = \begin{bmatrix} 4 & 6 & 2 \\ 6 & 16 & 3 \\ 2 & 3 & 9 \end{bmatrix} \cdot \begin{bmatrix} 0.927 \\ -0.361 \\ -0.102 \end{bmatrix} = \begin{bmatrix} 1.338 \\ -0.521 \\ -0.147 \end{bmatrix} = \begin{bmatrix} 0.927 \\ -0.361 \\ -0.102 \end{bmatrix} \cdot (1.443) = x_3 \cdot \lambda_3.$$

This can be expressed in matrix form with all three eigenvectors combined and with their associated eigenvalues in diagonal matrix form:

$$\mathbf{A} \cdot \mathbf{X} = \begin{bmatrix} 4 & 6 & 2 \\ 6 & 16 & 3 \\ 2 & 3 & 9 \end{bmatrix} \cdot \begin{bmatrix} 0.374 & -0.025 & 0.927 \\ 0.872 & -0.331 & -0.361 \\ 0.316 & 0.943 & -0.102 \end{bmatrix} = \begin{bmatrix} 7.360 & -0.197 & 1.338 \\ 17.144 & -2.611 & -0.521 \\ 6.205 & 7.449 & -0.147 \end{bmatrix}$$

$$= \begin{bmatrix} 0.374 & -0.025 & 0.927 \\ 0.872 & -0.331 & -0.361 \\ 0.316 & 0.943 & -0.102 \end{bmatrix} \cdot \begin{bmatrix} 19.662 & 0 & 0 \\ 0 & 7.895 & 0 \\ 0 & 0 & 1.443 \end{bmatrix} = \mathbf{X} \cdot \Lambda.$$

3.12 SPECTRAL DECOMPOSITION, TRIANGULAR DECOMPOSITION, AND SINGULAR VALUE DECOMPOSITION

Spectral decomposition, also called eigenvalue decomposition, and triangular decomposition, are two alternative methods for obtaining the square root of a matrix. Spectral decomposition is a matrix factoring method that also has important uses in finding powers of square matrices and expressing relationships among square matrices. The more general matrix factoring method of singular value decomposition has the same uses for any rectangular matrix.

"Decomposition" is the process of finding the roots of a matrix, or factoring it. It is the matrix analogue of finding the roots of a polynomial, such as:

$$a^2 + 2ab + b^2 = (a+b)(a+b) = (a+b)^2.$$

As everyone who has tried it knows, it is much easier to find the square of a polynomial such as $(a + b)^2$ than to find the square root of a polynomial such as $a^2 + 2ab + b^2$, particularly if the polynomial is a more difficult one such as $3a^2 - 5ab + 4b^2$. However, there are tried and true algorithms for such things.

So it is with matrix algebra. It is much easier to multiply matrices or square a matrix than to find the square roots of a matrix (matrices generally have multiple roots), but there are algorithms for such things. The two major methods for finding the square root of a matrix are *triangular decomposition* and *spectral decomposition*. The two give very different kinds of root matrices. The square root matrix that is found with spectral decomposition is symmetric, and, as the name implies, the square root matrix that is found with triangular decomposition is a triangular matrix. However, both qualify as being true square roots of the target matrix, in the sense that the root matrices created by either of the methods, when premultiplied or postmultiplied by their transpose (in order to square them), will yield the original target matrix.

3.12.1 Spectral Decomposition, Square Matrices, and Square Root Matrices

"Spectral decomposition" means finding the roots of a square matrix by the process of eigenvectors and eigenvalues. The eigenvectors in this case are analogous to "primary colors" from which all the colors of the rainbow can be constructed by mixing the primaries in additive combinations. In the same way, the entries of the original matrix can be reconstructed by combining the eigen-vectors of the matrix in additive or linear combinations.

Any power of a matrix can be obtained by the method of eigenvectors and eigenvalues. The foundational principle for this is given by Rencher in the middle of page 36. Eigenvectors are not only eigenvectors of matrix $\mathbf{A}$ (for example), but also eigenvectors of any power of matrix $\mathbf{A}$, including the 1/2th power of matrix $\mathbf{A}$ (square root of the matrix) and the -1 power of matrix $\mathbf{A}$ (inverse of the matrix). Therefore, either the *inverse* or the *square root* of matrix $\mathbf{A}$ can be obtained by the process of obtaining the eigenvectors and eigenvalues of the matrix. Let us again consider the matrix $\mathbf{A}$ that was used for the demonstration in the previous section:

$$\mathbf{A} = \begin{bmatrix} 4 & 6 & 2 \\ 6 & 16 & 3 \\ 2 & 3 & 9 \end{bmatrix}.$$

The fundamental equation for *spectral decomposition* is Equation 3.4 below. It shows that matrix $\mathbf{A}$ can be reconstructed by the product of $\mathbf{C}$, the *normalized* eigenvectors of $\mathbf{A}$, multiplied by $\mathbf{D}$, the diagonal matrix of eigenvalues, multiplied by the transpose of $\mathbf{C}$:

$$\mathbf{A} = \mathbf{CDC'} \tag{3.4}$$

Consider matrix $\mathbf{C}$ to be the matrix of *normalized* eigenvectors, meaning that they are normalized to 1.00. The matrix of normalized eigenvectors for matrix $\mathbf{A}$ is found (by the method of successive squaring demonstrated in Chapter 4, Sections 4.3 and 4.4) to be:

$$\mathbf{C} = \begin{bmatrix} 0.374 & -0.025 & 0.927 \\ 0.872 & -0.331 & -0.361 \\ 0.316 & 0.943 & -0.102 \end{bmatrix}.$$

and the diagonalized matrix of eigenvalues associated with these eigenvectors is:

$$\mathbf{D} = \begin{bmatrix} 19.662 & 0 & 0 \\ 0 & 7.895 & 0 \\ 0 & 0 & 1.443 \end{bmatrix}.$$

Once one has obtained the normalized eigenvectors matrix $\mathbf{C}$, and the diagonalized matrix of eigenvalues $\mathbf{D}$, they can be used to obtain any power of matrix $\mathbf{A}$ by a very simple process of matrix multiplication. This is because the eigenvectors $\mathbf{C}$ are not only eigenvectors of $\mathbf{C}$, but also of any power of $\mathbf{C}$. Therefore, various powers of A can be obtained from these variations on Equation 3.4:

$$\mathbf{A}^2 = \mathbf{C}\mathbf{D}^2\mathbf{C}'$$

$$\mathbf{A}^{1/2} = \mathbf{C}\mathbf{D}^{1/2}\mathbf{C}'$$

$$\mathbf{A}^{-1} = \mathbf{C}\mathbf{D}^{-1}\mathbf{C}'$$

To obtain any power of $\mathbf{A}$, we just enter the $\mathbf{D}$ matrix to that power in the midst of Equation 3.4. For example $\mathbf{A}^2$ is obtained from the formula:

$$\mathbf{A}^2 = \mathbf{C}\mathbf{D}^2\mathbf{C}' = \begin{bmatrix} 0.374 & -0.025 & 0.927 \\ 0.872 & -0.331 & -0.361 \\ 0.316 & 0.943 & -0.102 \end{bmatrix} \begin{bmatrix} 386.581 & 0 & 0 \\ 0 & 62.337 & 0 \\ 0 & 0 & 2.082 \end{bmatrix}.$$

$$\begin{bmatrix} 0.374 & 0.872 & 0.316 \\ -0.025 & -0.331 & 0.943 \\ 0.927 & -0.361 & -0.102 \end{bmatrix} = \begin{bmatrix} 56 & 126 & 44 \\ 126 & 301 & 87 \\ 44 & 87 & 94 \end{bmatrix}$$

Note how easily one obtains the square of the diagonal matrix of eigenvalues, $\mathbf{D}^2$. Since it is a diagonal matrix, the square of the matrix is obtained by merely squaring each diagonal element. The square of matrix $\mathbf{A}$ can be easily checked by directly calculating it, and we confirm that our obtained $\mathbf{A}^2$ matrix by the spectral decomposition method is correct:

$$\mathbf{A}^2 = \mathbf{A} \cdot \mathbf{A} = \begin{bmatrix} 4 & 6 & 2 \\ 6 & 16 & 3 \\ 2 & 3 & 9 \end{bmatrix} \cdot \begin{bmatrix} 4 & 6 & 2 \\ 6 & 16 & 3 \\ 2 & 3 & 9 \end{bmatrix} = \begin{bmatrix} 56 & 126 & 44 \\ 126 & 301 & 87 \\ 44 & 87 & 94 \end{bmatrix}.$$

In a similar manner, $\mathbf{A}^{-1}$, the inverse of matrix $\mathbf{A}$, is obtained from the formula:

$$\mathbf{A}^{-1} = \mathbf{C}\mathbf{D}^{-1}\mathbf{C}' = \begin{bmatrix} 0.374 & -0.025 & 0.927 \\ 0.872 & -0.331 & -0.361 \\ 0.316 & 0.943 & -0.102 \end{bmatrix} \cdot \begin{bmatrix} 0.0509 & 0 & 0 \\ 0 & 0.1267 & 0 \\ 0 & 0 & 0.6930 \end{bmatrix}.$$

$$\begin{bmatrix} 0.374 & 0.872 & 0.316 \\ -0.025 & -0.331 & 0.943 \\ 0.927 & -0.361 & -0.102 \end{bmatrix} = \begin{bmatrix} 0.6027 & -0.2143 & -0.0625 \\ -0.2143 & 0.1429 & 0.0000 \\ -0.0625 & 0.0000 & 0.1250 \end{bmatrix}.$$

Note how easily one obtains the inverse of the diagonal matrix of eigenvalues, $\mathbf{D}^{-1}$. Since it is a diagonal matrix, the inverse of the matrix is obtained by merely taking the reciprocal of each diagonal element. This inverse can be checked by showing that when $\mathbf{A}$ is multiplied by it, the result is an identity matrix:

$$\mathbf{A} \cdot \mathbf{A}^{-1} = \begin{bmatrix} 4 & 6 & 2 \\ 6 & 16 & 3 \\ 2 & 3 & 9 \end{bmatrix} \cdot \begin{bmatrix} 0.6027 & -0.2143 & -0.0625 \\ -0.2143 & 0.1429 & 0.0000 \\ -0.0625 & 0.0000 & 0.1250 \end{bmatrix} = \begin{bmatrix} 1 & 0 & 0 \\ 0 & 1 & 0 \\ 0 & 0 & 1 \end{bmatrix}.$$

The spectral square root of matrix A is calculated in a parallel manner:

$$\mathbf{A}^{1/2} = \mathbf{C}\mathbf{D}^{1/2}\mathbf{C}' = \begin{bmatrix} 0.374 & -0.025 & 0.927 \\ 0.872 & -0.331 & -0.361 \\ 0.316 & 0.943 & -0.102 \end{bmatrix} \cdot \begin{bmatrix} 4.434 & 0 & 0 \\ 0 & 2.810 & 0 \\ 0 & 0 & 1.201 \end{bmatrix}.$$

$$\begin{bmatrix} 0.374 & 0.872 & 0.316 \\ -0.025 & -0.331 & 0.943 \\ 0.927 & -0.361 & -0.102 \end{bmatrix} = \begin{bmatrix} 1.6553 & 1.0685 & 0.3440 \\ 1.0685 & 3.8351 & 0.3877 \\ 0.3440 & 0.3877 & 2.9549 \end{bmatrix}.$$

Note how easily one obtains the square root of the diagonal matrix of eigenvalues, $\mathbf{D}^{1/2}$. Since it is a diagonal matrix, the square root of the matrix is obtained by merely taking the square root of each diagonal element. The spectral square root obtained by this process can be checked by showing that when it is multiplied by itself, the result is the original $\mathbf{A}$ matrix.

$$\mathbf{A} = \mathbf{A}^{1/2} \cdot \mathbf{A}^{1/2} = \begin{bmatrix} 1.6553 & 1.0685 & 0.3440 \\ 1.0685 & 3.8351 & 0.3877 \\ 0.3440 & 0.3877 & 2.9549 \end{bmatrix} \cdot \begin{bmatrix} 1.6553 & 1.0685 & 0.3440 \\ 1.0685 & 3.8351 & 0.3877 \\ 0.3440 & 0.3877 & 2.9549 \end{bmatrix}$$

$$= \begin{bmatrix} 4 & 6 & 2 \\ 6 & 16 & 3 \\ 2 & 3 & 9 \end{bmatrix}.$$

3.12.2 Triangular Decomposition Compared to Spectral Decomposition

An alternative method for finding a square root of a square matrix is triangular decomposition. Although this method gives a very different kind of root matrix than spectral decomposition, it will also yield the original target matrix when premultiplied or postmultiplied by its transpose.

Consider **T**, for example, the triangular square root of matrix **A** obtained by the Cholesky algorithm in the demonstration above. Matrix **A** is a simple 3×3 square symmetric matrix (that could be, e.g., a covariance matrix):

$$\mathbf{A} = \begin{bmatrix} 4 & 6 & 2 \\ 6 & 16 & 3 \\ 2 & 3 & 9 \end{bmatrix}.$$

Using the Cholesky algorithm, we obtained the triangular square root **T** of this **A** matrix:

$$\mathbf{T} = \begin{bmatrix} 2 & 3 & 1 \\ 0 & \sqrt{7} & 0 \\ 0 & 0 & \sqrt{8} \end{bmatrix}.$$

When we premultiply this T matrix by its transpose in order to square it, we see that the squaring process yields matrix **A**:

$$\mathbf{T}'\mathbf{T} = \begin{bmatrix} 2 & 0 & 0 \\ 3 & \sqrt{7} & 0 \\ 1 & 0 & \sqrt{8} \end{bmatrix} \cdot \begin{bmatrix} 2 & 3 & 1 \\ 0 & \sqrt{7} & 0 \\ 0 & 0 & \sqrt{8} \end{bmatrix} = \begin{bmatrix} 4 & 6 & 2 \\ 6 & 16 & 3 \\ 2 & 3 & 9 \end{bmatrix} = \mathbf{A}.$$

thus demonstrating that **T** is a true square root of matrix **A**. This is the process of *triangular decomposition* of **A**, since the root matrix is a triangular matrix.

At the end of Section 3.12.1 above, the square root of this same **A** matrix was obtained by the eigenvectors/eigenvalues method, and was found to be:

$$\mathbf{A}^{1/2} = \begin{bmatrix} 1.6553 & 1.0685 & 0.3440 \\ 1.0685 & 3.8351 & 0.3877 \\ 0.3440 & 0.3877 & 2.9549 \end{bmatrix}.$$

Even though this $\mathbf{A}^{1/2}$ matrix is not triangular and seems to have little in common with matrix $\mathbf{T}$ (other than that both are 3×3), we see that it is also a root matrix of $\mathbf{A}$.

$$\mathbf{A}^{1/2} \cdot \mathbf{A}^{1/2} = \begin{bmatrix} 1.6553 & 1.0685 & 0.3440 \\ 1.0685 & 3.8351 & 0.3877 \\ 0.3440 & 0.3877 & 2.9549 \end{bmatrix} \cdot \begin{bmatrix} 1.6553 & 1.0685 & 0.3440 \\ 1.0685 & 3.8351 & 0.3877 \\ 0.3440 & 0.3877 & 2.9549 \end{bmatrix}$$

$$= \begin{bmatrix} 4 & 6 & 2 \\ 6 & 16 & 3 \\ 2 & 3 & 9 \end{bmatrix} = \mathbf{A}.$$

That is, when $\mathbf{A}^{1/2}$ is squared by multiplying it by itself, it yields matrix $\mathbf{A}$, thus demonstrating that it also is a true square root of matrix $\mathbf{A}$. Notice that matrix $\mathbf{A}^{1/2}$ is a symmetric square root matrix.

3.12.3 Singular Value Decomposition

A more general method for factoring matrices is sometimes required when the matrix of interest is not symmetric. This method is called singular value decomposition, and can be performed on any matrix whether it is square or rectangular, symmetric or nonsymmetric. Although the decomposition looks similar to Equation 3.4, the calculations are more complicated since the decomposition is not simply made up of the diagonal matrix of eigenvalues and the simple matrix of eigenvectors. This factoring method will not be required for our purposes and will therefore not be demonstrated.

3.13 NORMALIZATION OF A VECTOR

Sometimes, it will be useful to change the unit of scale of a vector (while, of course, preserving the proportional relationships of its elements, which is what makes it an eigenvector). This change in scale is referred to as *normalizing* the vector. One example of where this is useful is in the process of extracting the *eigenvectors* of a matrix. Eigenvectors are "defined only up to an arbitrary multiplier." For example, if the first eigenvector of matrix $\mathbf{S}$ were to come out to be:

$$x = \begin{bmatrix} 1 \\ 4 \\ 3 \\ 2 \end{bmatrix}.$$

then this eigenvector could also be expressed as:

$$y = \begin{bmatrix} 2 \\ 8 \\ 6 \\ 4 \end{bmatrix}.$$

The proportionality among the elements is preserved in this transformation of **x** into **y** by just, for example, doubling each element. If **x** is an eigenvector of matrix **S**, then **y** is also an eigenvector of matrix **S**, as is any other vector created by multiplying the elements of **x** by some constant. The way we specify which "version" of the eigenvector is wanted is by specifying what it is *normalized* to. By definition, its sum of squares is what it is normalized to. Vector **x** is thus normalized to 30:

$$\mathbf{SS(x)} = 1^2 + 4^2 + 3^2 + 2^2 = 1 + 16 + 9 + 4 = 30,$$

while vector **y** is normalized to 120:

$$\mathbf{SS(y)} = 2^2 + 8^2 + 6^2 + 4^2 = 4 + 64 + 36 + 16 = 120.$$

Often, we wish to normalize a vector to one. This is done by dividing each element of the vector by the **magnitude** of that vector, that is, by the square root of the value to which it is currently normalized. For example, if we wished to convert vector **x** (which is currently normalized to 30) to being normalized to one, we would divide each element of the vector by the magnitude of the vector, that is, the square root of 30.

$$\mathbf{x_{norm}} = \begin{bmatrix} \dfrac{1}{\sqrt{30}} \\[2mm] \dfrac{4}{\sqrt{30}} \\[2mm] \dfrac{3}{\sqrt{30}} \\[2mm] \dfrac{2}{\sqrt{30}} \end{bmatrix}.$$

The sum of squares of this *normalized* vector **x** is 1, so the vector has been successfully normalized to 1:

$$\mathbf{SS(x_{norm})} = \left(\frac{1}{\sqrt{30}}\right)^2 + \left(\frac{4}{\sqrt{30}}\right)^2 + \left(\frac{3}{\sqrt{30}}\right)^2 + \left(\frac{2}{\sqrt{30}}\right)^2$$

$$= \frac{1}{30} + \frac{16}{30} + \frac{9}{30} + \frac{4}{30} = \frac{30}{30} = 1.$$

Unless otherwise specified, when we say "*normalized*," we mean normalized to one. In other words, if you were instructed to normalize a vector, that would mean to normalize it to one. If we wish to have it normalized to some other value, such as 7, we say, "normalize the vector to 7."

3.14 CONCLUSION

This chapter has covered only the matrix algebra principles needed for this book. Additional information can be found in van de Geer (1971, chapters 1 to 7), Rencher (2002, chapter 2), Lattin, Carroll, and Green (2003, chapter 2), and Johnson and Wichern (2007, chapter 2).

STUDY QUESTIONS

A. Essay Questions

1. Explain what types of matrices are conformable for addition and subtraction, and explain how to add and subtract matrices.

2. Explain what makes matrices conformable for multiplication and explain how to multiply conformable matrices.

3. Explain what quadratic forms and bilinear forms are and how they are used.

4. Define the rank of a matrix and the terms singular and nonsingular. Relate these terms to the characteristics of the eigenvalues of the matrix.

5. Explain why an inverse matrix is useful.

6. What are orthogonal vectors and how can two vectors be shown to be orthogonal or nonorthogonal?

7. Explain eigenvectors and eigenvalues.

B. Calculation Questions

1. Use a **J** matrix to create a matrix of deviation scores corresponding to the following matrix of raw scores for six individuals:

$$\mathbf{X} = \begin{bmatrix} 8 & 6 \\ 6 & 3 \\ 10 & 4 \\ 7 & 1 \\ 8 & 5 \\ 5 & 6 \\ 7 & 5 \\ 9 & 2 \end{bmatrix}.$$

2. Use Equation 3.3 to calculate the CSSCP matrix for the data in problem 1.

3. Calculate the determinant of matrix **A** below using the matrix extension method.

$$\mathbf{A} = \begin{bmatrix} 6 & 3 & 1 \\ 3 & 5 & 4 \\ 1 & 4 & 2 \end{bmatrix}.$$

4. Calculate the triangular root of matrix **A** in problem 3 using the Cholesky method and confirm that it is a root by premultiplying by its transpose and showing that you get matrix **A**.

5. Use the method of cofactors to calculate the inverse of matrix **A** in problem 3. Verify that this is the inverse by multiplying $\mathbf{A}^{-1}$ by **A** and showing that you get the identity matrix.

6. Use the matrix obtained in question 5 and matrix **B** given below to solve the following matrix equation for **X**:

$$\mathbf{AX} = \mathbf{B}$$

$$\mathbf{B} = \begin{bmatrix} 2 & 3 & 1 \\ 5 & 1 & 3 \\ 4 & 6 & 0 \end{bmatrix}.$$

C. Data Analysis Questions

1. Use Microsoft Excel to set up a matrix multiplication structure for a 3×3 matrix multiplied by a 3×2 matrix. Use this structure to find the products: **AC**, **BD**, and **B(AC)**.

$$\mathbf{A} = \begin{bmatrix} 6 & 3 & 1 \\ 0 & 5 & 4 \\ 4 & 3 & 2 \end{bmatrix}$$

$$\mathbf{B} = \begin{bmatrix} 2 & 3 & 1 \\ 5 & 1 & 3 \\ 4 & 6 & 0 \end{bmatrix}$$

$$\mathbf{C} = \begin{bmatrix} 2 & 3 \\ 5 & 1 \\ 4 & 2 \end{bmatrix}$$

$$\mathbf{D} = \begin{bmatrix} 0 & 4 \\ 3 & 2 \\ 3 & 1 \end{bmatrix}.$$

2. Set up a structure in Excel that uses matrix multiplication to find each of the values below for a given matrix **X**.

$$\mathbf{X} = \begin{bmatrix} 10 & 6 & 13 \\ 16 & 8 & 15 \\ 5 & 4 & 12 \\ 8 & 3 & 10 \\ 12 & 9 & 16 \\ 2 & 10 & 9 \end{bmatrix}.$$

(a) Find $\mathbf{X}'\mathbf{X}$, the SSCP matrix.

(b) Find $\mathbf{x}$, the matrix of deviation scores.

(c) Find $\mathbf{X}'\mathbf{x}$, the CSSCP matrix.

(d) Add 3 to each of the elements in matrix **X** and explain what happens to each of the values in parts a, b and c.

REFERENCES

Box, G. E. P. 1954. Some theorems on quadratic forms applied in the study of analysis of variance problems: I. Effect of inequality of variance in the one-way classification. *Annals of Mathematical Statistics, 25*, 290–302.

Johnson, R. A., and Wichern, D. W. 2007. *Applied Multivariate Statistical Analysis, Sixth Edition.* Upper Saddle River, NJ: Pearson Prentice-Hall.

Lattin, J., Carroll, J. D., and Green, P. E. 2003. *Analyzing Multivariate Data.* Belmont, CA: Brooks/Cole.

Rencher, A. C. 2002. *Methods of Multivariate Analysis, Second Edition.* New York: Wiley.

Rencher, A. C., and Schaalje, G. B. 2008. *Linear Models in Statistics, Second Edtion.* New York: Wiley.

Van de Geer, J. P. 1971. *Introduction to Multivariate Analysis for the Social Sciences.* San Francisco, CA: W. H. Freeman and Company.

CHAPTER FOUR

FACTOR ANALYSIS AND RELATED METHODS: QUINTESSENTIALLY MULTIVARIATE

4.1 INTRODUCTION

In introducing factor analysis as the quintessential multivariate method, the etymology of the word "quintessential" is instructive. The medieval Latin term was "Quinta Essentia" or fifth element, the element beyond the four classical elements (earth, air, fire, and water). The *Oxford English Dictionary* captures the original meaning best, "The fifth essence of ancient and medieval philosophy, supposed to be the substance of which the heavenly bodies were composed, and to be actually latent in all things, the extraction of it by distillation or other methods being one of the great objects of alchemy." Similar to how alchemists sought to derive essential substances through distillation, the fundamental purpose of factor analysis is to find the latent variables that are presumed to underlie empirically observed variables. The search for latent structure remains a key component of twenty-first century behavioral science.

In 2007, Cudeck and McCallum published *Factor Analysis at 100: Historical Developments and Future Directions* to commemorate the 100-year history of factor analytic methods since Spearman's seminal 1904 paper. In the preface, they expressed their enthusiasm for all that factor analysis has accomplished in bringing immeasurable latent variables to life:

Multivariate Analysis for the Biobehavioral and Social Sciences: A Graphical Approach,
First Edition. Bruce L. Brown, Suzanne B. Hendrix, Dawson W. Hedges, Timothy B. Smith.
© 2012 John Wiley & Sons, Inc. Published 2012 by John Wiley & Sons, Inc.

Factor analysis is one of the great success stories of statistics in the social sciences because the primary focus of attention has always been on the relationships among fundamental traits such as intelligence, social class, or health status that are unmeasurable. Factor analysis provided a way to go beyond empirical variables, such as tests and questionnaires, to the corresponding latent variables that underlie them. The great accomplishment of Spearman was in advancing a method to investigate fundamental factors. It has proven its utility in hundreds of scientific studies, across dozens of disciplines, and over a hundred years (p. ix).

The popularity of factor analysis within psychology is attested by a count of the number of factor analysis articles in major American Psychological Association journals. Table 4.1 displays the frequency counts of the four most frequently used methods in each of six representative APA journals (counted since inception). In the first three of these journals, factor analysis is by far the most used method, even beating out such common methods as the *t*-test and analysis of variance (ANOVA). Even in basic science areas like comparative

TABLE 4.1. Frequencies of Occurrence of Four Most Common Statistical Methods in Each of Six APA Journals

	Factor Analysis	Multiple Regression	ANOVA	*t*-Test
Journal of Personality and Social Psychology	252	89	26	124
Journal of Consulting and Clinical Psychology	354	84	29	177
Journal of Counseling Psychology	168	168	24	69
Journal of Comparative Psychology	44	6	6	109
Journal of Experimental Psychology, General	41	11	1	106
Journal of Experimental Psychology: Learning, Memory, and Cognition	7	11	0	59

psychology and experimental psychology, factor analysis is in second place, only being exceeded by the *t*-test. In the *Journal of Experimental Psychology: Learning, Memory, and Cognition*, few statistical methods of any kind appear in the journal search, but factor analysis is third most frequent, behind only the *t*-test and multiple regression.

The development of factor analysis and principal component analysis (Pearson, 1901; Spearman, 1904) in the early part of the 20th century was closely connected to the rise of the psychological testing movement. It was central to a concerted attempt by scholars, such as Spearman (1904, 1934), Thurstone (1931,1935,1947), Holzinger (1930,1942), Burt (1909,1941), Pearson and Moul (1927), and others, to discover the structure of intellect. A good summary of the mathematical and conceptual foundations of this work is given in Harman's (1976) classic book, *Modern Factor Analysis*. Reading Harman's account now, about half a century after it was first written, one is struck with the quaint positivism implicit in the philosophical foundations of factor analysis. Many investigators seem to have viewed it as an ontological discovery procedure with near magical properties. The factor analytic literature is punctuated with an abundance of "scientific discovery" statements, such as "four factors emerged," with the implied presupposition that factor analytic methods can be trusted to answer the question of dimensionality. It should be added, however, that such overextended claims in the name of factor analysis are not restricted to the early studies. Factor analysis continues to be misunderstood well into the 21st century. The situation is well summarized by Kline's (1993, 95) comment that "factor analysis, although a complex technique, is not so difficult as to defy understanding and produce the regular misuse that is commonly found even in reputable journals, as has been shown by Cattell (1978)."

To minimize the perpetuation of error, it is essential to understand what factor analysis is—and what it is not. Factor analysis identifies the relationships among measured variables for the purposes of reducing data, such as collapsing several items on a test into subscales, and/or evaluating theoretical structures, such as hypothetical components of individuals' intellectual abilities. Factor analysis provides several key pieces of information about multivariate data: (1) identification of inferred latent variables referred to as factors, (2) estimates of the amount of variance explained by each factor, and (3) the relationship of the original data to each factor. In an example involving test item data, a factor analysis would, under optimal conditions, enable a researcher to estimate (1) how many statistically identified factors the items evaluate, (2) how related each of those factors are to one another and to the overall test score, and (3) which items best relate to the various factors identified (and which items relate poorly).

As will be demonstrated at the end of this chapter, it is easy to draw incorrect conclusions from factor analysis, especially when the measurements entering into the analysis have low reliability. This chapter could have been entitled "Factor Analysis: The Misunderstood Method," for truly it is a method that has been surrounded by controversy—one of the most seductive of quantitative procedures. It is often used in ways that draw unwarranted conclusions from data. Rencher (2002, 409), a well-known mathematically oriented multivariate statistician, comments that factor analysis "is considered controversial by some statisticians," and he gives as the reason that "in practice there are some data sets for which the factor analysis model does not provide a satisfactory fit."

Notwithstanding these misuses, factor analysis can be a valuable data analysis technique, particularly in connection with graphical applications, such as those presented in Chapter 5. When the strengths and weaknesses of the method are adequately understood, it is an important component in one's arsenal of data analytic strategies. Beyond this, there is an additional incentive for gaining a sound understanding of the concepts and methods of factor analysis—it *is* the quintessential multivariate method in the sense of being fundamental to all of the others. Not only is it the first of the multivariate methods to be developed, it also provides a conceptual foundation that makes the other methods easier to understand.

Clyde Coombs, one of the founding fathers of mathematical psychology (Coombs, 1964; Coombs et al., 1970), remarked in a seminar at the University of Michigan four decades ago that in the good old days (decades preceding the 1970s), one could ostensibly get a PhD in quantitative psychology simply by gaining a thorough understanding of factor analysis. However, he went on, in the decade of the 1970s, one was expected to be conversant with a wide variety of multivariate methods. He observed that it is not the case that that the PhD candidates of the 1970s were brighter than those of previous decades. Rather, there is sufficient redundancy in the principles underlying all of the multivariate methods, that once one thoroughly understands factor analysis, the conceptual basis of the others follows in a relatively clear and understandable manner.

For that reason, the factor analysis chapter in this book is the first and by far the largest of the substantive multivariate chapters. That is also why in the title we have referred to the variety of factor analysis and principal component methods as "quintessentially multivariate." As the rationale and computational basis for these methods becomes clear, they form a fitting backdrop for the study of the remaining multivariate methods in the book.

4.2 AN APPLIED EXAMPLE OF FACTORING: THE MENTAL SKILLS OF MICE

One of the prominent claims from a century of factor analyzing intelligence test results is that human intelligence is dominated by a single factor of general intelligence. Locurto et al. (2006) have used factor analysis to demonstrate a contrasting finding—that the "general factor" claim is not true of mice. Their paper is one of the 44 factor analytic studies enumerated in the table above for the *Journal of Comparative Psychology* since its inception. Table 4.2 pres-

TABLE 4.2. Correlation Matrix of Data Selected from Table 1 of Locurto et al. (2006)

Task	1	2	3	4
1. Olfactory	—			
2. Winshift	−0.21	—		
3. Detour	0.27	0.08	—	
4. Fear	−0.02	−0.2	−0.28	—
5. Operant	−0.04	0.06	0.03	0

TABLE 4.3. Factor Loadings and Eigenvalues for Data Selected from Table 2 of Locurto et al. (2006)

Task	Factor 1	Factor 2
Olfactory	0.40	0.74
Winshift	0.31	−0.74
Detour	0.81	0.19
Fear	−0.71	0.29
Operant	0.06	−0.25
Eigenvalue	1.41	1.29
% variance	28	26

ents the correlation matrix for the performance of 47 mice on five cognitive tasks (from table 1 of their paper). Table 4.3 gives the results of a factor analysis of this correlation matrix (from table 2 of their paper).

The factor loadings are the correlation coefficients between each of the five cognitive tasks (rows) and each of the two factors (columns). This 5 × 2 matrix of factor loadings is referred to as the *factor pattern*, the pattern of correlations between six variables and two factors. The largest factor loading for factor 1, 0.81, is the loading for "detour." The numerically smallest loading for factor 1 is "operant," 0.06. A small loading indicates that the variable is not very related to the factor in question. The largest negative loading for factor 1 is "fear," −0.71, indicating that the fear task is quite related to the first factor, but with an inverse relationship. The interpretation of each factor is found through the identity of its highly correlating variables, the ones with numerically high loadings whether positive or negative. Factor 1 in this analysis is accordingly defined as a "detour task versus fear task" dimension, with the positive end of the factor being defined by detour and the negative end being defined by fear. Similarly, factor 2 is seen to be an "olfactory versus winshift" factor, with the positive end defined by the olfactory task and the negative end defined by the winshift task. Thus the observed intercorrelations (Table 4.2) can be distilled to two latent factors (Table 4.3).

The eigenvalues at the bottom of the table indicate how much of the variance in the five variables is accounted for by each factor. Since there are five variables in this analysis, and since the variance for each standardized variable is 1,[1] there are five variance units to be accounted for. Accordingly, the first factor, with an eigenvalue of 1.41, accounts for 1.41 variance units out of five, or approximately 28% of the variance. Similarly, the second factor accounts for 1.29 variance units out of five, or approximately 26% of the variance. The two factors are about equal in the amount of variance they account for in the five cognitive tasks. On this basis, Lucurto et al. (2006) can argue that there is probably not a single general factor of intelligence in mice.

A *vector plot graph* can be constructed from the factor loadings of Table 4.2, as shown below. This is a graphical expression of the information in the factor pattern. Notice that this graph

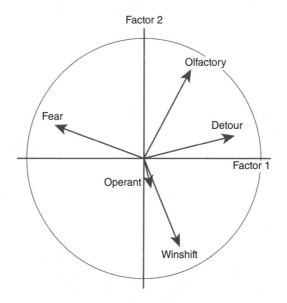

shows clearly that factor 1 is defined primarily by the detour task on the right (a positive loading) and the fear task on the left (a negative loading) as discussed above. The detour task involves blocking the mice from reaching a small collection of mouse toys, a kind of mouse playground, and the fear task involves avoiding foot shock. The structural organization of this first factor is telling us that the two are inversely related—play-motivated cognitive skill versus fear-motivated cognitive skill. That is, the kind of mice who do well in getting to the toys tend to not do so well in avoiding foot shock, and vice versa. The second factor is defined primarily by the olfactory task at the top (a positive loading) and the winshift task at the bottom (a negative loading). The

---

[1] The factoring is done on a correlation matrix, which is a standardized covariance matrix. The standardizing operation converts each covariance to a correlation coefficient, and each variance to 1.00, as explained in Chapter 2, Section 2.5.

olfactory task involves the attraction of hungry mice to chocolate, nutmeg, and cinnamon smells, and the winshift task involves water escape. Factor 2 therefore also involves an approach task on one end and an avoidance task at the other, but along a different positive versus negative dimension, one defined by sensory hedonism versus water aversion. The visual factor pattern does seem to provide some insight into the underlying latent structure of cognitive skills and motivation in mice, as a simple example of the broad methodological utility Cudeck and MacCallum (2007) are claiming for factor analysis.

How does one create a vector plot graph from the factor loadings? It is simply a matter of using the loadings of each variable on factor 1 and on factor 2 as coordinates for locating the endpoint to draw the vector for that variable within the two-factor space. Consider, for example, a variable that is correlated 0.60 with factor 1 and 0.80 with factor 2 as shown in the artificial "factor pattern" below. The circle for this factor pattern is created with a radius of 1. The endpoint for this vector is plotted at a coordinate of 0.60 on factor 1 and 0.80 on factor 2, as shown. Since correlation coefficients are cosines of the angle between the two variables, as explained in Chapter 2, Section 2.5, the angle between the variable and factor 1 is arccosine $(0.60) = 53.1°$, as shown in the figure. Similarly, the angle between the variable and factor 2 is arccosine $(0.80) = 36.9°$. The square of the correlation coefficient (R^2, the coefficient of determination) indicates the amount of variance accounted for, so factor 1 accounts for $0.60^2 = 0.36$, or 36% of the variance, and factor 2 accounts for $0.80^2 = 0.64$, or 64% of the variance. One hundred percent of the variance is therefore accounted for in the two-factor space. This vector plot graph shows the factor structure for a variable that is completely accounted for by two factors. Factor analysis is obviously built upon the geometry of correlation coefficients.

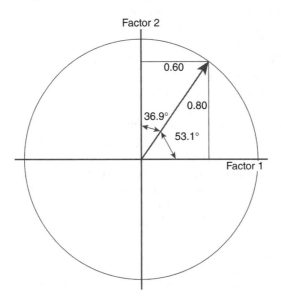

TABLE 4.4. Factor Analysis Summary Table for Locurto et al. Data (2006)

	Loadings		Communalities		
	Factor 1	Factor 2	Factor 1	Factor 2	Total
1. Olfactory	0.40	0.74	0.16	0.55	0.71
2. Winshift	0.31	−0.74	0.10	0.55	0.64
3. Detour	0.81	0.19	0.66	0.04	0.69
4. Fear	−0.71	0.29	0.50	0.08	0.59
5. Operant	0.06	−0.25	0.00	0.06	0.07
	Eigenvalues:		1.42	1.28	2.70
	Percents of eigenvalues:		28.4%	25.6%	54.0%

All of the information in the factor structure can be neatly summarized in a *factor analysis summary table* like the one shown in Table 4.4. The table is created by first entering the matrix of factor loadings on the left, as shown. Each entry in this 5×2 matrix of factor loadings is then squared to create the corresponding 5×2 matrix of *communalities* as shown in the right half of the table. Since these are squared loadings, that is, squared correlation coefficients, they indicate the proportion of variance in each variable accounted for by each factor. Communalities are therefore a particular type of coefficient of determination. These proportions of variance accounted for are additive, so they can be summed across the rows to get the *total communalities* as shown in the rightmost column in the table. To see how these are interpreted, consider the first variable, the olfactory task. Its loadings on factors 1 and 2 are 0.40 and 0.74, respectively. The corresponding communalities are 0.16 and 0.55, so factor 1 accounts for 16% of the variance in winshift, and factor 2 accounts for 55%, for a total of 71% of the variance in winshift accounted for within the two-factor space. Winshift has the largest total communality (0.71), and is therefore the variable best accounted for in this two-factor space. Variable 5, the operant task, has a very small total communality (0.07), and is therefore not at all well accounted for in this space.

The communalities can also be summated down each column in order to obtain the *eigenvalues* for each factor (column 1 and column 2) and also for the two factors combined (rightmost column). Obviously, the largest value an eigenvalue could take for five variables is five. That is, if the two-factor space accounted for all of the variance in each of the five variables, the final column would contain five ones and sum to five. The eigenvalue for the first factor of the Locurto et al. (2006) data is 1.42, meaning that 1.42 variance units are accounted for out of the five possible, which gives 28.4% of the variance in the five variables accounted for by factor 2 (1.42/5 = 0.284, which corresponds to 28.4%). Similarly, the second factor has an eigenvalue of 1.28, which indicates 25.6% of the variance accounted for. The cumulative eigenvalue for these two factors is 2.70, indicating that the two-dimensional factor space accounts for 54.0% of the variance in the five variables.

The vector plot graph below shows how the factor loadings are used as coordinates to plot the vectors for the five variables within the factor space. The factor analysis summary table shows the factor pattern and its properties numerically, and the vector plot graph shows the same information graphically. It is instructive to see the convergence between the numerical information and the graphical pattern. For example, the olfactory variable has the greatest proportion of variance accounted for within the two factor space (total communality of 0.71), and its vector in this factor pattern vector plot graph is the longest, the closest to the outer circle that has a radius of 1.00 (all variance accounted for). When the factor space accounts for all of the variance in a variable, the vector for that variable extends all the way out to the circle as was shown in the vector plot graph above for a variable with loadings of 0.60 and 0.80 on factor 1 and factor 2. The length of the vector within this two-factor space is in fact obtained as the square root of its total communality, so for olfactory, it is $\sqrt{0.71} = 0.84$. The lengths of the other four vectors are 0.80, 0.83, 0.77, and 0.25 for winshift, detour, fear, and operant, respectively. Of course, each of the five vectors actually has a length of 1.00, since the variance (standardized) of each is 1.00, but the full length vectors are not completely contained within this two-factor space. That is, each extends to some extent out beyond the two-factor plane into a third dimension, and the lengths given by the communalities are the projections of those vectors back into the two-factor space. It can be seen that the first four variables all have about the same magnitude of projection into this two-factor space (ranging from 0.77 to 0.84), but the operant task variable has little in common with this space, with a projection of 0.25.

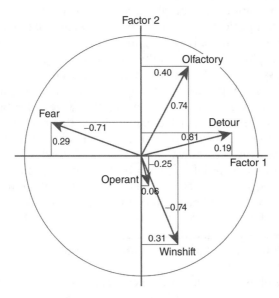

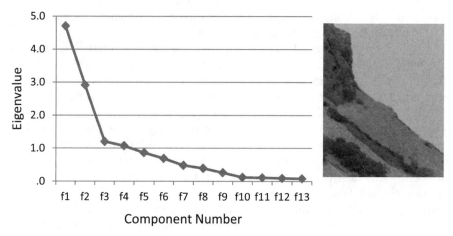

Figure 4.1   Scree plot for an example data set with 13 variables.

If a third factor were extracted, it would give a better idea of the location of each of these five variables, whether each extends upward from the page or recedes back below the page in the third dimension. One of the decisions to be made in doing factor analysis is how many factors should be included in the solution. Locurto et al. (2006) used the Kaiser (1960) criterion of an *eigenvalue greater than one* to decide when to stop factoring. This is a reasonable choice, since an eigenvalue of one indicates that we are accounting for only as much variance as is contained in one variable. A factor corresponding to less than one variable does not seem reasonable, so any factor with an eigenvalue less than one is not included in the analysis.

There are other criteria besides the Kaiser criterion for deciding how many factors to keep. Another commonly used method is the scree plot (Cattell, 1966). "Scree" is a geologic term. It refers to the accumulation of broken rock material that is found at the base of mountain cliffs. In factor analysis, a scree plot is simply a plot of the eigenvalues from successively extracted factors. Figure 4.1 shows a scree plot from a factor analysis of 13 variables. The first eigenvalue is nearly 5, the second nearly 3, the third a little above one, and the one for the fourth factor is ever so slightly above 1. By the eigenvector greater than one criterion, four factors would be retained, but by the use of the scree plot criterion, only three would be retained, which seems more reasonable in this case. As shown in the pictorial inset of geological scree (American Fork Canyon, Utah, photo by the authors), one can clearly see where the scree slope intersects with the nearly vertical mountain cliffs. In a similar way, a pronounced knee of the curve on the factor analytic scree plot (at factor 3, a value of 1.17) is taken as an indication of where the true variance due to factors ends and the scree of random error variance begins. All factors beyond the initial three in this analysis would, by the scree criterion, be disregarded as error.

Not all scree plots are so clearly delineated as this one for 13 variables. In the scree plot for five variables in the Locurto et al. (2006) study, Figure 4.2, there is

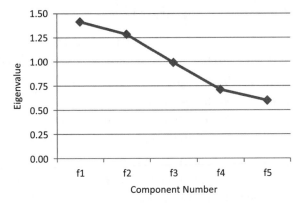

Figure 4.2   Scree plot for Locurto et al. factor analysis with five variables.

no clear delineation between factor structure and error. It therefore seems reasonable to go with the "eigenvalue greater than one" criterion as they did.

For a number of reasons, the Locurto et al. (2006) study is an ideal published paper to serve as an example in the initial explanation of factor analysis. It uses the principal component method of factor analysis, which is the simplest and most basic of factor analytic models. It also has a small dataset size (five variables), an easy interpretation, and an intrinsically interesting topic. The data are simple yet realistic.

4.3   CALCULATING FACTOR LOADINGS TO REVEAL THE STRUCTURE OF SKILLS IN MICE

Actually, what has already been covered in this chapter accounts for a major share of the mathematics and geometry of the principal component method of factor analysis. What is missing is an explanation of how factor loadings are obtained from the correlation matrix. The reader is at this point referred to the discussion of eigenvectors and eigenvalues in Chapter 3, Section 3.11, and vector normalization in Section 3.13, since *factor loadings are nothing more than eigenvectors normalized to their eigenvalues.*

As explained in Chapter 3, for a certain class of matrices including covariance matrices, correlation matrices, and SSCP matrices, there exist certain vectors $\mathbf{x}$ such that when the matrix is postmultiplied by them, the resultant vector is equal to vector $\mathbf{x}$ times a constant λ, as shown in Equation 4.1.

$$\mathbf{A}\,\mathbf{x} = \mathbf{x}\,\lambda \tag{4.1}$$

The vector $\mathbf{x}$ is an *eigenvector* of matrix $\mathbf{A}$, and λ is the *eigenvalue* corresponding to that eigenvector. When the correlation matrix $\mathbf{R}$ from the Locurto et al. (2006) data is postmultiplied by $\mathbf{f}_1$, the vector of loadings on factor 1, we see that the resultant vector is indeed equal to $\mathbf{f}_1$ multiplied by the first eigenvalue, which is $\lambda_1 = 1.414$.

$$\mathbf{Rf} = \mathbf{f}\,\lambda \tag{4.2}$$

$$\mathbf{Rf} = \begin{bmatrix} 1.000 & -0.210 & 0.270 & -0.020 & -0.040 \\ -0.210 & 1.000 & 0.080 & -0.200 & 0.060 \\ 0.270 & 0.080 & 1.000 & -0.280 & 0.030 \\ -0.020 & -0.200 & -0.280 & 1.000 & 0.000 \\ -0.040 & 0.060 & 0.030 & 0.000 & 1.000 \end{bmatrix} * \begin{bmatrix} 0.397 \\ 0.308 \\ 0.806 \\ -0.713 \\ 0.065 \end{bmatrix}$$

$$= \begin{bmatrix} 0.562 \\ 0.435 \\ 1.139 \\ -1.008 \\ 0.091 \end{bmatrix} = \begin{bmatrix} (0.397)*(1.414) \\ (0.308)*(1.414) \\ (0.806)*(1.414) \\ (-0.713)*(1.414) \\ (0.065)*(1.414) \end{bmatrix} = \mathbf{f}\,\lambda.$$

A more basic form of the equation, Equation 4.3, uses $\mathbf{k}$, the eigenvector normalized to one, rather than using $\mathbf{f}$.

$$\mathbf{Rk} = \mathbf{k}\,\lambda. \tag{4.3}$$

A vector is normalized to one by dividing it by its *magnitude*, which is the square root of its sum of squares. That is, it is divided by the square root of the value to which it is currently normalized, in this case λ:

$$\mathbf{k} = \frac{\mathbf{f}}{\sqrt{\lambda}} = \frac{\mathbf{f}}{\sqrt{1.414}} = \begin{bmatrix} 0.334 \\ 0.259 \\ 0.667 \\ -0.599 \\ 0.054 \end{bmatrix}.$$

The fundamental eigenvalue/eigenvector equation works equally well with the normalized eigenvectors of Equation 4.3.

$$\mathbf{Rk} = \begin{bmatrix} 1.000 & -0.210 & 0.270 & -0.020 & -0.040 \\ -0.210 & 1.000 & 0.080 & -0.200 & 0.060 \\ 0.270 & 0.080 & 1.000 & -0.280 & 0.030 \\ -0.020 & -0.200 & -0.280 & 1.000 & 0.000 \\ -0.040 & 0.060 & 0.030 & 0.000 & 1.000 \end{bmatrix} * \begin{bmatrix} 0.334 \\ 0.259 \\ 0.667 \\ -0.599 \\ 0.054 \end{bmatrix}$$

$$= \begin{bmatrix} 0.473 \\ 0.366 \\ 0.958 \\ -0.848 \\ 0.077 \end{bmatrix} = \begin{bmatrix} 0.334 \\ 0.259 \\ 0.667 \\ -0.599 \\ 0.054 \end{bmatrix} *1.414 = \mathbf{k}\,\lambda.$$

As explained in chapter three and as just demonstrated, eigenvectors have this interesting property of reproducing themselves (but multiplied by a constant) when a correlation matrix is postmultiplied by them.

How is the eigenvector obtained from the correlation matrix? The simplest way to state it is to say that we first raise the correlation matrix $\mathbf{R}$ to a particular power, reasonably high. The eigenvector is then found as *the vector of row sums of this $\mathbf{R}$ matrix that has been raised to a particular power*. We will illustrate how this is done.

Consider two matrices, $\mathbf{R}$ and $\mathbf{R}^{32}$. The $\mathbf{R}^{32}$ matrix is created by squaring $\mathbf{R}$ (multiplying it by itself using matrix multiplication), and then squaring its square, repeating this process five times to create the 32nd power of the $\mathbf{R}$ matrix.[2] For each successive matrix power, create a vector of row sums which will be symbolized as vector $\mathbf{m}_{(1)}$ for the $\mathbf{R}$ matrix, $\mathbf{m}_{(2)}$ for the $\mathbf{R}^2$ matrix, up to vector $\mathbf{m}_{(32)}$ for the $\mathbf{R}^{32}$ matrix. This is done by postmultiplying each successive matrix of $\mathbf{R}$ (to some power), by a vector of ones (symbolized by a boldface **1**).

$$\mathbf{m}_{(1)} = \mathbf{R} * \mathbf{1} \text{ and } \mathbf{m}_{(32)} = \mathbf{R}^{32} * \mathbf{1}$$

$$\mathbf{m}_{(1)} = \mathbf{R} * \mathbf{1} = \begin{bmatrix} 1.000 & -0.210 & 0.270 & -0.020 & -0.040 \\ -.210 & 1.000 & 0.080 & -0.200 & 0.060 \\ .270 & 0.080 & 1.000 & -0.280 & 0.030 \\ -.020 & -0.200 & -0.280 & 1.000 & 0.000 \\ -.040 & 0.060 & 0.030 & 0.000 & 1.000 \end{bmatrix} * \begin{bmatrix} 1 \\ 1 \\ 1 \\ 1 \\ 1 \end{bmatrix} = \begin{bmatrix} 1.000 \\ 0.730 \\ 1.100 \\ 1.500 \\ 1.0501 \end{bmatrix}.$$

$$\mathbf{m}_{(32)} = \mathbf{R}^{32}\mathbf{1} = \begin{bmatrix} 8590 & 4323 & 15088 & -12553 & 757 \\ 4323 & 5683 & 11089 & -10612 & 1343 \\ 15088 & 11089 & 29986 & -26330 & 2289 \\ -12553 & -10612 & -26330 & 23604 & -2284 \\ 757 & 1343 & 2289 & -2284 & 331 \end{bmatrix} * \begin{bmatrix} 1 \\ 1 \\ 1 \\ 1 \\ 1 \end{bmatrix} = \begin{bmatrix} 16206 \\ 11826 \\ 32123 \\ -28176 \\ 2436 \end{bmatrix}.$$

Notice how much larger the $\mathbf{R}^{32}$ matrix entries are than those in the $\mathbf{R}$ matrix, and also how much larger the row sums are in the $\mathbf{m}_{(32)}$ vector than those in the $\mathbf{m}_{(1)}$ vector. These row sum vectors $\mathbf{m}_{(1)}$ and $\mathbf{m}_{(32)}$ must now be normalized in such a way that the largest value in each vector will be set to one, in order to make them comparable with one another. This is done by dividing each vector by its own maximum value to create comparable $\mathbf{k}$ vectors.

$$\mathbf{k}_{(1)} = \frac{\mathbf{m}_{(1)}}{\max(\mathbf{m}_{(1)})} = \begin{bmatrix} 0.909 \\ 0.664 \\ 1.000 \\ 0.455 \\ 0.955 \end{bmatrix}, \quad \mathbf{k}_{(32)} = \frac{\mathbf{m}_{(32)}}{\max(\mathbf{m}_{(32)})} = \begin{bmatrix} 0.504 \\ 0.368 \\ 1.000 \\ -0.877 \\ 0.076 \end{bmatrix}.$$

[2] Five squarings of $\mathbf{R}$ yield $\mathbf{R}$ to the 32nd power:

These $\mathbf{k}_{(i)}$ vectors are referred to as *trial eigenvectors*, and up to an optimum point, the larger the power of the $\mathbf{R}$ matrix, the more precisely accurate will be the corresponding trial eigenvector $\mathbf{k}_{(i)}$. Consider the nine trial eigenvectors from $\mathbf{k}_{(1)}$ to $\mathbf{k}_{(256)}$ laid out side by side (in five-place accuracy) below. These correspond to each power of the $\mathbf{R}$ matrix from the unsquared $\mathbf{R}$ matrix up to the $\mathbf{R}$ matrix taken to the 256th power.

$\mathbf{k}_{(1)}=$	$\mathbf{k}_{(2)}=$	$\mathbf{k}_{(4)}=$	$\mathbf{k}_{(8)}=$	$\mathbf{k}_{(16)}=$	$\mathbf{k}_{(32)}=$	$\mathbf{k}_{(64)}=$	$\mathbf{k}_{(128)}=$	$\mathbf{k}_{(256)}=$
0.90909	0.82711	0.71300	0.61299	0.54470	0.50449	0.49394	0.49343	0.49343
0.66364	0.43261	0.23896	0.23797	0.31732	0.36814	0.38142	0.38207	0.38207
1.00000	1.00000	1.00000	1.00000	1.00000	1.00000	1.00000	1.00000	1.00000
0.45455	0.01970	-0.48142	-0.76384	-0.84742	-0.87713	-0.88448	-0.88484	-0.88484
0.95455	0.82340	0.51171	0.17171	0.06828	0.07583	0.07997	0.08017	0.08017

Notice that the two trial eigenvectors on the right, $\mathbf{k}_{(128)}$ and $\mathbf{k}_{(256)}$, are precisely identical to one another to five place accuracy. This is therefore the optimal power for the Locurto et al. (2006) correlation matrix, the power at which this iterative process converges.

This final vector $\mathbf{k}_{(256)}$ can now be normalized to one (as explained in Chapter 3, Section 3.13) to become the true eigenvector $\mathbf{k}$ of the correlation matrix $\mathbf{R}$. The sum of squares of vector $\mathbf{k}_{(256)}$ is:

$$SS = 0.493^2 + 0.382^2 + 1.000^2 + (-0.885)^2 + 0.080^2 = 2.179.$$

Using this value to normalize the vector to one, the resultant vector $\mathbf{k}$ is found to agree with the value of $\mathbf{k}$ calculated above from the loadings reported in the Locurto et al. (2006) paper.

$$\mathbf{k} = \frac{\mathbf{k}_{(256)}}{\sqrt{SS}} = \begin{bmatrix} 0.493/\sqrt{2.179} \\ 0.382/\sqrt{2.179} \\ 1.000/\sqrt{2.179} \\ -0.885/\sqrt{2.179} \\ 0.080/\sqrt{2.179} \end{bmatrix} = \begin{bmatrix} 0.334 \\ 0.259 \\ 0.677 \\ -0.559 \\ 0.054 \end{bmatrix}.$$

Although this process is somewhat tedious computationally, especially the successive squaring of the $\mathbf{R}$ matrix (the very kind of thing for which computers were invented), it is quite simple conceptually. *The eigenvector is nothing more than the vector of row sums of the optimal power of the $\mathbf{R}$ matrix. The usual form of this vector, $\mathbf{k}$, is normalized to one.* In the future, it might be possible with a new generation spreadsheet to directly specify matrix multiplication to a desired power rather than the laborious process of successively squaring. This reduces the problem of obtaining an eigenvector to one of determining what that optimal power might be, since it varies from one dataset to another. Of course, in actual practice, one uses a computer application to obtain eigenvalues and eigenvectors, but we have illustrated the "by-hand" calculations that could be used to get these same values in order to more fundamentally understand what eigenvalues and eigenvectors are.

The algorithm of the *method of successive squaring*[3] is only one of many methods for finding eigenvalues and their corresponding eigenvectors. It was chosen for this chapter, not because of its computational efficiency, but because it is pedagogically accessible and can be summarized in one reasonably simple sentence. It is not the algorithm used by statistical packages, such as SAS, SPSS, and Stata. They use more complex and also more computationally efficient algorithms for obtaining eigenvalues and their associated eigenvectors. Actually, the computing of eigenvalues/eigenvectors is something of an art within applied mathematics, in the subfield referred to as numerical analysis. See, for example, the book by Parlett (1998) entitled *The Symmetric Eigenvalue Problem.*

Now that we have accounted for the major steps in the process of obtaining a factor analysis of a correlation matrix by the principal component method, we will demonstrate the entire process from start (raw data) to finish (factor analysis summary table) using simplest case data.

4.4 SIMPLEST CASE MATHEMATICAL DEMONSTRATION OF A COMPLETE FACTOR ANALYSIS

In this section, the entire process of calculations for creating a factor analysis by the principal component method will be demonstrated, but using a small and artificial simplest case data set. In other words, the purpose in this section is to make the conceptual and mathematical basis of the method as clear as possible without being encumbered with realistic data.

Given below is a summary of the computational steps in factor analysis. Each of these steps will be demonstrated using a simplest case dataset.

COMPUTATIONAL STEPS IN FACTOR ANALYSIS

Step 1: Calculate the covariance matrix from raw data. $S = \left(\dfrac{1}{df}\right) X' X_c$

Step 2: Calculate the correlation matrix. $R = ESE$

Step 3: Extract the first eigenvector k_1 from the correlation matrix by the method of successive squaring.

Step 4: Obtain the first eigenvalue λ_1 corresponding to the first eigenvector k_1 using the characteristic equation. $Rk_1 = k_1\lambda_1$.

---

[3] In Chapter 3, Section 3.12, it is explained that if **k** is an eigenvector of matrix **R**, it is also an eigenvector of all powers of matrix **R**. The eigenvectors remain the same, but the eigenvalue λ is taken to the same power as **R**. For the 32nd power, for example, the equation would be $R^{32}k = k\,\lambda^{32}$. It is because of this property of eigenvectors that the method of successive squaring works.

Step 5: Obtain the first vector of factor loadings $\mathbf{f}_1$ by normalizing the eigenvector $\mathbf{k}_1$ to λ_1. $f_1 = k_1 \sqrt{\lambda_1}$

Step 6: Calculate the first factor product matrix. $G_1 = f_1 f_1'$

Step 7: Obtain the matrix of first factor residuals. $R_1 = R - G_1$

Step 8: Extract the second eigenvector $\mathbf{k}_2$ from the matrix of first factor residuals $\mathbf{R}_1$ by the method of successive squaring.

Step 9: Obtain the second eigenvalue λ_2 corresponding to the second eigenvector $\mathbf{k}_2$, using the characteristic equation. $Rk_2 = k_2 \lambda_2$

Step 10: Obtain the second vector of factor loadings $\mathbf{f}_2$ by normalizing the eigenvector $\mathbf{k}_2$ to λ_2. $f_2 = k_2 \sqrt{\lambda_2}$

Step 11: Calculate the second factor product matrix. $G_2 = f_2 f_2'$

Step 12: Obtain the matrix of second factor residuals. $R_2 = R - G_2$

Step 13: Repeat steps 8 through 12 for additional factors as necessary.

Step 14: Examine the diagonal of the final matrix of residuals to determine how much variance was not accounted for in each variable. Examine the off-diagonal elements to determine how well the correlations for each pair of variables are reproduced by the factor solution.

Step 15: Adjoin $\mathbf{f}_1$ and $\mathbf{f}_2$ (and any additional factors extracted) to create the matrix of factor loadings $\mathbf{F}$ and complete the factor analysis summary table.

Suppose that we have six subjects in a research study on cognitive development in adolescent males, and each of six males have been measured on three variables, a test of verbal ability, a test of mathematical ability, and a test of logical reasoning.

Al: verbal = 8, math = 12, logic = 16
Bill: verbal = 6, math = 10, logic = 20
Charlie: verbal = 5, math = 4, logic = 9
Dan: verbal = 5, math = 10, logic = 15
Ed: verbal = 4, math = 16, logic = 20
Frank: verbal = 2, math = 8, logic = 10

The data for these six persons are collected in data matrix $\mathbf{X}$.

$$\mathbf{X} = \begin{bmatrix} 8 & 12 & 16 \\ 6 & 10 & 20 \\ 5 & 4 & 9 \\ 5 & 10 & 15 \\ 4 & 16 & 20 \\ 2 & 8 & 10 \end{bmatrix}.$$

The first analytical step is to create a covariance matrix from the raw data. The means vector is calculated:

$$\bar{\mathbf{X}} = [5 \quad 10 \quad 15].$$

and the matrix of deviations away from these means, $\mathbf{X}_c$, is calculated to be:

$$\mathbf{X}_c = \begin{bmatrix} 3 & 2 & 1 \\ 1 & 0 & 5 \\ 0 & -6 & -6 \\ 0 & 0 & 0 \\ -1 & 6 & 5 \\ -3 & -2 & -5 \end{bmatrix}.$$

The subscript "c" on the symbol $\mathbf{X}_c$ for deviation score data stands for "centered." The CSSCP (centered sum of squares and cross products) matrix can be found by premultiplying the deviation score matrix $(\mathbf{X}_c)$ by the transpose of the raw score matrix $(\mathbf{X})$.[4]

$$\text{CSSCP} = \mathbf{X}'\mathbf{X}_c = \begin{bmatrix} 8 & 6 & 5 & 5 & 4 & 2 \\ 12 & 10 & 4 & 10 & 16 & 8 \\ 16 & 20 & 9 & 15 & 20 & 10 \end{bmatrix} \begin{bmatrix} 3 & 2 & 1 \\ 1 & 0 & 5 \\ 0 & -6 & -6 \\ 0 & 0 & 0 \\ -1 & 6 & 5 \\ -3 & -2 & -5 \end{bmatrix} = \begin{bmatrix} 20 & 6 & 18 \\ 6 & 80 & 78 \\ 18 & 78 & 112 \end{bmatrix}.$$

The covariance matrix is obtained by dividing the CSSCP matrix $\mathbf{X}'\mathbf{X}_c$ by the degrees of freedom value, which is 5 for this data matrix:

---

[4] It is curious that one may obtain CSSCP, the centered form of the SSCP matrix, by multiplying a centered raw score matrix by an uncentered one. When there is rounding error in creating the centered matrix, as often is the case, it is advantageous to have only one centered matrix in the equation to avoid squaring and multiplying the error. It is true of course, that SSCP = $\mathbf{X}'\mathbf{X}$ and that CSSCP = $\mathbf{X}'_c\mathbf{X}_c$, but it is also true that CSSCP = $\mathbf{X}'_c\mathbf{X}$. This could, of course, be verified numerically by running the calculations, but it may also be of interest to know why. The reason it works is explained in an old multivariate statistics book by Lunneborg and Abbott (1983, 55–60). They introduce the concept of a centering matrix, C, which is a square matrix of dimensions $n \times n$, where n is the number of observations in the data matrix. When the $n \times m$ square and symmetric matrix of raw data is premultiplied by C, the result is centered data matrix: $\mathbf{X}_c = \mathbf{CX}$. But the centering matrix has a special property. It is *idempotent*, which means that when it squared, that is multiplied by itself, the result is equal to the original C matrix: $\mathbf{C}' \mathbf{C} = \mathbf{C}$. That is, the square (or for that matter, any power) of the matrix is equal to the matrix itself. Substituting this into the equation for a CSSCP matrix, we can see the reason for the anomaly.

$$\text{CSSCP} = \mathbf{X}'_c\mathbf{X}_c = (\mathbf{X}'\mathbf{C}')(\mathbf{CX}) = \mathbf{X}'\mathbf{C}'\mathbf{CX} = \mathbf{X}'\mathbf{CX} = \mathbf{X}'(\mathbf{CX}) = \mathbf{X}'\mathbf{X}_c$$

$$\mathbf{S} = \left(\frac{1}{df}\right)\mathbf{X}'\mathbf{X}_c = \left(\frac{1}{5}\right)\begin{bmatrix} 20 & 6 & 18 \\ 6 & 80 & 78 \\ 18 & 78 & 112 \end{bmatrix} = \begin{bmatrix} 4 & 1.2 & 3.6 \\ 1.2 & 16 & 15.6 \\ 3.6 & 15.6 & 22.4 \end{bmatrix}.$$

The covariance matrix is transformed into a correlation matrix by premultiplying and also postmultiplying it by a standardizing matrix $\mathbf{E}$ (which is a diagonal matrix of reciprocals of standard deviations).

$$\mathbf{R} = \mathbf{ESE} = \begin{bmatrix} \dfrac{1}{2} & 0 & 0 \\ 0 & \dfrac{1}{4} & 0 \\ 0 & 0 & \dfrac{1}{\sqrt{22.4}} \end{bmatrix}\begin{bmatrix} 4 & 1.2 & 3.6 \\ 1.2 & 16 & 15.6 \\ 3.6 & 15.6 & 22.4 \end{bmatrix}\begin{bmatrix} \dfrac{1}{2} & 0 & 0 \\ 0 & \dfrac{1}{4} & 0 \\ 0 & 0 & \dfrac{1}{\sqrt{22.4}} \end{bmatrix}$$

$$= \begin{bmatrix} 1.000 & 0.150 & 0.380 \\ 0.150 & 1.000 & 0.824 \\ 0.380 & 0.824 & 1.000 \end{bmatrix}.$$

This process of using matrix algebra to create a correlation matrix from a raw data matrix was demonstrated in Chapter 2, Section 2.5, but is reviewed again here for completeness in demonstrating this process.

Now we are ready to factor the correlation matrix using the method of successive squaring. A row-sum vector $\mathbf{s}_{(1)}$ is created for matrix $\mathbf{R}$ (by postmultiplying $\mathbf{R}$ by a vector of ones), and this $\mathbf{s}_{(1)}$ vector is normalized to create a $\mathbf{k}_{(1)}$ trial eigenvector, whose largest value is one. R is then squared to obtain $\mathbf{R}^2$, and $\mathbf{R}^2$ is postmultiplied by a vector of ones to get the $\mathbf{s}_{(2)}$ vector, which is normalized to have one as its largest value, to become trial eigenvector $\mathbf{k}_{(2)}$. The trial eigenvectors $\mathbf{k}_{(1)}$ and $\mathbf{k}_{(2)}$ are compared to see whether they converge within an acceptable tolerance. They do not, so the correlation matrix is squared again. This is repeated for successive powers of $\mathbf{R}$ until convergence is reached.

$$\mathbf{s}_{(1)} = \mathbf{R}1 = \begin{bmatrix} 1.000 & 0.150 & 0.380 \\ 0.150 & 1.000 & 0.824 \\ 0.380 & 0.824 & 1.000 \end{bmatrix}\begin{bmatrix} 1 \\ 1 \\ 1 \end{bmatrix} = \begin{bmatrix} 1.530 \\ 1.974 \\ 2.204 \end{bmatrix} \qquad \mathbf{k}_{(1)} = \frac{\mathbf{s}_{(1)}}{\max(\mathbf{s}_{(1)})} = \begin{bmatrix} 0.694 \\ 0.896 \\ 1.000 \end{bmatrix}$$

$$\mathbf{s}_{(2)} = \mathbf{R}^2 1 = \begin{bmatrix} 1.167 & 0.613 & 0.884 \\ 0.613 & 1.702 & 1.705 \\ 0.884 & 1.705 & 1.824 \end{bmatrix}\begin{bmatrix} 1 \\ 1 \\ 1 \end{bmatrix} = \begin{bmatrix} 2.665 \\ 4.020 \\ 4.413 \end{bmatrix} \qquad \mathbf{k}_{(2)} = \frac{\mathbf{s}_{(2)}}{\max(\mathbf{s}_{(2)})} = \begin{bmatrix} 0.604 \\ 0.911 \\ 1.000 \end{bmatrix}$$

$$\mathbf{s}_{(4)} = \mathbf{R}^4 1 = \begin{bmatrix} 2.520 & 3.267 & 3.690 \\ 3.267 & 6.179 & 6.553 \\ 3.690 & 6.553 & 7.015 \end{bmatrix}\begin{bmatrix} 1 \\ 1 \\ 1 \end{bmatrix} = \begin{bmatrix} 9.478 \\ 15.999 \\ 17.259 \end{bmatrix} \qquad \mathbf{k}_{(4)} = \frac{\mathbf{s}_{(4)}}{\max(\mathbf{s}_{(4)})} = \begin{bmatrix} 0.549 \\ 0.927 \\ 1.000 \end{bmatrix}$$

$$\mathbf{s}_{(8)} = \mathbf{R}^8 1 = \begin{bmatrix} 30.647 & 52.607 & 56.602 \\ 52.607 & 91.797 & 98.519 \\ 56.602 & 98.519 & 105.774 \end{bmatrix} \begin{bmatrix} 1 \\ 1 \\ 1 \end{bmatrix} = \begin{bmatrix} 139.856 \\ 242.923 \\ 260.894 \end{bmatrix}$$

$$\mathbf{k}_{(8)} = \frac{\mathbf{s}_{(8)}}{\max(\mathbf{s}_{(8)})} = \begin{bmatrix} 0.536 \\ 0.931 \\ 1.000 \end{bmatrix}$$

$$\mathbf{s}_{(16)} = \mathbf{R}^{16}1 = \begin{bmatrix} 6910 & 12018 & 12905 \\ 12018 & 20900 & 22442 \\ 12905 & 22442 & 24098 \end{bmatrix} \begin{bmatrix} 1 \\ 1 \\ 1 \end{bmatrix} = \begin{bmatrix} 31833 \\ 55360 \\ 59444 \end{bmatrix}$$

$$\mathbf{k}_{(16)} = \frac{\mathbf{s}_{(16)}}{\max(\mathbf{s}_{(16)})} = \begin{bmatrix} 0.536 \\ 0.931 \\ 1.000 \end{bmatrix}.$$

The last two trial eigenvectors, $\mathbf{k}_{(8)}$ and $\mathbf{k}_{(16)}$, agree to three place accuracy, so trial eigenvector $\mathbf{k}_{(16)}$ is normalized to 1, and becomes the true eigenvector, $\mathbf{k}_1$. The **magnitude** of a vector (as explained in Chapter 3, Section 3.13) is the square root of its sum of squares, that is, the square root of the value to which it is currently normalized. In this case, its sum of squares is $0.536^2 + 0.931^2 + 1.000^2 = 2.154$, so the vector is currently normalized to 2.154. The final trial eigenvector, $\mathbf{k}_{(16)}$, is therefore normalized to one by dividing each entry in the vector by the magnitude of that vector, $\sqrt{2.154} = 1.467$, to become the true eigenvector $\mathbf{k}_1$.

$$\mathbf{k}_1 = \frac{\mathbf{k}_{(16)}}{\sqrt{2.154}} = \begin{bmatrix} \dfrac{0.536}{1.4677} \\ \dfrac{0.931}{1.4677} \\ \dfrac{1.000}{1.4677} \end{bmatrix} = \begin{bmatrix} 0.365 \\ 0.635 \\ 0.681 \end{bmatrix}.$$

By definition, the vector of factor loadings is the eigenvector normalized to its eigenvalue. The eigenvalue is found (as described above) by using the fundamental eigenvalue/eigenvector equation.

$$\mathbf{Rk}_1 = \mathbf{k}_1 \lambda \tag{4.3}$$

$$\mathbf{Rk}_1 = \begin{bmatrix} 1.000 & 0.150 & 0.380 \\ 0.150 & 1.000 & 0.824 \\ 0.380 & 0.824 & 1.000 \end{bmatrix} \begin{bmatrix} 0.365 \\ 0.635 \\ 0.681 \end{bmatrix} = \begin{bmatrix} 0.719 \\ 1.251 \\ 1.343 \end{bmatrix} = \begin{bmatrix} 0.365 \\ 0.635 \\ 0.681 \end{bmatrix} (1.971) = \mathbf{k}_1 \lambda_1.$$

The eigenvalue for the first factor is therefore found to be 1.971, and the corresponding desired magnitude for the first factor is $\sqrt{\lambda_1} = \sqrt{1.971} = 1.4039$.

The normalized eigenvector $\mathbf{k}_1$ is transformed into $\mathbf{f}_1$, the vector of factor loadings on factor 1, by multiplying each value in the $\mathbf{k}_1$ vector by the desired magnitude, that is, by the square root of the eigenvalue. This is the process explained in Chapter 3, Section 3.13, for normalizing a vector to a particular value.

$$\mathbf{f}_1 = \mathbf{k}_1 \sqrt{\lambda_1} = \mathbf{k}_1(1.4039) = \begin{bmatrix} (0.365)(1.4039) \\ (0.635)(1.4039) \\ (0.681)(1.4039) \end{bmatrix} = \begin{bmatrix} 0.512 \\ 0.891 \\ 0.957 \end{bmatrix}.$$

Having obtained factor loadings, we have now successfully extracted factor 1 from the correlation matrix. To now extract factor 2, we must first remove all of the variance and covariance in the original $\mathbf{R}$ matrix that has been accounted for by factor 1. This is done by creating a $\mathbf{G}_1$ matrix, the *matrix of first factor products*, by postmultiplying the vector of factor loadings $\mathbf{f}_1$ by its own transpose.

$$\mathbf{G}_1 = \mathbf{f}_1\mathbf{f}_1' = \begin{bmatrix} 0.512 \\ 0.891 \\ 0.957 \end{bmatrix} \begin{bmatrix} 0.512 & 0.891 & 0.957 \end{bmatrix} = \begin{bmatrix} 0.262 & 0.456 & 0.490 \\ 0.456 & 0.794 & 0.852 \\ 0.490 & 0.852 & 0.915 \end{bmatrix}.$$

The diagonal of this $\mathbf{G}_1$ matrix is the square of each of the loadings on factor 1—in other words, the *communalities*, the amount of variance that factor one accounts for in each of the three variables. The off-diagonal values in this $\mathbf{G}_1$ matrix give the amount of *covariance* in each *pair* of variables accounted for by each factor. We subtract this $\mathbf{G}_1$ matrix from the $\mathbf{R}$ matrix to find out how much variance and covariance is left over after factor 1 is extracted—in other words, the *amount of variance/covariance not accounted for by factor 1*. The resultant matrix is the *matrix of first factor residuals*, symbolized as $\mathbf{R}_1$. The diagonal entries in this residual matrix indicate what proportion of variance in each variable is *not* accounted for by factor 1, and the off-diagonal entries indicate how much *covariance* in each pair of variables is not accounted for by factor 1.

$$\mathbf{R}_1 = \mathbf{R} - \mathbf{G}_1 = \begin{bmatrix} 1.000 & 0.150 & 0.380 \\ 0.150 & 1.000 & 0.824 \\ 0.380 & 0.824 & 1.000 \end{bmatrix} - \begin{bmatrix} 0.262 & 0.456 & 0.490 \\ 0.456 & 0.794 & 0.852 \\ 0.490 & 0.852 & 0.915 \end{bmatrix}$$

$$= \begin{bmatrix} 0.738 & -0.306 & -0.110 \\ -0.306 & 0.206 & -0.028 \\ -0.110 & -0.028 & 0.085 \end{bmatrix}.$$

This matrix indicates how much more factoring needs to be done, that is, how much in each of the variances and how much in each of the covariances remains to be accounted for. Variable 3, "reasoning," has 0.915 variance units out of 1.000, 91.5% accounted for, leaving only 8.5% to be explained. It is very well accounted for by the first factor. Variable 1, "math," on the other hand, has only 26.2% of the variance accounted for, so there is still 73.8% left to be explained.

We are now ready to extract factor two by the method of successive squaring. The same algorithm applied to the original correlation matrix is now applied to the matrix of first factor residuals to extract a second factor from the variance and covariance that is left after the extraction of factor 1. The trial eigenvector is obtained for each successively squared $\mathbf{R}_1$ matrix, until convergence is reached.

$$\mathbf{s}_{(1)} = \mathbf{R}_1 1 = \begin{bmatrix} 0.738 & -0.306 & -0.110 \\ -0.306 & 0.206 & -0.028 \\ -0.110 & -0.028 & 0.085 \end{bmatrix} \begin{bmatrix} 1 \\ 1 \\ 1 \end{bmatrix} = \begin{bmatrix} 0.322 \\ -0.128 \\ -0.053 \end{bmatrix}$$

$$\mathbf{k}_{(1)} = \frac{\mathbf{s}_{(1)}}{\max(\mathbf{s}_{(1)})} = \begin{bmatrix} 1.000 \\ -0.398 \\ -0.164 \end{bmatrix}$$

$$\mathbf{s}_{(2)} = \mathbf{R}^2 1 = \begin{bmatrix} 0.650 & -0.286 & -0.082 \\ -0.286 & 0.137 & 0.025 \\ -0.082 & 0.025 & 0.020 \end{bmatrix} \begin{bmatrix} 1 \\ 1 \\ 1 \end{bmatrix} = \begin{bmatrix} 0.282 \\ -0.123 \\ -0.036 \end{bmatrix}$$

$$\mathbf{k}_{(2)} = \frac{\mathbf{s}_{(2)}}{\max(\mathbf{s}_{(2)})} = \begin{bmatrix} 1.000 \\ -0.437 \\ -0.128 \end{bmatrix}$$

$$\mathbf{s}_{(4)} = \mathbf{R}^4 1 = \begin{bmatrix} 0.511 & -0.227 & -0.062 \\ -0.227 & 0.101 & 0.027 \\ -0.062 & 0.027 & 0.008 \end{bmatrix} \begin{bmatrix} 1 \\ 1 \\ 1 \end{bmatrix} = \begin{bmatrix} 0.222 \\ -0.099 \\ -0.027 \end{bmatrix}$$

$$\mathbf{k}_{(4)} = \frac{\mathbf{s}_{(4)}}{\max(\mathbf{s}_{(4)})} = \begin{bmatrix} 1.000 \\ -0.445 \\ -0.121 \end{bmatrix}$$

$$\mathbf{s}_{(8)} = \mathbf{R}^8 1 = \begin{bmatrix} 0.317 & -0.141 & -0.038 \\ -0.141 & 0.063 & 0.017 \\ -0.038 & 0.017 & 0.005 \end{bmatrix} \begin{bmatrix} 1 \\ 1 \\ 1 \end{bmatrix} = \begin{bmatrix} 0.137 \\ -0.061 \\ -0.017 \end{bmatrix}$$

$$\mathbf{k}_{(8)} = \frac{\mathbf{s}_{(8)}}{\max(\mathbf{s}_{(8)})} = \begin{bmatrix} 1.000 \\ -0.445 \\ -0.121 \end{bmatrix}.$$

The sum of squares of the last trial eigenvector $\mathbf{k}_{(8)}$ is $1.000^2 + (-0.445)^2 + (-0.121)^2 = 1.213$, so the magnitude of the last trial eigenvector is $\sqrt{1.213} = 1.1012$.

The second true eigenvector of matrix **R**, the $\mathbf{k}_2$ vector, is calculated by normalizing the last trial eigenvector to 1 by dividing each entry in that vector by the magnitude of the vector.

$$\mathbf{k}_2 = \frac{\mathbf{k}_{(8)}}{\sqrt{1.213}} = \begin{bmatrix} \dfrac{1.000}{1.1012} \\ \dfrac{-0.445}{1.1012} \\ \dfrac{-0.121}{1.1012} \end{bmatrix} = \begin{bmatrix} 0.908 \\ -0.404 \\ -0.110 \end{bmatrix}.$$

The corresponding eigenvalue is obtained from the fundamental eigenvalue/eigenvector equation. Note that this is to be an eigenvalue of the original **R** matrix, so the original **R** matrix is what is used in the equation to obtain the eigenvalue.

$$\mathbf{R}\,\mathbf{k}_2 = \mathbf{k}_2 \lambda \tag{4.3}$$

$$\mathbf{R}\mathbf{k}_2 = \begin{bmatrix} 1.000 & 0.150 & 0.380 \\ 0.150 & 1.000 & 0.824 \\ 0.380 & 0.824 & 1.000 \end{bmatrix} \begin{bmatrix} 0.908 \\ -0.404 \\ -0.110 \end{bmatrix} = \begin{bmatrix} 0.806 \\ -0.358 \\ -0.098 \end{bmatrix} = \begin{bmatrix} 0.908 \\ -0.404 \\ -0.110 \end{bmatrix}(0.887) = \mathbf{k}_2\lambda_2.$$

The eigenvalue for the first factor is therefore found to be $\lambda_2 = 0.887$, and the corresponding desired magnitude for the first factor is $\sqrt{\lambda_2} = \sqrt{0.887} = 0.9419$. The normalized eigenvector $\mathbf{k}_2$ is converted into the vector of factor loadings, $\mathbf{f}_2$, by multiplying each value in the $\mathbf{k}_2$ vector by the desired magnitude, 0.9419.

$$\mathbf{f}_2 = \mathbf{k}_2 \sqrt{\lambda_2} = \mathbf{k}_2 \sqrt{0.887} = \begin{bmatrix} (0.908)(0.9419) \\ (-0.404)(0.9419) \\ (-0.110)(0.9419) \end{bmatrix} = \begin{bmatrix} 0.855 \\ -0.381 \\ -0.104 \end{bmatrix}.$$

With the calculation of $\mathbf{f}_2$, the factor loadings on factor two, we have successfully extracted two factors from the correlation matrix **R**. The *matrix of second factor products* is obtained by postmultiplying $\mathbf{f}_2$ by its transpose.[5]

$$\mathbf{G}_2 = \mathbf{f}_2\mathbf{f}_2' = \begin{bmatrix} 0.855 \\ -0.381 \\ -0.104 \end{bmatrix} \begin{bmatrix} 0.855 & -0.381 & -0.104 \end{bmatrix} = \begin{bmatrix} 0.732 & -0.326 & -0.089 \\ -0.326 & 0.145 & 0.039 \\ -0.089 & 0.039 & 0.011 \end{bmatrix}.$$

---

[5] Occasionally, one or two of the products reported here will differ from a calculator result. For example, in this $\mathbf{G}_2$ matrix, the first entry is 0.732, but 0.855^2 calculated to three place accuracy is 0.731. In fact, all of these calculations are from a spreadsheet and are therefore more precise than what can be obtained from the rounded values on the page. They are transcribed here to fit three-place accuracy, but the actual calculations on the spreadsheet are carried out to several more places.

This matrix of second factor products is subtracted from the matrix of first factor residuals to obtain the *matrix of second factor residuals.* Since we are only extracting two factors, this is the final matrix of residuals. This matrix contains the amount of variance and covariance *not* accounted for after the factor analysis is completed. It is an important matrix for interpreting the results of the factor analysis.

$$\mathbf{R}_2 = \mathbf{R}_1 - \mathbf{G}_2 = \begin{bmatrix} 0.738 & -0.306 & -0.110 \\ -0.306 & 0.206 & -0.028 \\ -0.110 & -0.028 & 0.085 \end{bmatrix} - \begin{bmatrix} 0.732 & -0.326 & -0.089 \\ -0.326 & 0.145 & 0.039 \\ -0.089 & 0.039 & 0.011 \end{bmatrix}$$

$$= \begin{bmatrix} 0.006 & 0.019 & -0.021 \\ 0.019 & 0.062 & -0.068 \\ -0.021 & -0.068 & 0.074 \end{bmatrix}.$$

In interpreting the factor analysis, there are two kinds of information in this matrix of final residuals—information in the diagonal elements (variance information), and information in the off-diagonal elements (covariance information). With respect to the first kind of information, variance, the two-factor solution does very well in accounting for the variance in variable 1, math scores, with less than 1% variance not accounted for. It does not do quite as well in accounting for the variance in variables 2 and 3, where the proportion of variance not accounted for is 0.062 and 0.074, respectively.

The second kind of information, the covariance information, tells two things—how far off the factor analysis is in accounting for the correlations, and whether the factor solution over- or underestimates those correlations. The entry for the correlation between variable 1 and variable 2 is 0.019, indicating that we have *underestimated* the actual correlation, but by a small amount. The entry for the correlation between variable one and variable three is about the same size, but with a negative sign (−0.021), indicating that we have *overestimated* the correlation this time but not by very much. The entry for the correlation between variable 2 and variable 3 indicates a larger amount of error, and the correlation is *overestimated* by that amount (almost 7%).

Both the "variance accounted for criterion" of completeness of factoring, and also the "residual correlations criterion" provide useful information. The variance criterion tends to be used more by those who favor a principal component approach, and the residual correlations criterion is used more by those who lean toward a more traditional factor analytic approach. The reasons for this will become apparent toward the end of the chapter. One method for evaluating whether enough factors have been extracted is to test the matrix of residuals (after all factors are extracted) for statistically significant deviation from random error.

The major information for interpreting a factor analysis is contained in a factor analysis summary table, like the one given in Table 4.4 for the Locuro

TABLE 4.5. Unrotated Factor Analysis Summary Table for Simplest Case Data

	Loadings		Communalities			Uniqueness
	Factor 1	Factor 2	Factor 1	Factor 2	Total	
Verbal test	0.512	0.855	0.262	0.732	0.994	0.006
Math test	0.891	−0.381	0.794	0.145	0.938	0.062
Logic test	0.957	−0.104	0.915	0.011	0.926	0.074
Eigenvalues:			1.971	0.887	2.858	0.142
Percents of eigenvalues:			65.7%	29.6%	95.3%	4.7%

et al. data, where factor loadings, communalities, eigenvalues, and percents of eigenvalues are placed in relation to one another. Table 4.5 is the factor analysis summary table for the results of the present analysis. Once the factor loadings are entered into this table, the other entries can be calculated from them. The communalities for the two factors are the squared factor loadings. The total communalities are the sums of the individual factor communalities by rows. The eigenvalues are the sums of the communalities by columns. The percents of eigenvalues are the eigenvalues divided by the number of variables in the analysis, and then multiplied by 100 to convert them to percents. The uniqueness values for each variable are calculated by subtracting each of the total communalities from 1.000.

The uniqueness column of the summary table is a new one introduced in this section. It is an indication of the proportion of variance in each of the variables *not* accounted for by the factor solution. For example, the total communality for the reasoning test is 0.926, indicating that 92.6% of the variance in that variable is accounted for by the two-factor solution. The amount of variance not accounted for is 1 − 0.926 = 0.074, the uniqueness. When using Stata to do a factor analysis, uniqueness is one of the default statistics reported.

The factor loadings can be used to interpret the identity of the factors in terms of the common elements that load highly on each factor. However, the interpretation will usually be clearer after rotation of the factor pattern. In Table 4.5, it can be seen that the first factor is most highly related to the math test and the logic test, but the verbal test also loads quite strongly, so the factor seems to reflect general intelligence. The second factor has a high loading on verbal ability, and negative loadings on math and logic. It appears to be a bipolar factor of verbal ability versus math/logic. However, there is another—much clearer—interpretation that will be obvious once we have rotated the factor pattern. In the vector plot (Fig. 4.3) of these data, we can see that by rotating the factor pattern about 24°, factor 1 coincides very closely with the math test, and factor 2 coincides very closely with the verbal test. We could dub factor 1 a quantitative ability factor, and factor 2 a verbal

TABLE 4.6. Rotated Factor Analysis Summary Table for Simplest Case Data

	Loadings		Communalities			Uniqueness
	Factor 1	Factor 2	Factor 1	Factor 2	Total	
Verbal test	0.122	0.989	0.015	0.979	0.994	0.006
Math test	0.969	0.013	0.938	0.000	0.938	0.062
Logic test	0.917	0.293	0.840	0.086	0.926	0.074
Variance accounted for:			1.793	1.065	2.858	0.142
Percents of total variance:			59.8%	35.5%	95.3%	4.7%

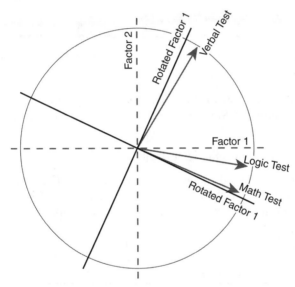

Figure 4.3 Vector plot for the factor pattern of the three variables in the simplest case dataset of Table 4.5 (unrotated) and Table 4.6 (rotated). The axes for unrotated factors are shown with dashed lines and the axes for rotated factors are shown with bold lines.

intelligence factor. Logic loads strongly on the quantitative factor, but also is slightly related to the verbal factor, as seen in the rotated vector plot, and also in Table 4.6, the factor analysis summary table for the rotated factor solution. Rotation will be explained more fully in Section 4.7.

To run the factor analysis on this simple data set in Stata, the first step is to get the data into Stata. One of the easiest ways is to copy the 6 × 3 matrix of data (with labels included at the top) and paste it directly into the Stata data editor, as shown.

Once the data have been pasted (or otherwise entered into the editor), the three variables, verbal performance, math performance, and logic, will appear

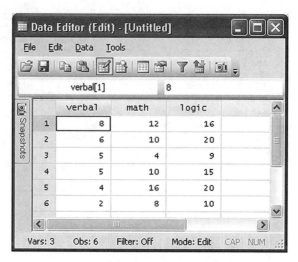

(StataCorp, 2009. Stata: Release 11. Statistical Software. College Station, TX: StataCorp LP.)

in the "variables" window of Stata. The following code is then entered in the "command" window of Stata, followed by the "return" key which runs the program.

factor verbal math logic, mineigen (0.80) pcf

Two tables will then appear in the output window. The first table contains eigenvalue information (not shown). The second table contains factor loadings, and a column of uniqueness values as shown in the table below.

Factor loadings (pattern matrix) and unique variances

Variable	Factor 1	Factor 2	Uniqueness
verbal	**0.5123**	**0.8553**	**0.0060**
math	**0.8909**	**−0.3806**	**0.0615**
logic	**0.9566**	**−0.1036**	**0.0742**

(StataCorp, 2009. Stata: Release 11. Statistical Software. College Station, TX: StataCorp LP.)

These are the three factor analysis results selected by Stata to form the default condition for output: eigenvalue information, loadings, and uniqueness values. The loadings are all one actually needs in order to create the complete factor analysis summary table as demonstrated above.

Notice how simple the Stata commands are. The first command word, "factor," indicates the statistical method to be used. The next three words, "math verbal reasoning," indicate names of the three variables from the entered data to be included in the analysis. With just these four words of instruction, the factor analysis would run complete, if the default settings were sufficient for one's purposes. Following the comma, special conditions are given. The first condition specified here is to set "mineigen," the minimum eigenvalue criterion, to something lower than the usual 1.00. This is necessary because with this unrealistically small data set, the second factor only has an eigenvalue of 0.887, and the second factor would therefore be excluded by the default criterion. The second condition, "pcf," indicates to Stata that the principal component method of factor analysis is to be used. (The default method for Stata is the principal factors method, explained in Section 4.8 of this chapter.)

To obtain the rotated factor pattern, a second Stata command is necessary, such as this one for obtaining a varimax rotation.

rotate, varimax

The output from this additional instruction includes first a table of "variance accounted for" by each factor, analogous to eigenvalues (not shown), and then a table of rotated loadings and uniqueness values (the table below). Note that the uniqueness values do not change from the unrotated analysis to this one. That, of course, will always be true, because rotation does not change the amount of variance in each variable accounted for in the space of m factors, it just changes the rotation of axes within that space.

Rotated factor loadings (pattern matrix) and unique variances

Variable	Factor 1	Factor 2	Uniqueness
verbal	**0.1221**	**0.9895**	**0.0060**
math	**0.9687**	**0.0127**	**0.0615**
logic	**0.9166**	**0.2926**	**0.0742**

(StataCorp, 2009. Stata: Release 11. Statistical Software. College Station, TX: StataCorp LP.)

It will be instructive to now show the comparable code for factor analysis in SAS and SPSS. The SAS code for accomplishing this same analysis is as follows:

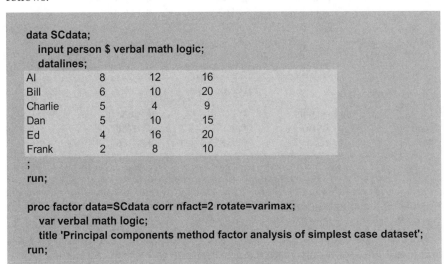

```
data SCdata;
    input person $ verbal math logic;
    datalines;
Al        8      12      16
Bill      6      10      20
Charlie   5      4       9
Dan       5      10      15
Ed        4      16      20
Frank     2      8       10
;
run;

proc factor data=SCdata corr nfact=2 rotate=varimax;
    var verbal math logic;
    title 'Principal components method factor analysis of simplest case dataset';
run;
```

The first 11 lines of this code constitute the SAS "data statement," that is, the SAS module that enters the data. In the first line, we have chosen the name of the data set to be "SCdata," followed by a semicolon. (Every individual instruction in SAS is terminated with a semicolon, and failure to do so is one of the most common SAS coding errors.) The second line, "input," indicates the identity and sequential order of each of the four variables being entered. The first variable, "person," is followed by a $ to indicate that it is text, a label, rather than quantitative data. The instruction "datalines" indicates that the input that follows will be successive rows of data, until they are terminated by another semicolon. The "run" instruction completes and executes the data step.

The last four lines of code constitute the actual SAS procedure, in this case PROC FACTOR. The first line calls that procedure, indicates the data set to be analyzed, calls for a correlation matrix to be included in output, sets the number of factors (nfact) to be two, and specifies that a varimax rotation be used. The next line names the variables to be factored ("var" stands for variables). The second to last line, the title, is for convenience in labeling output. It may contain any text you desire, entered between the two single quotes. The final line is the "run" instruction to complete and execute the FACTOR procedure. The SAS output created by these instructions includes a correlation matrix, eigenvalues, communalities, and so on, but the two tables of central interest are the factor pattern and the rotated factor pattern.

Factor Pattern

	Factor 1	Factor 2
verbal	0.51225	0.85534
math	0.89086	−0.38059
logic	0.95658	−0.10360

Rotated Factor Pattern

	Factor 1	Factor 2
verbal	0.12738	0.98883
math	0.96872	0.00750
logic	0.91817	0.28766

In SPSS, one first enters the 6 × 3 matrix of data in the spreadsheet format of the "data view" (similar to Stata). One can then use the "point and click" menu item selection to accomplish the analysis, or alternately, one can use command line code, both of which are shown here.

SPSS "POINT AND CLICK" INSTRUCTIONS FOR FACTOR

1. Click **Analyze**, click **Dimension Reduction**, and click **Factor**.
2. A "Factor Analysis" window will appear, with your variables listed in a pane on the left. Use the arrow to move them to the right, into the **Variables** pane.
3. Click **Extraction**. A window entitled "Factor Analysis: Extraction" will appear. In the area labeled "Analyze," click the radio button for Correlation matrix, and in the area labeled "Display," check Unrotated factor solution (so you can compare unrotated with rotated), and also check Scree plot. In the area labeled "Extract," click the radio button for **Fixed number of factors**, and put "2" in the box for the number to extract.
4. In that window, there is a "Method:" drop down menu at the top. Select **Principal components** from that menu. Click **Continue**. You will be returned to the "Factor Analysis" window.
5. On the right of that window, click the third button down, **Rotation**. This will cause a "Factor Analysis: Rotation" window to appear. Click **Varimax** in the "Method" area and click **Loading plot(s)** and **Rotated solution** in the "Display" box. Click **Continue**. You will be returned to the "Factor Analysis" window.
6. If you would like descriptive statistics on your variables, click the button on the top right, **Descriptives,** and when the "Factor Analysis: Descriptives" window appears, select the ones you would like. Click **Continue**, and you will be returned to the "Factor Analysis" window.
7. Click the **OK** button on the bottom left and SPSS will run your analysis. An output window of results will fill your screen, containing all of the things you have requested (table and graph of rotated loadings, prerotation loadings, communalities, eigenvalues, descriptive statistics that have been requested, etc.).
8. At the top of your output of results, you will see a listing of the syntax statements SPSS has created, corresponding to your selections.

The output from this particular set of selections contains a number of useful tables and graphs. The most important tables are the two sets of factor loadings, rotated and unrotated. From these loadings, one can construct the factor analysis summary table,[6] which will place in context the many other summary statistics reported (such as communalities, eigenvalues) One particularly useful graph included in this output (invoked by the PLOT ROTATION command) is the plot of the vectors for verbal, math, and logic variables in the

---

[6] When comparing output across statistical analysis applications, the factor analysis summary table format is a convenient way to see all of the essential factor analytic information in context.

rotated 2D factor space. This shows clearly the spatial orientation of the variables to one another. When there are three factors, a three-dimensional cube will display the location of each variable.)

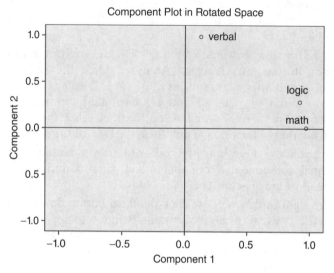

Reprinted courtesy of International Business Machines Corporation, © SPSS, Inc., an IBM Company.

You can also save the syntax for this analysis (the command line statements) in a file for future use. To do so, one follows steps 1 through 6 in the instructions just given, and then for step 7, rather than clicking the **OK** button to run the analysis, click the **Paste** button right next to it. A window labeled "Syntax1" will appear with a listing of the command line code.

```
DATASET ACTIVATE DataSet0.
FACTOR
  /VARIABLES verbal math logic
  /MISSING LISTWISE
  /ANALYSIS verbal math logic
  /PRINT UNIVARIATE INITIAL EXTRACTION ROTATION
  /PLOT EIGEN ROTATION
  /CRITERIA FACTORS(2) ITERATE(25)
  /EXTRACTION PC
  /CRITERIA ITERATE(25)
  /ROTATION VARIMAX
  /METHOD=CORRELATION.
```

You can select the **File** menu item, then the **Save As** item, and follow instructions to save this code as a file that can later be called up to rerun the analysis with any desired changes.

4.5 FACTOR SCORES: THE RELATIONSHIP BETWEEN LATENT VARIABLES AND MANIFEST VARIABLES

Many applications of factor analysis, perhaps the majority of those found in published papers, are concerned only with identifying the factor structure, the relationships among the variables expressed in the factor analysis summary tables, and in the vector plot graphs of the factor pattern. However, in some applications, the investigator wishes to obtain *factor scores*, estimates of the underlying scores on the latent variables for each observation. Factor scores are sometimes used as input to another analysis, such as a multivariate analysis of variance (MANOVA, described in Chapter 8). In this case, the factor analysis is functioning primarily as a data simplification tool in the initial phase of the data analysis process.

The calculation of factor score is reasonably simple for the principal component method of factor analysis. To obtain the matrix of factor scores, $\mathbf{Y}$, one postmultiplies the matrix of raw input data in Z-score form, $\mathbf{Z}$, by a transformation matrix $\mathbf{T}$ as shown in Equation 4.4.

$$\mathbf{Y} = \mathbf{ZT} \tag{4.4}$$

The $\mathbf{T}$ matrix is a matrix of eigenvectors, but they are the eigenvectors normalized to the reciprocals of their respective eigenvalues. This is the third form of the eigenvectors that we have encountered in the factor analysis model.

4.5.1 The Three Types of Eigenvector in Factor Analysis

It is helpful in understanding factor analysis to identify the three types of eigenvectors that play a part in the factor analysis formulas. The first is the $\mathbf{K}$ matrix (or rather vector $\mathbf{k}_1$ and vector $\mathbf{k}_2$, which are concatenated to create the $\mathbf{K}$ matrix), and as you will remember, these vectors are normalized to 1. It follows that matrix $\mathbf{K}$, when premultiplied by its transpose, will have an identity matrix as its product.

$$\mathbf{K'K} = \mathbf{I}$$

$$\mathbf{K'K} = \begin{bmatrix} 0.365 & 0.635 & 0.681 \\ 0.908 & -0.404 & -0.110 \end{bmatrix} \begin{bmatrix} 0.365 & 0.908 \\ 0.635 & -0.404 \\ 0.681 & -0.110 \end{bmatrix} = \begin{bmatrix} 1.000 & 0.000 \\ 0.000 & 1.000 \end{bmatrix} = \mathbf{I}.$$

Since they are normalized to 1, the diagonal entries will be equal to 1. The off-diagonal entries are zero because the two eigenvectors are orthogonal to (independent of) one another.

The second form of the eigenvectors that we have encountered, $\mathbf{F}$, is the factor loadings matrix, which consists of eigenvectors normalized to their eigenvalues. When a similar product matrix is created by premultiplying $\mathbf{F}$ by its transpose, the resultant matrix is a diagonal matrix of eigenvalues, Λ.

$$\mathbf{F'F} = \Lambda$$

$$\mathbf{F'F} = \begin{bmatrix} 0.512 & 0.891 & 0.957 \\ 0.855 & -0.381 & -0.104 \end{bmatrix} \begin{bmatrix} 0.512 & 0.855 \\ 0.891 & -0.381 \\ 0.957 & -0.104 \end{bmatrix} = \begin{bmatrix} 1.971 & 0.000 \\ 0.000 & 0.887 \end{bmatrix} = \Lambda.$$

The third form of the eigenvectors used in factor analysis is this transformation matrix $\mathbf{T}$, which consists of eigenvectors normalized to the reciprocals of their eigenvalues. When the product matrix is created by premultiplying $\mathbf{T}$ by its transpose, the resultant matrix is a diagonal matrix of reciprocals of eigenvalues.

$$\mathbf{T'T} = \Lambda^{-1}$$

$$\mathbf{T'T} = \begin{bmatrix} 0.260 & 0.452 & 0.485 \\ 0.964 & -0.429 & -0.117 \end{bmatrix} \begin{bmatrix} 0.260 & 0.964 \\ 0.452 & -0.429 \\ 0.485 & -0.117 \end{bmatrix} = \begin{bmatrix} 0.507 & 0.000 \\ 0.000 & 1.127 \end{bmatrix} = \Lambda^{-1}.$$

The transformation matrix $\mathbf{T}$ can be created by postmultiplying the matrix of factor loadings $\mathbf{F}$ by the matrix Λ^{-1}, a diagonal matrix of reciprocals of the eigenvalues as given in Equation 4.5.

$$\mathbf{T} = \mathbf{F}\Lambda^{-1} \tag{4.5}$$

$$\mathbf{T} = \mathbf{F}\Lambda^{-1} = \begin{bmatrix} 0.512 & 0.855 \\ 0.891 & -0.381 \\ 0.957 & -0.104 \end{bmatrix} \begin{bmatrix} \dfrac{1}{1.971} & 0.000 \\ 0.000 & \dfrac{1}{0.887} \end{bmatrix} = \begin{bmatrix} 0.260 & 0.964 \\ 0.452 & -0.429 \\ 0.485 & -0.117 \end{bmatrix}.$$

Similar equations can be specified for transforming any of these forms of the eigenvector matrix into one of the other forms, such as Equation 4.6 for obtaining the loadings matrix $\mathbf{F}$ from matrix $\mathbf{K}$, or Equation 4.7 for obtaining matrix $\mathbf{K}$ from the loadings matrix $\mathbf{F}$.

$$\mathbf{F} = \mathbf{K}\Lambda^{1/2}. \tag{4.6}$$

$$\mathbf{K} = \mathbf{F}\Lambda^{-1/2}. \tag{4.7}$$

4.5.2   Factor Scores Demonstration Using Simplest Case Data from Section 4.4

Since the factor scores are created in standardized score form, the original data must also be converted to standardized form for the calculating of factor scores. The **Z** matrix can be created from the original data by postmultiplying the matrix of deviation scores **x** by a standardizing matrix **E**, that is, a diagonal matrix of reciprocals of standard deviations.

$$\mathbf{Z} = \mathbf{X}_c \mathbf{E} \tag{4.8}$$

$$\mathbf{Z} = \mathbf{X}_c \mathbf{D} = \begin{bmatrix} 3 & 2 & 1 \\ 1 & 0 & 5 \\ 0 & -6 & -6 \\ 0 & 0 & 0 \\ -1 & 6 & 5 \\ -3 & -2 & -5 \end{bmatrix} \begin{bmatrix} \dfrac{1}{\sqrt{4}} & 0 & 0 \\ 0 & \dfrac{1}{\sqrt{16}} & 0 \\ 0 & 0 & \dfrac{1}{\sqrt{22.4}} \end{bmatrix} = \begin{bmatrix} 1.5 & 0.5 & 0.211 \\ 0.5 & 0 & 1.056 \\ 0 & -1.5 & -1.268 \\ 0 & 0 & 0.000 \\ -0.5 & 6 & 1.056 \\ -1.5 & -2 & -1.056 \end{bmatrix}.$$

The matrix of factor scores **Y** may now be calculated by Equation 4.4, as the matrix product of the **Z** matrix and the **T** matrix. The entries in the **Y** matrix are the latent variable scores within the two-factor space.

$$\mathbf{Y} = \mathbf{ZT} \tag{4.4}$$

$$\mathbf{Y} = \mathbf{ZT} = \begin{bmatrix} 1.5 & 0.5 & 0.211 \\ 0.5 & 0 & 1.056 \\ 0 & -1.5 & -1.268 \\ 0 & 0 & 0.000 \\ -0.5 & 6 & 1.056 \\ -1.5 & -2 & -1.056 \end{bmatrix} \begin{bmatrix} 0.260 & 0.964 \\ 0.452 & -0.429 \\ 0.485 & -0.117 \end{bmatrix} = \begin{bmatrix} 0.718 & 1.207 \\ 0.643 & 0.359 \\ -1.293 & 0.792 \\ 0.000 & 0.000 \\ 1.061 & -1.249 \\ -1.129 & -1.108 \end{bmatrix}.$$

On the one side of the equation, the **Z** scores are manifest variable scores. On the other side of the equation, the **Y** scores are the latent variable scores, and the **T** eigenvectors are the coefficients for transforming the manifest variable scores into latent variable scores. The latent variable scores **Y** are also in Z-score form, with means of zero and standard deviations of one.

Matrix **F** can be used as linear combination weights to go the other direction, to *transform latent variable scores back into manifest variable score*, as expressed in Equation 4.9. Notice that transforming data in this direction requires that the eigenvectors be in the form of loadings, matrix **F**. Also, the **F** matrix must be in transposed form to make the matrices conformable.

$$\mathbf{Z} = \mathbf{YF}' \tag{4.9}$$

$$\mathbf{Z} = \mathbf{YF'} = \begin{bmatrix} 0.718 & 1.207 \\ 0.643 & 0.359 \\ -1.293 & 0.792 \\ 0.000 & 0.000 \\ 1.061 & -1.249 \\ -1.129 & -1.108 \end{bmatrix} \begin{bmatrix} 0.512 & 0.891 & 0.957 \\ 0.855 & -0.381 & -0.104 \end{bmatrix}$$

$$= \begin{bmatrix} 1.400 & 0.181 & 0.562 \\ 0.636 & 0.436 & 0.578 \\ 0.015 & -1.453 & -1.319 \\ 0.000 & 0.000 & 0.000 \\ -0.525 & 1.420 & 1.144 \\ -1.526 & -0.584 & -0.965 \end{bmatrix}.$$

As we examine the "reconstructed" Z-scores in the $\mathbf{Z}$ matrix from this calculation, we notice that the values do not come out quite right. For example, on variable one, the math test scores, the actual z scores are 1.5, 0.5, 0, 0, –0.5, and –1.5, but the corresponding reconstructed Z-scores come out to 1.400, 0.636, 0.015, 0.000, –0.525, and –1.526. This is because the two factors extracted have not accounted for all of the variance in the original variables. As shown in the factor analysis summary table in Section 4.4 above, this factor analysis only accounts for 95.3% of the variance in the original data. The factor scores are therefore only approximate.

For comparison now we will consider the case where the factor scores *are* exact, where the factors account for all of the variance in the original data.

4.5.3 Factor Analysis and Factor Scores for Simplest Case Data with a Rank of 2

It will be instructive to consider now this same analysis on a data matrix that is *not* full rank, a matrix with a rank of 2, where we account for all of the variance in the three-variable data set within a reduced two-factor space. To do this, we will make a small change in the simple data set just used.

$$\mathbf{X} = \begin{bmatrix} 8 & 12 & 20 \\ 6 & 10 & 16 \\ 5 & 4 & 9 \\ 5 & 10 & 15 \\ 4 & 16 & 20 \\ 2 & 8 & 10 \end{bmatrix}.$$

Notice that this 6×3 matrix of data is almost the same as the one for which factor scores were demonstrated in the preceding section. The only difference is in the two upper values of the third variable, where the 16 and the 20 exchange places. Notice also that now that this change has been made, on every row, the value of the third variable is the sum of the values in the first two variables. Variable 3 is therefore a linear combination of the other two variables, and the rank of the matrix is therefore only two. The covariance matrix is calculated as outlined above, and only the circled cells change, those that involve the covariance between variable 3 and each of the other two. When the covariance matrix is not full rank, that is, when there is a linear dependency, the third column of covariances is the sum of the other two columns as is seen in the shaded entries of the **S** matrix below. The same is true of rows.

$$\mathbf{S} = \begin{bmatrix} 4 & 1.2 & 5.2 \\ 1.2 & 16 & 17.2 \\ 5.2 & 17.2 & 22.4 \end{bmatrix}.$$

The correlation matrix also differs by the two values in the same locations above and below the diagonal.

$$\mathbf{R} = \begin{bmatrix} 1.000 & 0.150 & 0.549 \\ 0.150 & 1.000 & 0.909 \\ 0.549 & 0.909 & 1.000 \end{bmatrix}.$$

As we factor this correlation matrix by the process given in the preceding section, we see in the factor analysis summary table (Table 4.7) that the three communalities are all equal to 1.000, signifying that all of the variance is accounted for in each variable. Also, the two eigenvalues sum to 3.00, indicating that all of the variance in the entire data set is accounted for by two factors.

The factor scores for this rank-of-two data set are obtained from Equation 4.4.

TABLE 4.7. Factor Analysis Summary Table for Simplest Case Data with a Rank of Two

	Loadings		Communalities			Uniqueness
	Factor 1	Factor 2	Factor 1	Factor 2	Total	
Verbal test	0.601	0.799	0.362	0.638	1.000	0.000
Math test	0.880	−0.475	0.775	0.225	1.000	0.000
Logic test	0.998	−0.063	0.996	0.004	1.000	0.000
	Eigenvalues:		2.132	0.868	3.000	0.000
	Percents of eigenvalues:		71.1%	28.9%	100.0%	0%

$$\mathbf{Y} = \mathbf{ZT} \tag{4.4}$$

$$\mathbf{Y} = \mathbf{ZT} = \begin{bmatrix} 1.5 & 0.5 & 1.056 \\ 0.5 & 0 & 0.211 \\ 0 & -1.5 & -1.268 \\ 0 & 0 & 0.000 \\ -0.5 & 6 & 1.056 \\ -1.5 & -2 & -1.056 \end{bmatrix} \begin{bmatrix} 0.282 & 0.921 \\ 0.413 & -0.547 \\ 0.468 & -0.073 \end{bmatrix} = \begin{bmatrix} 1.124 & 1.031 \\ 0.240 & 0.445 \\ -1.213 & 0.913 \\ 0.000 & 0.000 \\ 0.973 & -1.358 \\ -1.124 & -1.031 \end{bmatrix}.$$

The factor scores in this case reproduce the information in the original data set completely, since the two factors in this case account for all of the variance. When Equation 4.9 is used to reconstruct the original manifest variable Z scores from the factor scores, we see that the process is completely reversible.

$$\mathbf{Z} = \mathbf{YF'} \tag{4.9}$$

$$\mathbf{Z} = \mathbf{YF'} = \begin{bmatrix} 1.124 & 1.031 \\ 0.240 & 0.445 \\ -1.213 & 0.913 \\ 0.000 & 0.000 \\ 0.973 & -1.358 \\ -1.124 & -1.031 \end{bmatrix} \begin{bmatrix} 0.601 & 0.880 & 0.998 \\ 0.799 & -0.475 & -0.063 \end{bmatrix} = \begin{bmatrix} 1.5 & 0.5 & 1.056 \\ 0.5 & 0 & 0.211 \\ 0 & -1.5 & -1.268 \\ 0 & 0 & 0.000 \\ -0.5 & 6 & 1.056 \\ -1.5 & -2 & -1.056 \end{bmatrix}.$$

This same simple data set with a rank of two can now be used to illustrate additional fundamental equations of factor analysis.

4.5.4   Factor Analysis as Data Transformation

The first principle to be illustrated with this data set is Equation 4.10, which shows that the matrix of factor loadings **F** is in fact a "root" matrix of the correlation matrix from which it is derived. That is, when **F** is postmultiplied by its own transpose, the resultant matrix is the "reproduced correlation matrix," symbolized by $\mathbf{R}^+$, which will approximate the original **R** matrix as closely as is possible given the data configuration. It is precisely equivalent to the original correlation matrix with this particular demonstration data set, since we have accounted for all of the variance with the two factors. Thurstone (1935), one of the founding fathers of factor analysis referred to this equation as the "fundamental factor theorem."

$$\mathbf{R}^+ = \mathbf{FF'} \tag{4.10}$$

$$\mathbf{R}^+ = \mathbf{FF'} = \begin{bmatrix} 0.601 & 0.799 \\ 0.880 & -0.475 \\ 0.998 & -0.063 \end{bmatrix} \begin{bmatrix} 0.601 & 0.880 & 0.998 \\ 0.799 & -0.475 & -0.063 \end{bmatrix} = \begin{bmatrix} 1.000 & 0.150 & 0.549 \\ 0.150 & 1.000 & 0.909 \\ 0.549 & 0.909 & 1.000 \end{bmatrix}.$$

Just as eigenvectors can be used to transform observed data (manifest variables) into factor scores (latent variables), they can also be used to transform $\mathbf{R}$, the variance/covariance matrix[7] among the manifest variables, into a variance/covariance matrix among the latent variables. Equation 4.11 shows that Λ, the variance/covariance matrix for the latent variables, is equal to the original correlation matrix $\mathbf{R}$ both pre-multiplied and also post-multiplied by its matrix of normalized eigenvectors $\mathbf{K}$.

$$\Lambda = \mathbf{K'RK} \tag{4.11}$$

$$\Lambda = \mathbf{K'RK} = \begin{bmatrix} 0.282 & 0.413 & 0.468 \\ 0.921 & -0.547 & -0.073 \end{bmatrix} \begin{bmatrix} 1.000 & .150 & 0.549 \\ 0.150 & 1.000 & 0.909 \\ 0.549 & .909 & 1.000 \end{bmatrix} \begin{bmatrix} 0.282 & 0.921 \\ 0.413 & -0.547 \\ 0.468 & -0.073 \end{bmatrix}$$

$$= \begin{bmatrix} 2.132 & 0 \\ 0 & 0.868 \end{bmatrix}.$$

Equation 4.11 shows clearly what is accomplished by an eigenvector transformation of a *covariance structure*. First, it reduces the number of variables. With this simple demonstration it only reduces from three manifest variables down to two latent variables, but with actual data, it is not unusual to reduce a dozen variables down to three or four. The second thing accomplished is that the resultant latent variables are orthogonal to one another. That is, they are independent of one another; their intercorrelations are zero. This orthogonal simplification of the covariance structure is also an important property. It often facilitates further statistical analyses performed on the data in that the variables can be dealt with individually, without concern for the covariances among them.

The transformation from raw data to factor scores and back was shown in the last section to be a completely reversible process using this "errorless" data (i.e., data with all the variance accounted for). The transformation between manifest variable *covariance structures* and latent variable *covariance structures* is also completely reversible with "errorless" data. The variance/covariance matrix for latent variables, Λ, can be transformed back into the original correlation matrix by the operation of premultiplying and also post-multiplying it by the matrix of eigenvectors $\mathbf{K}$, as expressed in Equation 4.12.

$$\mathbf{R} = \mathbf{K}\Lambda\mathbf{K'} \tag{4.12}$$

---

[7] The correlation matrix is, of course, also a variance/covariance matrix. It is the variance/covariance matrix of standardized variables.

$$\mathbf{R} = \mathbf{K}\Lambda\mathbf{K}' = \begin{bmatrix} 0.282 & 0.921 \\ 0.413 & -0.547 \\ 0.468 & -0.073 \end{bmatrix} \begin{bmatrix} 2.132 & 0 \\ 0 & 0.868 \end{bmatrix} \begin{bmatrix} 0.282 & 0.413 & 0.468 \\ 0.921 & -0.547 & -0.073 \end{bmatrix}$$

$$= \begin{bmatrix} 1.000 & 0.150 & 0.549 \\ 0.150 & 1.000 & 0.909 \\ 0.549 & 0.909 & 1.000 \end{bmatrix}.$$

Both Equation 4.11 and also Equation 4.12 are examples of data transformation by the method of quadratic and bilinear forms, explained in Chapter 3, Section 3.10. Actually, Equation 4.10, the fundamental factor theorem, is also an example of covariance matrix transformation by quadratic and bilinear forms, in that it is just a derived form of Equation 4.12. Remembering that the matrix of factor loadings, $\mathbf{F}$, is the product of the matrix of normalized eigenvectors $\mathbf{K}$ postmultiplied by $\Lambda^{1/2}$, as given in Equation 4.6, $\mathbf{F} = \mathbf{K}\Lambda^{1/2}$, the derivation is:

$$\mathbf{R} = \mathbf{K}\Lambda\mathbf{K}'$$
$$= \mathbf{K}\Lambda^{1/2}\Lambda^{1/2}\mathbf{K}'$$
$$= (\mathbf{K}\Lambda^{1/2})(\Lambda^{1/2}\mathbf{K}')$$
$$= \mathbf{F}\mathbf{F}.$$

4.6 PRINCIPAL COMPONENT ANALYSIS: SIMPLIFIED FACTORING OF COVARIANCE STRUCTURE

Three years before Spearman's (1904) landmark paper that marked the beginning of the development of factor analysis, Karl Pearson (1901) published the paper that marked the beginning of factor analysis's twin method, principal component analysis (PCA). The two methods have much in common, but they also differ in important ways (Jolliffe, 2002).

What is a principle component? The first principal component is defined as the linear combination of the variables that has maximal variance. The second principal component is the linear combination with maximal variance in a direction orthogonal to the first principal component, and so on to as many components as are extracted. Consider the bivariate Gaussian distribution of points shown in the figure on the left below (courtesy of Wikimedia Commons). If we create an ellipse corresponding to the 97% confidence interval for these points, as shown in the figure at the right, the first principal component is the major axis (the principal axis) of that ellipse, and the second principal component is the minor axis of the ellipse. If the swarm of points were located in a three-dimensional space, the three principal components would be the three orthogonal axes of the enclosing ellipsoid. The same logic extends to higher dimensional spaces.

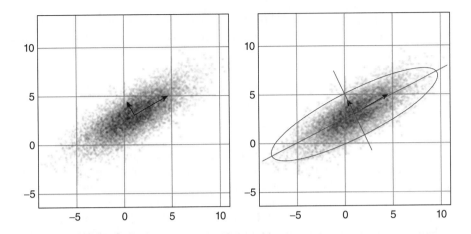

The mathematical procedure for PCA corresponds closely to the process demonstrated in Section 4.4 for factor analysis by the principal components method. Two characteristics of PCA differentiate it from factor analysis. First, the mathematics of principal components is simpler and more direct. Second, it usually involves factoring the entire covariance matrix (rather than the correlation matrix), so that variances of the variables are taken into account. It should be emphasized, however, in order for this process to make sense, the variables must be *commensurate*. That is, they must be measured on the same scale; otherwise, it makes no sense to say that one variable has twice as much variance as another. Suppose we are comparing IQ scores (with an expected population standard deviation of 16) to GRE scores (with an expected population standard deviation of 100). It would make no sense to conduct a principal components analysis of a covariance matrix with these two variables included since they are not comparable.

To illustrate PCA, we will go back to the rank-of-two simplest case data of Section 4.5.3, but we will now factor it differently. We will factor the *covariance matrix*, **S**, rather than the correlation matrix. For illustration purposes, we will assume that the three variables, a math test, a verbal test, and a reasoning test, are all measured on the same scale. Notice that the three variances in this covariance matrix differ substantially from one another. Variable 2 (math test) has a variance of 16, and variable 3 (logic test) has a variance of 22.4, while variable 1 (verbal test), has a variance of only 4. Since the three tests are commensurate, that is, measured on the same scale, these differences in variance are of interest and PCA will incorporate them into the analysis.

$$\mathbf{S} = \begin{bmatrix} 4 & 1.2 & 5.2 \\ 1.2 & 16 & 17.2 \\ 5.2 & 17.2 & 22.4 \end{bmatrix}.$$

When this **S** matrix is factored by the same method demonstrated in Section 4.4 (for a factor analysis), the first eigenvector and the first eigenvalue are obtained as:

$$\mathbf{k}_1 = \begin{bmatrix} 0.142 \\ 0.625 \\ 0.767 \end{bmatrix} \qquad \lambda_1 = 37.379.$$

Vector $\mathbf{k}_1$ is in fact the vector of transformation coefficients for obtaining component scores (analogous to factor scores) on the first principal component. The vector $\mathbf{k}_1$ (the eigenvector normalized to one) functions in PCA as the vector $\mathbf{t}_1$ (the eigenvector normalized to $1/\lambda$) did in obtaining factor scores. This is one of the senses in which PCA is simpler than factor analysis.

Next, the *matrix of first component products*, $\mathbf{G}_1$, is created by postmultiplying $\mathbf{k}_1$ by its transpose, and then multiplying the resulting matrix by the eigenvalue.

$$\mathbf{G}_1 = \mathbf{k}_1\mathbf{k}_1' * \lambda_1 = \begin{bmatrix} 0.142 \\ 0.625 \\ 0.767 \end{bmatrix} [0.142 \quad 0.625 \quad 0.767] * 37.379$$

$$= \begin{bmatrix} 0.754 & 3.320 & 4.074 \\ 3.320 & 14.616 & 17.936 \\ 4.074 & 17.936 & 22.009 \end{bmatrix}.$$

This matrix indicates the amount of variance and covariance accounted for by the first principal component. It is subtracted from the covariance matrix **S** to obtain matrix $\mathbf{S}_1$, the *matrix of first component residuals*, which indicates the amount of variance and covariance yet to be accounted for.

$$\mathbf{S}_1 = \mathbf{S} - \mathbf{G}_1 = \begin{bmatrix} 4 & 1.2 & 5.2 \\ 1.2 & 16 & 17.2 \\ 5.2 & 17.2 & 22.4 \end{bmatrix} - \begin{bmatrix} 0.754 & 3.320 & 4.074 \\ 3.320 & 14.616 & 17.936 \\ 4.074 & 17.936 & 22.009 \end{bmatrix}$$

$$= \begin{bmatrix} 3.246 & -2.120 & 1.126 \\ -2.120 & 1.384 & -0.736 \\ 1.126 & -0.736 & 0.391 \end{bmatrix}.$$

Matrix $\mathbf{S}_1$ is now factored in the same way to obtain the second eigenvector and the second eigenvalue.

$$\mathbf{k}_2 = \begin{bmatrix} 0.804 \\ -0.525 \\ 0.279 \end{bmatrix} \qquad \lambda_2 = 5.021$$

The *matrix of second component products*, $\mathbf{G}_2$, is created.

$$\mathbf{G}_2 = \mathbf{k}_2\mathbf{k}_2' *\lambda_2 = \begin{bmatrix} 0.804 \\ -0.525 \\ 0.279 \end{bmatrix} [0.804 \quad -0.525 \quad 0.279]*5.021$$

$$= \begin{bmatrix} 3.246 & -2.120 & 1.126 \\ -2.120 & 1.384 & -0.736 \\ 1.126 & -0.736 & 0.391 \end{bmatrix}$$

Not surprisingly, this product matrix $\mathbf{G}_2$ is exactly equivalent to the matrix of first component residuals. The *matrix of second component residuals*, $\mathbf{S}_2 = \mathbf{S}_1 - \mathbf{G}_2$ is therefore a matrix of zeros. This is, of course, because we created this demonstration matrix to have a rank of two so that two components would extract all of the variance and covariance from the covariance matrix $\mathbf{S}$.

The two normalized eigenvectors $\mathbf{k}_1$ and $\mathbf{k}_2$ are adjoined to one another to create the matrix of PCA coefficients, $\mathbf{K}$.

$$\mathbf{K} = \begin{bmatrix} 0.142 & 0.804 \\ 0.625 & -0.525 \\ 0.767 & 0.279 \end{bmatrix}.$$

This is the matrix of PCA transformation coefficients. When the original raw data matrix $\mathbf{X}$ is postmultiplied by this matrix, component scores (the analogue of factor scores) are created as expressed in Equation 4.13a.

$$\mathbf{Y} = \mathbf{XK} \tag{4.13a}$$

$$\mathbf{Y} = \mathbf{XK} = \begin{bmatrix} 8 & 12 & 20 \\ 6 & 10 & 16 \\ 5 & 4 & 9 \\ 5 & 10 & 15 \\ 4 & 16 & 20 \\ 2 & 8 & 10 \end{bmatrix} \begin{bmatrix} 0.142 & 0.804 \\ 0.625 & -0.525 \\ 0.767 & 0.279 \end{bmatrix} = \begin{bmatrix} 23.99 & 5.71 \\ 19.38 & 4.04 \\ 10.12 & 4.43 \\ 18.47 & 2.96 \\ 25.92 & 0.40 \\ 12.96 & 0.20 \end{bmatrix}.$$

These are the scores of each of the six persons (Al, Bill, Charlie, Dan, Ed, and Frank) on the two principal components that were extracted from the covariance matrix. Let us now consider geometrically what we have just done. Figure 4.4 shows the input matrix of data of the six male subjects and their scores on each test (math test, verbal test, and logic reasoning test). To the right of that matrix, we have plotted the point for each subject in the two

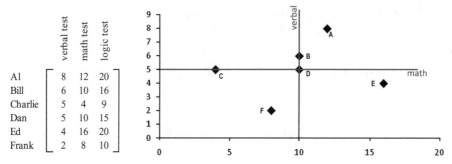

	verbal test	math test	logic test
Al	8	12	20
Bill	6	10	16
Charlie	5	4	9
Dan	5	10	15
Ed	4	16	20
Frank	2	8	10

Figure 4.4 Simplest case dataset of six men on three tests of mental abilities, with a scatterplot of the first two variables.

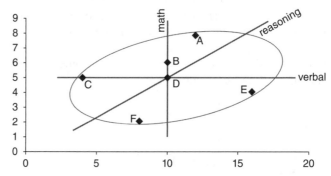

Figure 4.5 Vector plot from unscaled coefficients for the three cognitive test variables of the simplest case data superimposed on the scatterplot of raw data from Figure 4.4.

dimensional space with math test score on the horizontal axis and verbal test score on the vertical axis.[8]

Since these demonstration data were created to have linear dependency, that is, created to have a rank of 2, the third variable, the logic test, is located within this same two dimensional space, as shown in Figure 4.5.

This two-dimensional space is easy to comprehend, but the same principles demonstrated here generalize to the larger data sets of actual practice. Whereas we have reduced three manifest variables to two latent components, it is not unusual to account for a high percentage of the variance while reducing a dozen or more variables down to several orthogonal components. The twin advantages of principal components are (1) they are orthogonal to one another, and (2) they account for largest amount of variance possible (in descending order from the first factor, to the second, to the third, and so on.)

The maximization of variance by PCA can be shown with this simple demonstration data. As shown in Figure 4.6, the first principal component is in fact

---

[8] We have violated the usual convention of having the first variable on the horizontal axis and the second on the vertical axis for esthetic reasons.

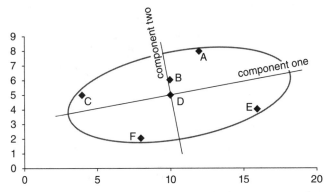

Figure 4.6   Principal component axis one and principal component axis two superimposed upon the scatterplot of raw data from Figure 4.4.

the major axis of the ellipse enclosing the scatterplot of the data. It is the longest possible line passing through the ellipse and its center. This is another way of seeing that it maximizes the variance. Successive principal components are minor axes of an ellipsoid within a many-dimensional space, with each accounting for maximum amount of variance orthogonal to all previous components.

Notice in Figure 4.7 that if lines perpendicular to the first principal component (i.e., isoquants[9]) are drawn through each data point, the metric location of each intersection corresponds to the component scores calculated above: 25.92 for Ed (E), 12.96 for Frank (F), and so on. The same would be true for projections onto the axis for component two.

The information from PCA can be collected into a summary table as was demonstrated for factor analysis. However, the table will be more informative if the coefficient matrix $\mathbf{K}$ is first rescaled by postmultiplying it by $\mathbf{\Lambda}^{1/2}$ the square root of the diagonal matrix of eigenvalues. This normalizes the eigenvectors to their eigenvalues, as was done in creating factor loadings in factor analysis. However, it should be noted that whereas this process applied to factor analysis (with a correlation matrix—standardized covariances) creates factor loadings that can be interpreted as correlation coefficients between the manifest variables and the factors, the same is not true of these scaled PCA coefficients. Therefore, rather than labeling them as $\mathbf{F}$ (the symbol used in this chapter for factor loadings), we will designate them as $\mathbf{K}_{sc}$, where the subscript sc stands for scaled coefficients.

$$\mathbf{K}_{sc} = \mathbf{K}\mathbf{\Lambda}^{1/2} \qquad (4.14)$$

[9] Isoquants are lines marking equal quantitative value, even as isobars are lines connecting locations of equal barometric pressure. Each of the six isoquants in Figure 4.7 marks the locations that have equal value projections onto component 1. Similar isoquants could be constructed for component 2 or any of the variables. Since these data have a rank of 2, all five sets of isoquants (three variables and two components) would have perfect metric properties.

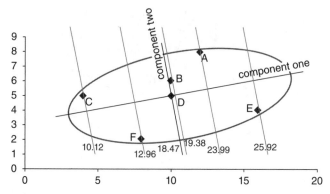

Figure 4.7 Isoquant projection of the six multivariate data points of Figure 4.4 on the first principal component.

TABLE 4.8. PCA Summary Table for Simplest Case Data with a Rank of Two

	Scaled Coefficients		Squared Coefficients		
	Factor 1	Factor 2	Factor 1	Factor 2	Total
Verbal test	0.868	1.802	0.754	3.246	4.000
Math test	3.823	−1.176	14.616	1.384	16.000
Logic test	4.691	0.625	22.009	0.391	22.400
Eigenvalues:			37.379	5.021	42.400
Percents of eigenvalues:			88.2%	11.8%	100.0%

$$\mathbf{K}_{sc} = \mathbf{K}\mathbf{\Lambda}^{1/2} = \begin{bmatrix} 0.142 & 0.804 \\ 0.625 & -0.525 \\ 0.767 & 0.279 \end{bmatrix} \begin{bmatrix} \dfrac{1}{\sqrt{37.379}} & 0.000 \\ 0.000 & \dfrac{1}{\sqrt{5.021}} \end{bmatrix} = \begin{bmatrix} 0.868 & 1.802 \\ 3.823 & -1.176 \\ 4.691 & 0.625 \end{bmatrix}.$$

Using scaled coefficients for the summary table (Table 4.8), the sum of the squared coefficients (by columns) are equal to the eigenvalues, as they should be. To obtain the percents of eigenvalues, the eigenvalues are divided by the trace of the **S** matrix, the sum of all the variance units, which is equal to $tr(\mathbf{S}) = 42.40$.

It is important to note from this summary table that the row totals of the squared coefficients are the variances of the three manifest variables (math test, verbal test, and reasoning test). The two column totals (37.379 and 5.021) are in fact the eigenvalues,[10] the variances of the latent variables, the compo-

---

[10] The reason the row and column totals are equivalent to the variances of the three variables and the eigenvalues of the two components respectively is because we have created data that fit the two component PCA model perfectly, that is, data with a rank of two. With empirically derived data, where the rank is greater than two, these statistics for a two-component PCA model would indicate the number of variance units accounted for by the model rather than the total amount of variance in the data.

nents. The sum of sums of all of the squared coefficients (42.400, in the bottom right) is the total amount of variance in the data set, and therefore the trace of both the covariance matrix **S** and also the matrix of eigenvalues Λ.

In SPSS, PCA is run (point and click method) by essentially the same steps given at the end of Section 4.4 for factor analysis, with successive menu selections of "analyze," "dimension reduction," and "factor." The difference is that when this process brings up the window entitled "Factor Analysis" for selection of options, one should click on the "Extraction" button and select "Covariance matrix" (under "Analyze") and "Unrotated factor solution" (under "Display"). This will produce a PCA of a covariance matrix.

Stata and SAS, on the other hand, each use two completely different modules or "procedures" for doing PCA and for doing factor analysis. In Stata, once the data have been entered as described in Section 4.4, the command line instruction for PCA is:

```
pca verbal math logic, cov
```

The "cov" option after the comma indicates that the matrix to be factored is the covariance matrix rather than the correlation matrix. Notice that the first command word in this Stata code is "pca," not "factor." Likewise, SAS uses a separate procedure, PROC PRINCOMP, for PCA rather than using PROC FACTOR.

The SAS code for accomplishing this PCA is as follows:

```
data DSA;
  input person $ verbal math logic;
  datalines;
Al          8    12   20
Bill        6    10   16
Charlie     5    4    9
Dan         5    10   15
Ed          4    16   20
Frank       2    8    10
;
run;

proc princomp data=DSA out=DSAout cov n=2;
  var verbal math logic;
  title 'Principal Component Analysis of Dataset DSA';
run;

proc print data=DSAout;
  title 'Printout of input data augmented by the component scores.';
run;

proc plot;
  plot prin2*prin1=person / vpos=19;
  title2 'Plot of Principal Components';
run;
```

The first 11 lines of this SAS code constitute the data statement. They are used to create the SAS data set for analysis. The next four lines of code are PROC PRINCOMP, which actually carries out the PCA. The "$n = 2$" option specifies that two components will be extracted. The "cov" option is included to instruct PROC PRINCOMP to calculate the PCA on the covariance matrix (rather than the correlation matrix) and to print out the covariance matrix. The "out" option is an important one. It specifies that component scores should be calculated and then appended to the input data. The entire structure is then written to the file that is specified after "out=." The following section of code is PROC PRINT, to print out the input data augmented by component scores. Last, PROC PLOT is used to print a scatter plot of each observation in the two-dimensional space of component 1 and component 2.

The three crucial elements of output are the PCA coefficients (the **K** matrix, analogous to factor loadings), the eigenvalues, and the component scores (analogous to factor scores). Both the coefficients and also the eigenvalues are included in the default display of results from all three statistical packages. However, the coefficients are reported in two different ways. Both SAS and Stata report unscaled coefficients (the **K** matrix), but SPSS reports scaled coefficients (the $\mathbf{K}_{sc}$ matrix).

The SAS code given above creates the third kind of crucial information, the component scores, with an "out=<filename>" option in the first line of PROC PRINCOMP. Component scores are also easy to obtain in SPSS. From the window entitled "Factor Analysis," one simply clicks on the "Scores" button and selects one of the three methods for calculating scores. In Stata, the code for component scores is somewhat tedious and will not be explained here, but it is obviously a simple matter to calculate the component scores on a spreadsheet using $\mathbf{Y} = \mathbf{XK}$ (Eq. 4.13a) once one has obtained the coefficients matrix **K**.

The output file generated by the SAS code above gives the component scores appended to the input data, as shown below. These component scores differ from those calculated above using Equation 4.13a, $\mathbf{Y} = \mathbf{X}\,\mathbf{K}$. They differ because the default condition in SAS is to calculate *centered component scores*, that is, component scores as deviations from their respective means. The component scores calculated above by Equation 4.13a are *uncentered component scores*. Both kinds of component scores are useful, but for different purposes.

Obs	person	verbal	math	logic	Prin1	Prin2
1	Al	8	12	20	5.51339	2.75726
2	Bill	6	10	16	0.90936	1.08308
3	Charlie	5	4	9	-8.35593	1.47593
4	Dan	5	10	15	0.00000	0.00000
5	Ed	4	16	20	7.44657	-2.55901
6	Frank	2	8	10	-5.51339	-2.75726

The equation for centered component scores is the same as Equation 4.13a for uncentered ones, except that a centered data matrix $\mathbf{X_c}$ is entered into the formula rather than $\mathbf{X}$. As we calculate centered component scores using Equation 4.13b, we see that they correspond to the SAS output.

$$\mathbf{Y_c} = \mathbf{X_c K} \tag{4.13b}$$

$$
\mathbf{Y_c} = \mathbf{X_c K} =
\begin{bmatrix}
3 & 2 & 5 \\
1 & 0 & 1 \\
0 & -6 & -6 \\
0 & 0 & 0 \\
-1 & 6 & 5 \\
-3 & -2 & -5
\end{bmatrix}
\begin{bmatrix}
0.142 & 0.804 \\
0.625 & -0.525 \\
0.767 & 0.279
\end{bmatrix}
=
\begin{bmatrix}
5.51 & 2.76 \\
0.91 & 1.08 \\
-8.36 & 1.48 \\
0 & 0 \\
7.45 & -2.56 \\
-5.51 & -2.76
\end{bmatrix}.
$$

If one were to directly calculate the variance of the component scores in either the centered or the un-centered $\mathbf{Y}$ matrix, the variance in the first component is 37.379, which is equivalent to eigenvalue one. The variance of the second is 5.021, which is equivalent to eigenvalue two. We see that the eigenvalues are, in fact, the variances of the component scores.

The scaled coefficients in the PCA summary table above can be used to create a vector plot of the three variables within the principal components space (Fig. 4.8). The values in that matrix are coordinates of the three variables with regard to the two orthogonal axes, component 1 and component 2. Consider the coordinates for the verbal vector, 0.868 and 1.802. The vector for the verbal cognitive test is plotted according to these coordinates as shown in the vector plot below. The coordinates for the math test vector are 3.823 and −1.176, and for the logic test vector are 4.691 and 0.625. The lengths of each of these three vectors can be calculated from the coordinates using the Pythagorean theorem, and they come out to 2.00, 4.00, and 4.73, for verbal, math, and logic respectively. These are, of course, the standard deviations for each of the three variables. The standard deviation of each variable is the length, or magnitude, of the vector representing that variable. We see that when we refer to the square root of the sum of squares of a vector array of numbers as the *magnitude* of that vector, geometrically, it is indeed quite literal.

We will conclude this section by reviewing the major equations for PCA in comparison with the comparable equations for factor analysis. It was shown in Sections 4.5.3 and 4.5.4 (using Eqs. 4.4, 4.9–4.12) that factor analysis consists essentially of a set of data transformations using linear combinations. Equation 4.4, $\mathbf{Y} = \mathbf{ZT}$, for example, is the factor analysis transformation from manifest variable $\mathbf{Z}$ scores into factor scores. For PCA, the analogous transformation is given in Equations 4.13a and 4.13b, the former transforming uncentered raw data into uncentered component scores, and the latter transforming centered raw data into centered component scores, as demonstrated.

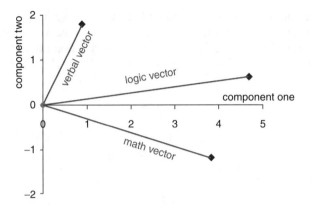

Figure 4.8 Vector plot from scaled coefficients of the three mental test variables from a principal component analysis of the simplest case data of Figure 4.4.

$$\mathbf{Y} = \mathbf{X}\mathbf{K} \tag{4.13a}$$

$$\mathbf{Y} = \mathbf{X}\mathbf{K} = \begin{bmatrix} 8 & 12 & 20 \\ 6 & 10 & 16 \\ 5 & 4 & 9 \\ 5 & 10 & 15 \\ 4 & 16 & 20 \\ 2 & 8 & 10 \end{bmatrix} \begin{bmatrix} 0.142 & 0.804 \\ 0.625 & -0.525 \\ 0.767 & 0.279 \end{bmatrix} = \begin{bmatrix} 23.99 & 5.71 \\ 19.38 & 4.04 \\ 10.12 & 4.43 \\ 18.47 & 2.96 \\ 25.92 & 0.40 \\ 12.96 & 0.20 \end{bmatrix}.$$

$$\mathbf{Y_c} = \mathbf{X_c}\mathbf{K} \tag{4.13b}$$

$$\mathbf{Y_c} = \mathbf{X_c}\,\mathbf{K} = \begin{bmatrix} 3 & 2 & 5 \\ 1 & 0 & 1 \\ 0 & -6 & -6 \\ 0 & 0 & 0 \\ -1 & 6 & 5 \\ -3 & -2 & -5 \end{bmatrix} \begin{bmatrix} 0.142 & 0.804 \\ 0.625 & -0.525 \\ 0.767 & 0.279 \end{bmatrix} = \begin{bmatrix} 5.51 & 2.76 \\ 0.91 & 1.08 \\ -8.36 & 1.48 \\ 0 & 0 \\ 7.45 & -2.56 \\ -5.51 & -2.76 \end{bmatrix}.$$

This transformation is completely and precisely reversible when enough components are extracted to account for all of the variance. Since the artificial data matrix for this demonstration has a rank of 2, the two components do account for all the variance, and the process is reversible, as expressed in Equations 4.15a and 4.15b. This is obviously an idealized condition for demonstration purposes that would rarely occur in actual practice.

$$\mathbf{X} = \mathbf{Y}\mathbf{K}' \tag{4.15a}$$

$$\mathbf{X} = \mathbf{YK'} = \begin{bmatrix} 23.99 & 5.71 \\ 19.38 & 4.04 \\ 10.12 & 4.43 \\ 18.47 & 2.96 \\ 25.92 & 0.40 \\ 12.96 & 0.20 \end{bmatrix} \begin{bmatrix} 0.142 & 0.625 & 0.767 \\ 0.804 & -0.525 & 0.279 \end{bmatrix} = \begin{bmatrix} 8 & 12 & 20 \\ 6 & 10 & 16 \\ 5 & 4 & 9 \\ 5 & 10 & 15 \\ 4 & 16 & 20 \\ 2 & 8 & 10 \end{bmatrix}.$$

$$\mathbf{X_c} = \mathbf{Y_c\, K'} \tag{4.15b}$$

$$\mathbf{X_c} = \mathbf{Y_c\, K'} = \begin{bmatrix} 5.51 & 2.76 \\ 0.91 & 1.08 \\ -8.36 & 1.48 \\ 0 & 0 \\ 7.45 & -2.56 \\ -5.51 & -2.76 \end{bmatrix} \begin{bmatrix} 0.142 & 0.625 & 0.767 \\ 0.804 & -0.525 & 0.279 \end{bmatrix} = \begin{bmatrix} 3 & 2 & 5 \\ 1 & 0 & 1 \\ 0 & -6 & -6 \\ 0 & 0 & 0 \\ -1 & 6 & 5 \\ -3 & -2 & -5 \end{bmatrix}.$$

The fundamental factor theorem, $\mathbf{R}^+ = \mathbf{FF'}$ in factor analysis, Equation 4.10, reveals that the matrix of factor loadings $\mathbf{F}$ is the root matrix of the correlation matrix (or of the reproduced correlation matrix $\mathbf{R}^+$ when all of variance is not accounted for). For PCA, the scaled coefficients $\mathbf{K}_{sc}$ are the matrix square root of the covariance matrix $\mathbf{S}$, as shown in the analogous Equation 4.16.

$$\mathbf{S} = \mathbf{K}_{sc}\mathbf{K}'_{sc}. \tag{4.16}$$

$$\mathbf{S} = \mathbf{K}_{SC}\mathbf{K}'_{SC} = \begin{bmatrix} 0.868 & 1.802 \\ 3.823 & -1.176 \\ 4.691 & 0.625 \end{bmatrix} \begin{bmatrix} 0.868 & 3.823 & 4.691 \\ 1.802 & -1.176 & 0.625 \end{bmatrix} = \begin{bmatrix} 4 & 1.2 & 5.2 \\ 1.2 & 16 & 17.2 \\ 5.2 & 17.2 & 22.4 \end{bmatrix}.$$

This set of Equations 4.13a–4.18 is simpler than the comparable set of formulas for factor analysis (4.4 to 4.12), because factor analysis uses three forms of the eigenvector matrix ($\mathbf{K}$, $\mathbf{F}$, and $\mathbf{T}$), whereas PCA uses only one ($\mathbf{K}$), although for some purposes, such as the summary table, it is desirable to use the scaled form of the coefficients ($\mathbf{K}_{sc}$).

Coefficients matrix $\mathbf{K}$, the normalized eigenvectors, can also be used to transform the covariance matrix of manifest variables, $\mathbf{S}$, into the covariance matrix of latent variables, $\mathbf{\Lambda}$ (the diagonal matrix of eigenvalues), as shown in Equation 4.17.

$$\mathbf{\Lambda} = \mathbf{K'SK} \tag{4.17}$$

$$\mathbf{\Lambda} = \mathbf{K'SK} = \begin{bmatrix} 0.142 & 0.625 & 0.767 \\ 0.804 & -0.525 & 0.279 \end{bmatrix} \begin{bmatrix} 4 & 1.2 & 5.2 \\ 1.2 & 16 & 17.2 \\ 5.2 & 17.2 & 22.4 \end{bmatrix} \begin{bmatrix} 0.142 & 0.804 \\ 0.625 & -0.525 \\ 0.767 & 0.279 \end{bmatrix}$$

$$= \begin{bmatrix} 37.379 & 0 \\ 0 & 5.021 \end{bmatrix}.$$

Of course, when all of the variance is accounted for by the extracted components, this process is also completely reversible. One can use matrix $\mathbf{K}$ to transform the covariance matrix of latent variables, Λ, back into the covariance matrix of manifest variables, $\mathbf{S}$, as shown in Equation 4.18.

$$\mathbf{S} = \mathbf{K}\Lambda\mathbf{K}' \tag{4.18}$$

$$\mathbf{S} = \mathbf{K}\Lambda\mathbf{K}' = \begin{bmatrix} 0.142 & 0.804 \\ 0.625 & -0.525 \\ 0.767 & 0.279 \end{bmatrix} \begin{bmatrix} 37.379 & 0 \\ 0 & 5.021 \end{bmatrix} \begin{bmatrix} 0.142 & 0.625 & 0.767 \\ 0.804 & -0.525 & 0.279 \end{bmatrix}$$

$$= \begin{bmatrix} 4 & 1.2 & 5.2 \\ 1.2 & 16 & 17.2 \\ 5.2 & 17.2 & 22.4 \end{bmatrix}.$$

These two transformation, Equations 4.17 and 4.18, demonstrate clearly what is accomplished by a PCA. It transforms a larger set of correlated manifest variables into a smaller set of orthogonal latent variables that account for the maximum amount of variance possible, subject to the constraint of orthogonality. This simplifies the data, both in terms of enabling one to account for most of the variance with a reduced set of variables, and also in terms of setting the covariances among the latent variables to zero.

We've seen that PCA is simpler and more direct computationally than factor analysis. It also usually involves factoring a covariance matrix rather than a correlation matrix. It is a straightforward application of factoring logic (eigenvectors and eigenvalues) to a covariance structure, with the intent of creating linear combinations of the variables that are orthogonal to one another and that account for the maximum amount of variance in the original variables. It is particularly useful as a data simplification technique to prepare data for additional statistical analysis (such as a MANOVA, explained in Chapter 8).

4.7   ROTATION OF THE FACTOR PATTERN

One of the most important uses of both factor analysis and PCA is dimension reduction, simplifying the data down to a smaller set of variables. The key to doing so is to be able to extract factors that maximize the amount of variance accounted for, subject to the constraint of orthogonality. One of the characteristics of the process is that the amount of variance accounted for becomes substantially smaller with each additional factor. This leaves the last few factors with very little variance, so that they can be disregarded in the final solution, thus reducing the dimensionality. However, this particular variance structure of the latent variables, with the first factor accounting for a large amount of variance, then dropping substantially with each succeeding factor, is not always the desired or optimal configuration for understanding the data. Often, the data are more meaningful with a reconfiguration, a rotation, of the latent variables.

The process of rotating the factor pattern to a scientifically or theoretically more meaningful configuration has for the better part of a century been an integral part of the factor analytic landscape. Suppose, for example, that an organizational psychologist has created a management profiling instrument consisting of 12 question, six related to effectiveness in working with people ("maintenance"), and six that relate to task effectiveness.[11] The data are factor analyzed and the following factor pattern is obtained.

Unrotated Factor Pattern

	Factor 1	Factor 2
Maintenance variable 1	0.696	−0.712
Maintenance variable 2	0.712	−0.695
Maintenance variable 3	0.752	−0.652
Maintenance variable 4	0.769	−0.631
Maintenance variable 5	0.792	−0.604
Maintenance variable 6	0.798	−0.595
Task variable 1	0.711	0.695
Task variable 2	0.711	0.697
Task variable 3	0.656	0.748
Task variable 4	0.663	0.742
Task variable 5	0.609	0.786
Task variable 6	0.571	0.816

Using the factor loadings as coordinates for plotting in the space of factor 1 and factor 2, a vector plot is obtained for this unrotated factor pattern (Fig. 4.9). Factor 1 is located squarely in the middle of the two bundles of vectors, the bundle of six vectors for the task variables, and the bundle of six variables for the maintenance variables. This is the usual state of affairs in the initial factor analysis solution. The first factor (or component) is always in the geometric center of the collection of variables. It is like a least squares midpoint vector of all of the other vectors.[12]

Given that the six task variables all indicate positive performance, and given that the six maintenance variables also indicate positive performance, the first factor could be interpreted as overall positivity of the ratings—a kind of G factor like Spearman's (1904) hypothesized general intelligence. Factor 2 in this configuration is more difficult to interpret, with the task variables all loading on the positive side of that factor, and the maintenance variables

[11] This is actually simulated data in which the underlying structure is created to be two factors, with individual reliabilities of the 12 variables set at 0.90. It appears again in Section 4.9 of this chapter, together with comparable simulations at other reliability levels.

[12] Components and factors are not exactly a least squares fit to the variables, but they use a similar criterion. Whereas a least squares fit, as for example in multiple regression, is obtained by minimizing the squared deviations of data points on a single criterion variable Y from the plane or hyperplane of prediction, eigenvectors as expressed in principal components analysis or factor analysis are obtained by minimizing the squared deviations of perpendicular distances from the data points to the line of the eigenvector. See Rencher (2002, 387–389) for an explanation of a principal component as a "perpendicular regression line."

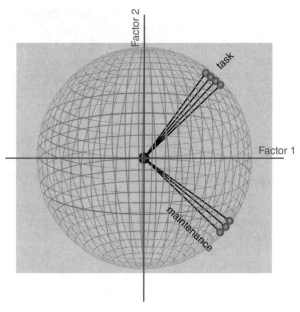

Figure 4.9   Vector plot for unrotated factor pattern of 12 management-profiling variables. (Image courtesy of Metrika (metrika.org), © Appzilla, LLC.)

loading on the negative side. Perhaps it could be interpreted as a kind of polar factor of task versus maintenance, but it is not a very clear factor. A rotation of the pattern of approximately 40 degrees counter-clockwise would make it easier to interpret, as shown in the rotated factor pattern below and its accompanying vector plot, where factor 1 is defined by the cluster of six maintenance variables, and factor 2 is defined by the cluster of six task variables (Fig. 4.10).

Rotated Factor Pattern

	Factor 1	Factor 2
Maintenance variable 1	0.992	−0.081
Maintenance variable 2	0.994	−0.058
Maintenance variable 3	0.995	0.000
Maintenance variable 4	0.995	0.027
Maintenance variable 5	0.993	0.063
Maintenance variable 6	0.993	0.074
Task variable 1	0.081	0.991
Task variable 2	0.080	0.992
Task variable 3	0.005	0.995
Task variable 4	0.014	0.995
Task variable 5	−0.055	0.993
Task variable 6	−0.103	0.990

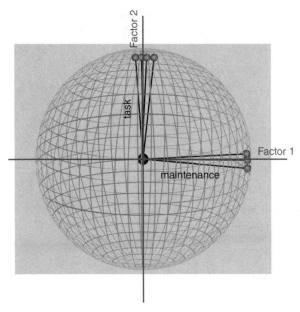

Figure 4.10 Vector plot for rotated factor pattern of twelve management-profiling variables. (Image courtesy of Metrika (metrika.org), © Appzilla, LLC.)

This example was selected to have a simple and clear pattern, but often they are neither simple nor clear, and a plan of rotation is not always obvious. Subjective graphical rotation is plagued with the problem of "indeterminacy." To the extent that persons have varying esthetic preferences for rotating complex configurations, scientific replication would be difficult. Thurstone (1947) proposed a criterion, rotation to "simple structure," as a way of solving the indeterminacy problem in factor analysis. The fundamental idea of simple structure is to seek factors that have mainly zero or near zero loadings with just a few high loadings, which places the factors as much as possible within the center of clusters of variables (as shown in the post-rotational vector plot). At first, graphical methods were used to rotate a factor pattern to simple structure, but computational rotation methods (such as varimax, oblimin, promax) were needed, and were soon forthcoming.

In the graphical approach, one can rotate the matrix of factor loadings by first estimating the desired angle of rotation. From a visual inspection of the prerotation plot, for example, it appears that a rotation counter-clockwise of about 40° would give us the desired orientation, the one closest to simple structure. We carefully measure the angle of rotation and find it to be −41°, that is, 41° in a counter-clockwise direction.

To avoid the confusion of a negative angle of rotation, we will subtract 41° from 360° (since a 41° rotation counter-clockwise, a negative direction, is equivalent to a 319° rotation in a positive direction, clockwise). Matrix **V**, a 2×2 transformation matrix, is obtained by placing the cosine of α (where

$\alpha = 319°$) in the diagonal locations of the matrix (cosine $(319°) = 0.755$). The sine of α is placed in the location below the diagonal (sine $(319°) = -0.656$), and the negative sine of α ($-$sine $(319°) = 0.656$) in the location above the diagonal as shown.

$$V = \begin{bmatrix} \cos\alpha & -\sin\alpha \\ \sin\alpha & \cos\alpha \end{bmatrix}$$
$$= \begin{bmatrix} 0.755 & 0.656 \\ -0.656 & 0.755 \end{bmatrix} \tag{4.19}$$

When the unrotated factor pattern $\mathbf{F}$ is postmultiplied by this rotation matrix $\mathbf{V}$, the result is indeed the rotated factor pattern $\mathbf{F_r}$ that has the desired geometrical arrangement as shown in the vector plot above.

$$\mathbf{F_r} = \mathbf{FV} = \begin{bmatrix} 0.696 & -0.712 \\ 0.712 & -0.695 \\ 0.752 & -0.652 \\ 0.769 & -0.631 \\ 0.792 & -0.604 \\ 0.798 & -0.595 \\ 0.711 & 0.695 \\ 0.711 & 0.697 \\ 0.656 & 0.748 \\ 0.663 & 0.742 \\ 0.609 & 0.786 \\ 0.571 & 0.816 \end{bmatrix} \begin{bmatrix} 0.755 & 0.656 \\ -0.656 & 0.755 \end{bmatrix} = \begin{bmatrix} 0.992 & -0.081 \\ 0.994 & -0.058 \\ 0.995 & 0.000 \\ 0.995 & 0.027 \\ 0.993 & 0.063 \\ 0.993 & 0.074 \\ 0.081 & 0.991 \\ 0.080 & 0.992 \\ 0.005 & 0.995 \\ 0.014 & 0.995 \\ -0.055 & 0.993 \\ -0.103 & 0.990 \end{bmatrix}.$$

The graphical approach to rotation is reasonable when there are only two factors, or even three, but becomes difficult if not impossible with four or more. The search for computational methods for rotating factors began early in the 20th century, and has resulted in a proliferation of methods, some of which produce orthogonal factors (uncorrelated) and some oblique (correlated). The most commonly used method is the varimax rotational procedure, published in mid-century by Henry Kaiser (1958). Varimax seeks a rotated factor loadings matrix $\mathbf{F}$ that maximizes the squared loadings within each column (each factor). It provides a fairly good approximation to the results of graphical attempts to achieve simple structure.

The SAS command code option for a varimax solution is "rotate=varimax," as shown in the first line of the PROC FACTOR procedure below.

```
proc factor data=Mdata nfact=2 rotate=varimax;
  var maint1-maint6 task1-task6;
  title 'Factor analysis of management profiling data with rotation';
run;
```

The SAS output resulting from this code gives both the unrotated factor pattern (matrix of factor loadings) and also the rotated factor pattern. It does not indicate the angle of rotation, but it does report the V matrix, the transformational matrix that is used to create a rotated pattern, as shown in the SAS output fragment below. Clearly, the varimax solution results in the same sine and cosine values in the transformation matrix as those obtained by choosing a rotational angle of –41 degrees and calculating the transformation matrix using Equation 4.19, thus illustrating the similarity of varimax rotation to a graphical rotation according to simple structure. The creation of mathematical algorithms of rotation over the past century and their availability in statistical packages has greatly increased the variety of ways in which one can factor data.

Orthogonal Transformation Matrix		
	1	2
1	0.75532	0.65536
2	−0.65536	0.75532

The art of selecting an optimal rotational method has become well-practiced within the factor analytic community. The possibilities are many. Chapter 32 of the SAS manual (SAS Institute Inc, 2009), entitled "The FACTOR Procedure," lists nine primarily orthogonal methods and 15 oblique choices under the "ROTATE=" option, and there are a number of additional options available under each of these. Examples of oblique rotation will not be given in this chapter, but will be given in Chapter 5.

Although rotational methods are used extensively with factor analysis, they are not commonly used with PCA of a covariance matrix. There are several reasons for this. For one, the factor analytic approach tends to take latent variables more seriously in an ontological sense. Factor analytic latent variables are often a central part of theory building and modeling, and rotation is often needed to make the latent variables more clear and meaningful.

On the other hand, PCA is much less concerned with theory. It seems to be the method of choice when one's primary goal is data simplification or reduction, perhaps in preparation for additional analyses by other methods, such as multivariate analysis of variance. There are also reasons for approaching the rotation of principal component coefficients with caution. Rotation negates one of the chief simplifying characteristics of PCA—its power to give orthogonal structure to the latent variables. That is, when one rotates PCA coefficients (to anything other than 90, 180, or 270°) the component scores are no longer orthogonal to one another—they become correlated.

4.8 THE RICH VARIETY OF FACTOR ANALYSIS MODELS

Just as there is a wide variety of possible rotational methods to choose from in the factor analytic art, there are also a number of estimation methods for accomplishing the prerotational factoring in the first place. Kline (1993) refers to them as "condensation methods." Chapter 32 of the SAS (2009) manual lists nine choices under the "METHOD=" option. There have been, however, many unique condensation methods, a number of which have fallen by the wayside. The major watershed in the factor analytic methods is between the principal component method and the multiple varieties of *common factor* methods. So far in this chapter, we have relied upon the principal component method (because of its simplicity) to explain the conceptual basis of factoring, but to factor-analytic purists, only the common factor methods count as being "real" factor analyses. We turn now to a consideration of the additional properties of these common factor methods that distinguish them from the principal component method.

The simplest of the common factor methods is the *principal factor* method, also sometimes called the *principal axis* method. Mathematically, the principal factor method is very similar to the principal component method, but philosophically, they differ substantially. That difference turns on the concept of *common factors*, the idea that we are searching for underlying latent variables that transcend any method devised to measure them. These common factors may be manifested in several observed variables. The search for common factors is focused upon finding those properties that are common to several, many, or sometimes all of the observed variables.

It may be remembered that in Chapter 1, Section 1.5, factor analysis and PCA were diagrammed differently. In PCA, the arrows go from the observed variables to the components, indicating that components are linear functions derived from the observed variables. In factor analysis, the "causal" arrows go from the factors (latent variables) to the observed (manifest) variables, indicating that the observed variables are caused by or an expression of the latent ones. The observed variables are considered as mere surface manifestations of actually existing underlying variables. To re-quote Cudeck and McCallum (2007, ix) from the second paragraph of this chapter, the genius of factor analysis is that it provides "a way to go beyond empirical variables, such as tests and questionnaires, to the corresponding latent variables that underlie them." In other words, our interest does not lie with responses to any particular questionnaire, but with the fundamental but unobservable traits that lead to particular responses on any of a variety of questionnaires. That is why Kline (1993) can say that factor analysis "lies at the heart of psychometrics" (p. 93). The quest of true factor analysis is to find a way to get beyond the observable surface variables to measure the unobservable source traits that are the real focus of psychological theorizing.

In PCA, there is no such ontological quest. PCA is merely a method for creating more elegant, simple, parsimonious, and useful expressions of observed data. In some ways, the distinction is much like the historically important distinction made by MacCorquodale and Meehl (1948) more than half a century ago between intervening variables and hypothetical constructs. Intervening variables are more general explanatory variables that may be useful in prediction, but are not considered to be "true" in any fundamental sense or to have any ontological significance. This is the nature of components in PCA. On the other hand, hypothetical constructs are fundamental explanatory constructs that are hypothesized to be the real underlying elements that give theory its substance. This is the fundamental idea of the *common factors model*, that common factors are real, not just useful. The argument has some rhetorical force, maintaining that verbal intelligence, for example, is deeper than and transcends any particular way we may devise to measure it. Verbal intelligence is in some sense more real and more central to our understanding than any particular method devised to measure it.

This major philosophical divide is expressed mathematically through the most simple, perhaps even trivial, of differences in computational methods. Consider the following possibility. If we were to go back to Section 4.4 and change nothing whatsoever in the factor analysis computations other than to replace the ones in the diagonal of the correlation matrix with estimates of the communalities, those same calculations would yield a *common factor* method of factor analysis rather than merely a principal component factor analysis. More precisely, those calculations would yield a factor solution by what is called the *principal factor* method, and it would now qualify as a true *common factor* analysis.

Since communalities are not known until after running the factor analysis, they must be estimated. There are a number of ways of estimating communalities. One of the most popular methods is to use squared multiple correlations (SMCs) as estimates of the communalities. They are in fact lower bounds to the communalities. This, of course requires prior calculation of several multiple correlations, one for each variable in the data set, and these calculations are in fact integrated into data analysis software systems like SAS, Stata, and SPSS.

The SAS code for calculating a principal factor method consists of two parts, specifying the method as "principal," and setting the communality estimate with a "priors" option, as shown in the code fragment below. It is important to note that in SAS, the "method=PRINCIPAL" option can lead to either a principal component method or a principal factor method, depending on whether or not the "priors" option is invoked. If this option does not appear in the code line, the analysis will be principal component factor analysis. If the "priors" option does appear, the analysis will be principal factor method of factor analysis, with the communality estimates being whatever has been specified in the priors option (SMC, ASMC, INPUT, MAX, ONE, or RANDOM). The priors option is set to "SMC" in the following code fragment, which is to indicate that squared multiple correlations will be used as the communality estimates.

```
proc factor data=Mdata method=PRINCIPAL priors=SMC nfact=2;
```

How does using communalities or communality estimates in the diagonal produce common factors, that is, factors that better reflect the common variance? As can be seen in the presentation of PCA (Section 4.6), it is a truism that the variables in the covariance matrix having larger variances exert more influence on the final factor solution than those with smaller variances. Placing communality estimates in the diagonal of the matrix (where variances are located) gives more influence to those variables that have greater communalities, that is, are more related to, and have more variance in common with, all of the other variables.

The obvious question is why should one not employ actual communalities in the diagonal rather than estimating them from SMCs or in some other way? One could simply run the factor analysis twice, once to obtain the communalities, and then a second time with the actual communalities from the first run being placed in the diagonal elements of the correlation matrix. Consider what would happen. This second run would produce a new solution with communalities that differ somewhat from those in the first run, and so the process would have to be repeated with these new communalities. But then this would produce yet a third group of communalities different from both of the other two, leading to continued iterations. The iterative process just described is in fact one of the most popular methods of common factor analysis. It is referred to as the *iterated principal factor* method. The process of iteration does not continue indefinitely, but rather will converge after several such iterations. Iterated principal factor analysis generally produces a solution superior to that of the principal factor method itself. It is specified in SAS with the instruction "method=PRINIT."

Another important variety of common factor method is the maximum likelihood (MLFA) method (Lawley, 1940). Although the principal factor method (particularly in its iterated form) is probably the most often used common factor method, statisticians tend to prefer the maximum likelihood method (Lawley and Maxwell, 1971), presumably because of its inferential and hypothesis testing properties. The maximum likelihood criterion is inferential in nature in that rather than explaining as much variance as possible in observed data (as does principal components), it explains as much variance as possible in the population correlation matrix as inferred from the sample correlation matrix. Because of this, the mathematical basis of maximum likelihood is substantially more complex than that of the principal components method, and MLFA assumes that the sample is selected at random from a population. It also requires that the correlation matrix be nonsingular, that is, of full rank, whereas most other methods do not (unless one is doing factor scores, see Eq. 4.20).

Because MLFA is accomplished by solving equations through an iterative estimation process, the eigenvalues are not always found to be in descending order across successive factors, and in practice the MLFA procedure may occasionally fail to converge. Nevertheless, as we shall see in Section 4.9, the maximum likelihood approach is superior in some ways in dealing with some kinds of data. Perhaps its most important application, however, is not in exploratory factor analysis, but in confirmatory factor analysis.

One other variety of factor analysis belonging to the common factor model should be mentioned here, alpha factor analysis (Kaiser & Caffery, 1965). This method is particularly tailored for psychometric applications of factor analysis. The criterion by which the alpha method is derived is to find uncorrelated common factors that have maximum generalizability in the sense measured by Cronbach's (1951) alpha. Cronbach's alpha is a prominent index of reliability[13] that is often used as a measure of internal consistency. Clearly, factor analysis is relevant to the pursuit of internal consistency, in that when items making up a scale are characterized by multiple factors, internal consistency will often suffer. Given the centrality of factor analysis to psychometric work, this general idea of combining dimensionality (factoring) and reliability considerations is salutary whether or not this method fully achieves it purposes. One of the distinguishing characteristics of alpha factor analysis is that it gives reliability estimates for each of the factors.

The major contrast between the maximum likelihood factor analytic method (MLFA) and the alpha method is that alpha factoring is a psychometric method, while MLFA is primarily a statistical method. The MLFA method assumes multivariate normality, whereas there are no distributional assumptions for alpha factoring. To illustrate how alpha factor analysis deals with the problem of deciding how many factors to extract, consider again the data set (from a simulation exercise) that was used in Section 4.7 above to illustrate rotation. As mentioned in footnote 8, the underlying structure for these simulated data was in fact based upon two factors, with the reliability of the 12 individual variables set at 0.90. When SAS is used to conduct an alpha factor analysis of these data, using the "method=ALPHA" option, two factors are extracted by the default setting. However, when we override the default by specifying "nfact=3," the following table of coefficient alphas for each of the three factors appears in the SAS output. The first two factors obviously have rather good reliabilities. The third variable, which was excluded in SAS's default analysis, in fact should have been excluded. It would have been

---

[13] Nunnaly (1978) considers Cronbach's alpha to be the most important index of test reliability.

excluded by reliability considerations[14] had it not first been excluded by the minimum eigenvalues criterion.

Coefficient Alpha for Each Factor		
Factor1	Factor2	Factor3
0.90484911	0.88221137	-0.1077113

Several of the common factor models, because of their use of iterated estimation methods, will sometimes produce communalities greater than one. This peculiarity should, in theory, not be possible. Since communalities are squared correlation coefficients, they should fall between zero and one, but estimation by iterative methods can and does produce "impossible" communalities. If an iterative solution produces a communality equal to one, it is referred to as a *Heywood case* (Heywood, 1931), and if one of the communalities is greater than one, it is referred to as an *ultra-Heywood* case. There are methods in SAS and the other statistical packages for dealing with these problems. The maximum likelihood method is particularly susceptible to the Heywood case, perhaps occurring in practice as much as half the time, even when the correlation matrix is positive definite (nonsingular). The alpha method is not as susceptible to the Heywood case (Kaiser and Derflinger, 1990).

We have discussed four ways of conducting a factor analysis using the common factor model: the principal factor method, the iterated principle factor method, the maximum likelihood method, and the alpha method of factor analysis. One additional method that grows out of the common factor model, and one that has to some extent eclipsed these methods in popular interest, is *confirmatory factor analysis* (CFA). All four of the common factor methods discussed here, and also the principal component method, have in the past several decades come to be referred to collectively as *exploratory factor analysis* (EFA) to distinguish them from the confirmatory approach. Many are of the opinion (implicit within the title EFA) that the main function of old-fashioned factor analysis methods is to investigate data structure in an exploratory mode as a prelude to constructing a reasonable factor analytic model that can be tested (and perhaps confirmed) with CFA. However, notwithstand-

---

[14] It should be noted that alpha factor analysis does not always identify spurious factors as well as it does here. See, for example, the observation in footnote 19 of this chapter, and the accompanying text. These coefficient alphas that are reported by the SAS PROC FACTOR alpha method factor analysis are an indication of the reliability of the factor scores. In Section 4.10, where we examine the effects of reliability on factor analysis, the focus is on a different (but related) reliability—the reliability of the individual items or variables entering into the factor analysis.

ing the developments in confirmatory factor analysis, exploratory factor analysis continues to thrive as one of the primary methods of choice (as shown in Table 4.1), often as an end in and of itself. Judging from the papers that appear in the top journals, exploratory factor analysis still substantially exceeds confirmatory factor analysis in actual usage.

Nunnaly (1978) tersely summarizes the contrast between the common factor method approach and the principal component approach to factoring. The principal component method of factor analysis deals with empirically real factors, in the sense that they are linear combinations of observed variables. The factor loadings likewise have a clear interpretation. They are simply correlation coefficients between the observed variables and the calculated factors. Common factors are theoretical or hypothetical and have to be estimated using any of a multitude of methods that define the factors in terms of common variance with uniqueness removed. Another major difference is that the calculation of factor scores in the common factors model is more complex. Since the factors are estimated rather than directly calculated, the factor scores must also be estimated. There are several estimation procedures available. The most popular approach to estimating factor scores for the common factor model is the regression method (Thomson, 1951), shown in Equation 4.20.[15] For comparison, the direct factor score calculation formula used with the principal component method, Equation 4.4, is shown below it.

$$\mathbf{Y} = \mathbf{Z}\mathbf{R}^{-1}\mathbf{F}. \qquad (4.20)$$

$$\mathbf{Y} = \mathbf{Z}\mathbf{T}. \qquad (4.4)$$

Both of these formulas have $\mathbf{Z}$ scores of the manifest variables as the premultiplying matrix, and eigenvectors as the postmultiplying matrix (either $\mathbf{F}$ or $\mathbf{T}$). The regression-based formula, Equation 4.20, has $\mathbf{R}^{-1}$ inserted between $\mathbf{Z}$ and the eigenvector matrix, which is not surprising, since the computational basis of multiple regression (Chapter 9) is the inverse of a correlation matrix. The other difference in the two formulas is that the regression estimation equation, Equation 4.20, has matrix $\mathbf{F}$, factor loadings (the eigenvectors normalized to their eigenvalues), as the postmultiplying matrix, whereas the direct factor score formula, Equation 4.4, has matrix $\mathbf{T}$ (the eigenvectors normalized to the inverse of their eigenvalues) in the postmultiplying position.

SPSS offers a choice of three factor score estimation procedures. When one selects the **Scores** button from within the FACTOR procedure (as demonstrated at the end of Section 4.4), one is given the choice between: (1) the regression method, (2) the Bartlett method, and (3) the Anderson–Rubin method. There is some controversy concerning which estimation method is most appropriate in particular situations. This is another example of the

---

[15] For a detailed account of this and other approaches to the estimation of factor scores, see chapter 16 of Harman (1976).

relative simplicity of the principal component method where factor scores have only one method and a straightforward interpretation. They are simply linear combinations of the manifest variables, like a weighted sum.

We will hold until the end of the chapter the question of whether the common factor method has compensating advantages for all of the added complexity, but there is a prior question. Does using the common factor model rather than principal component factor analysis actually make a difference in the final factor solution one obtains? That depends. With unstable data sets (those having low reliability coefficients for the individual variables), or with data sets that have a small number of variables, the condensation method chosen can have a sizeable effect upon the results. However, as Harman (1976), Nunnaly (1978), Kline (1993), Rencher (2002),[16] and others have pointed out, with data matrices that provide a good fit to the factor analysis model, the results of the common factor methods differ little from one another and also differ little from the results produced by the principal component factor analysis method. Although maximum likelihood is considered to be much more sophisticated statistically, Cattell (1978) has argued that with large data matrices, the MLFA results are so similar to those obtained from the principal component approach, that the additional complexity of maximum likelihood is probably not justified. The next question of course is whether the more complex common factor methods give in some sense better results when applied to data sets with a small number of variables, or when applied to those with low-item reliabilities. The two final sections of the chapter deal with those questions. Section 4.9 examines the performance characteristics of the various methods with a small set of variables (the Locurto et al. data), and Section 4.10 examines the performance characteristics of the various methods, with data sets having a variety of levels of item reliability.

4.9   FACTOR ANALYZING THE MENTAL SKILLS OF MICE: A COMPARISON OF FACTOR ANALYTIC MODELS

We will now close the discussion of common factor methods by comparatively applying four of them to a data set with a small number of variables, the simple five-variable data from the Locurto et al. (2006) study presented in Sections 4.2 and 4.3. We do this in order to compare the operating characteristics of the common factor methods in comparison with the principal component approach when applied to simple but useful and informative empirical data. The following is the correlation matrix **R** for these data.

---

[16] Rencher (2002) mentions in particular that the principal factors methods and the iterated principal factors method give results very similar to the principal components method when *either* of two things is true: (1) fairly large values in the correlation matrix, such that a small number of factors can account for most of the variance, (2) the number of variables is large.

$$\mathbf{R} = \begin{bmatrix} 1.000 & -0.210 & 0.270 & -0.020 & -0.040 \\ -0.210 & 1.000 & 0.080 & -0.200 & 0.060 \\ 0.270 & 0.080 & 1.000 & -0.280 & 0.030 \\ -0.020 & -0.200 & -0.280 & 1.000 & 0.000 \\ -0.040 & 0.060 & 0.030 & 0.000 & 1.000 \end{bmatrix}.$$

We will first demonstrate the principal factor method, the simplest of the common factor models in comparison with the principal component method. As mentioned above, it is mathematically equivalent to principal component factoring (demonstrated in Section 4.4), except that the diagonal of the **R** matrix is replaced with estimated communalities before factoring. We will use squared multiple correlation (SMC) values as the communality estimates. Guttman (1956) has shown that SMC is in fact the lower bound for communalities. Individual item reliabilities are the upper bound, which seems reasonable since one would not expect a variable to have a higher correlation with other variables than with itself. We have chosen to use the lower bound estimate, likely to be an underestimate, to amplify the contrast between the principal component solution and the principal factor solution.

The SMC values for each of the five variables (the olfactory task, the win-shift task, the detour task, the fear task, and the operant task) can be obtained from the inverse of the correlation matrix.

$$\mathbf{R}^{-1} = \begin{bmatrix} 1.000 & -0.210 & 0.270 & -0.020 & -0.040 \\ -0.210 & 1.000 & 0.080 & -0.200 & 0.060 \\ 0.270 & 0.080 & 1.000 & -0.280 & 0.030 \\ -0.020 & -0.200 & -0.280 & 1.000 & 0.000 \\ -0.040 & 0.060 & 0.030 & 0.000 & 1.000 \end{bmatrix}^{-1}$$
$$= \begin{bmatrix} 1.147 & 0.262 & -0.337 & -0.019 & 0.040 \\ 0.262 & 1.106 & -0.102 & 0.198 & -0.053 \\ -0.337 & -0.102 & 1.186 & 0.305 & -0.043 \\ -0.019 & 0.198 & 0.305 & 1.125 & -0.022 \\ 0.040 & -0.053 & -0.043 & -0.022 & 1.006 \end{bmatrix}.$$

The equation[17] for calculating squared multiple correlation coefficients (R_i^2) from the inverse of the correlation matrix ($\mathbf{R}^{-1}$) is:

$$\text{SMC} = R_i^2 = 1 - \frac{1}{r^{ii}}. \tag{4.21}$$

[17] This formula works because the inverse of a correlation matrix contains truly multivariate information in a way that the correlation matrix itself does not. Each element in the $\mathbf{R}^{-1}$ matrix contains information about *all* of the elements in the **R** matrix. If any element in **R** changes, it will change all of the elements in $\mathbf{R}^{-1}$.

where r^{ii} is the ith diagonal entry of the inverse matrix $\mathbf{R}^{-1}$. For the Locurto et al. 5×5 correlation matrix, the five SMC values are calculated as:

$$SMC_1 = R_1^2 = 1 - \frac{1}{1.147} = 0.1283.$$

$$SMC_2 = R_2^2 = 1 - \frac{1}{1.106} = 0.0957.$$

$$SMC_3 = R_3^2 = 1 - \frac{1}{1.186} = 0.1567.$$

$$SMC_4 = R_4^2 = 1 - \frac{1}{1.125} = 0.1107.$$

$$SMC_5 = R_5^2 = 1 - \frac{1}{1.006} = 0.0060.$$

This correlation matrix is particularly notable for its lack of common variance. The SMC values are remarkably low. The actual communalities (0.7076, 0.6437, 0.6922, 0.5882, and 0.0661) are substantially higher than the SMCs. This is an illustration of the fact that there are some data sets for which setting the priors at SMC values (as a lower bound estimate) may not always be a good practice.

If we were to apply the equations and processes demonstrated in Section 4.4 for extracting factor loadings, with these five SMC entries substituted into the diagonal of the correlation matrix, it would indeed give us the principal factor results. However, rather than demonstrating the obvious with matrix algebra, we will let SAS do the work.

Because we do not have the raw data matrix for the Locurto et al. data, we will have to enter the correlation matrix directly. The SAS data step for entering the correlation matrix directly is shown below. Notice that the process is to use a parenthetical expression (type=CORR) adjoining the data set name (Locurto) in the first line of the data statement.

```
data Locurto(type=CORR);
input olfactory winshift detour fear operant;
datalines;
   1.000    -.210    .270    -.020    -.040
   -.210    1.000    .080    -.200    .060
   .270     .080     1.000   -.280    .030
   -.020    -.200    -.280    1.000   .000
   -.040    .060     .030     .000    1.000
;
run;
```

Now that the correlation matrix has been directly entered, we can run the factor analysis in the usual way. The PROC FACTOR code for doing the principal factor analysis of these data is:

```
proc factor data=Locurto method=principal priors=smc nfact=2;
   var olfactory winshift detour fear operant;
   title 'Locurto factor analysis of mouse skills, principal factor method';
run;
```

In the first line of this procedure, the method is set at "principal," and the priors are set at "SMC," which is how one calls for a principal factor method of analysis with SMC as the chosen priors.

Notice in the side-by-side summary tables (Fig. 4.9) how the summary statistics for the principal factor method compare with the statistics for the principal component method. The principle component method accounts for almost three times as much variance with two factors (54.0% compared with 18.8%). Obviously, this profound failure of the principal factor method, in variance terms, is because of the unusually low values of the squared multiple correlations (SMCs) due to the spread of the variables in the Locurto et al. data.

It should, however, be acknowledged that maximizing the amount of variance accounted for is not the goal of common factor analysis. The goal is to account for the correlations among the variables. This is done by separating the unique variance for each variable from the common variance component. Common factor analysis thus revises the fundamental factor theorem $R^+ = FF'$ by adding a diagonal matrix of uniqueness values as a separate component (diagonal matrix U), as shown in the schematic diagram of Equation 4.22.

$$R^+ = \boxed{\begin{matrix}F\end{matrix}}\ \boxed{\begin{matrix}F'\end{matrix}}\ +\ \diagbox{U}$$

$$R^+ = F\,F' + U. \tag{4.22}$$

It is assumed that the common factors are independent of the uniqueness values (in the population), and the two can thus be considered separately. Using the factor loadings and the uniqueness values shown in the principal factor summary table above, the common factor model can be used to reconstruct the original correlation matrix as closely as possible, according to Equation 4.22.

$$\mathbf{R^+} = \mathbf{F\,F'} + \mathbf{U} = \begin{bmatrix} 0.279 & 0.403 \\ 0.151 & -0.420 \\ 0.523 & 0.070 \\ -0.417 & 0.199 \\ 0.025 & -0.101 \end{bmatrix} \begin{bmatrix} 0.279 & 0.151 & 0.523 & -0.417 & 0.025 \\ 0.403 & -0.420 & 0.070 & 0.199 & -0.101 \end{bmatrix}$$

$$+ \begin{bmatrix} 0.760 & 0 & 0 & 0 & 0 \\ 0 & 0.801 & 0 & 0 & 0 \\ 0 & 0 & 0.722 & 0 & 0 \\ 0 & 0 & 0 & 0.787 & 0 \\ 0 & 0 & 0 & 0 & 0.989 \end{bmatrix}.$$

$$= \begin{bmatrix} 0.240 & -0.127 & 0.174 & -0.036 & -0.034 \\ -0.127 & 0.199 & 0.050 & -0.146 & 0.046 \\ 0.174 & 0.050 & 0.278 & -0.204 & 0.006 \\ -0.036 & -0.146 & -0.204 & 0.213 & -0.031 \\ -0.034 & 0.046 & 0.006 & -0.031 & 0.011 \end{bmatrix}$$

$$+ \begin{bmatrix} 0.760 & 0 & 0 & 0 & 0 \\ 0 & 0.801 & 0 & 0 & 0 \\ 0 & 0 & 0.722 & 0 & 0 \\ 0 & 0 & 0 & 0.787 & 0 \\ 0 & 0 & 0 & 0 & 0.989 \end{bmatrix}$$

$$= \begin{bmatrix} 1 & -0.127 & 0.174 & -0.036 & -0.034 \\ -0.127 & 1 & 0.050 & -0.146 & 0.046 \\ 0.174 & 0.050 & 1 & -0.204 & 0.006 \\ -0.036 & -0.146 & -0.204 & 1 & -0.031 \\ -0.034 & 0.046 & 0.006 & -0.031 & 1 \end{bmatrix}.$$

This reconstructed **R+** matrix is subtracted from the original correlation matrix **R** (shown below) to evaluate how well the principal factor analysis has accomplished its purpose of reconstructing the correlation matrix. The largest discrepancy value is 0.096, in cell 1,3 of the residual matrix. The original correlation matrix value in this cell is 0.270, and the reconstructed correlation matrix value underestimates it, with a value of 0.174. If we construct the same kind of model according to Equation 4.22 for the results of the principal components analysis from Table 4.9, the largest discrepancy value is 0.240, more than twice as large as the maximum discrepancy for the principal factor analysis.

TABLE 4.9. Factor Analysis Summary Tables Comparing Principal Component and Principal Factor Methods for Locurto et al. Data

| | Principal Component Method | | | | | | Principal Factor Method (Using SMSc as Communalities) | | | | | |
| | Factor Loadings | | Communalities | | | Uniqueness | Factor Loadings | | Communalities | | | Uniqueness |
	Factor 1	Factor 2	Factor 1	Factor 2	Total		Factor 1	Factor 2	Factor 1	Factor 2	Total	
Olfactory	0.397	0.744	0.158	0.553	0.711	0.289	0.279	0.403	0.078	0.163	0.240	0.760
Winshift	0.308	−0.746	0.095	0.556	0.651	0.349	0.151	−0.420	0.023	0.176	0.199	0.801
Detour	0.806	0.190	0.649	0.036	0.685	0.315	0.523	0.070	0.274	0.005	0.278	0.722
Fear	−0.713	0.285	0.508	0.081	0.589	0.411	−0.417	0.199	0.174	0.040	0.213	0.787
Operant	0.065	−0.242	0.004	0.058	0.063	0.937	0.025	−0.101	0.001	0.010	0.011	0.989
Eigenvalues =	1.414	1.285			2.699	2.301			0.548	0.394	0.942	4.058
Percents of eigenvalues =	28.3%	25.7%			54.0%	46.0%			11.0%	7.9%	18.8%	81.2%

205

$$\mathbf{R_{res} = R - R^+}$$

$$
= \begin{bmatrix}
1 & -0.210 & 0.270 & -0.020 & -0.040 \\
-0.210 & 1 & 0.080 & -0.200 & 0.060 \\
0.270 & 0.080 & 1 & -0.280 & 0.030 \\
-0.020 & -0.200 & -0.280 & 1 & 0.000 \\
-0.040 & 0.060 & 0.030 & 0.000 & 1
\end{bmatrix}
$$

$$
- \begin{bmatrix}
1 & -0.127 & 0.174 & -0.036 & -0.034 \\
-0.127 & 1 & 0.050 & -0.146 & 0.046 \\
0.174 & 0.050 & 1 & -0.204 & 0.006 \\
-0.036 & -0.146 & -0.204 & 1 & -0.031 \\
-0.034 & 0.046 & 0.006 & -0.031 & 1
\end{bmatrix}
$$

$$
= \begin{bmatrix}
0 & -0.083 & 0.096 & 0.016 & -0.006 \\
-0.083 & 0 & 0.030 & -0.054 & 0.014 \\
0.096 & 0.030 & 0 & -0.076 & 0.024 \\
0.016 & -0.054 & -0.076 & 0 & 0.031 \\
-0.006 & 0.014 & 0.024 & 0.031 & 0
\end{bmatrix} .
$$

This demonstration makes clear why we say that the goal of common factor analysis is to account for the correlations, while the goal of PCA is to account for as much variance as possible in the observed variables. For this data set, each method seems to do better than the other in its avowed task. However, even though the common factor structure does well in reconstructing the correlation matrix, it would also be desirable to have more than 18.8% of the variance accounted for in the common factors ($\mathbf{FF'}$) part of the model. In fact, several of the other common factor methods do considerably better in this regard than SMC principal factor analysis. For example, when actual communalities from the principal component method are used in the diagonal rather than SMCs, 40.3% of the variance is accounted for, as shown in the summary table on the left of Table 4.10. The results from the iterated principal factor method are shown on the right in Table 4.10. These results are about mid-way between the other two principal factors methods, with 28.3% of the variance accounted for. It should be mentioned that the low variance accounted for is not some kind of failure of the methods, but rather an expression of different goals. In common factor condensation methods, the goal is to separate out the variance that is unique to each variable from the variance that is common to all (in order to account for the correlation matrix as well as possible), which has the effect here of reducing the total amount of communality.

Figure 4.11 compares these three principal factor methods with one another and also with three additional methods: maximum likelihood, alpha, and principal component factoring. The six methods are compared in three ways: percent of variance accounted for, percent not accounted for, and discrepancy

TABLE 4.10. Factor Analysis Summary Tables Comparing Principal Factor (Actual Communalities) and Iterated Principal Factor Methods for Locurto et al. Data

| | Principal Factor Method (Using Actual Communalities) | | | | | | Iterated Principal Factor Method | | | | | |
| | Factor Loadings | | Communalities | | | Uniqueness | Factor Loadings | | Communalities | | | Uniqueness |
	Factor 1	Factor 2	Factor 1	Factor 2	Total		Factor 1	Factor 2	Factor 1	Factor 2	Total	
Olfactory	0.500	0.517	0.250	0.333	0.583	0.417	0.433	0.463	0.187	0.214	0.401	0.599
Winshift	0.129	-0.681	0.017	0.464	0.480	0.520	0.080	-0.535	0.006	0.287	0.293	0.707
Detour	0.746	-0.019	0.557	0.000	0.557	0.443	0.681	-0.054	0.464	0.003	0.467	0.533
Fear	-0.502	0.370	0.252	0.137	0.389	0.611	-0.384	0.311	0.148	0.097	0.245	0.755
Operant	0.010	-0.074	0.000	0.005	0.006	0.994	0.010	-0.087	0.000	0.008	0.008	0.992
Eigenvalues =	1.076	0.939	2.015			2.985	0.805	0.608	1.413			3.587
Percents of eigenvalues =	21.5%	18.8%	40.3%			59.7%	16.1%	12.2%	28.3%			71.7%

207

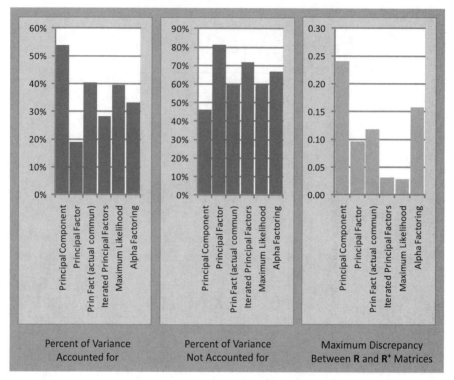

Figure 4.11   Comparison of five common factor models to the principal component model in the amount of variance accounted for and not accounted for, and in the size of discrepancies between the **R** matrix and the reconstructed **R**$^+$ matrix.

between actual correlations and reconstructed correlations (the maximum value of entries in the residual matrix, $\mathbf{R}_{res} = \mathbf{R} - \mathbf{R}^+$).

It is clear from this figure that for this particular data set two of the common factor methods are notably better than the other four models in reproducing correlations—the iterated principal factor method and the maximum likelihood method. On the other hand, no other condensation method rivals the principal component method in the amount of variance accounted for in the data. This is consistent with Kline's (1993, 105–106) summary of 10 mathematical properties of the principal component method, in which he says that principal components "will account for more or as much variance in the matrix as the same number of factors derived by any other method."

Table 4.11 displays the factor analysis summary tables for the maximum likelihood method and also for the alpha factoring method. The first run of the maximum likelihood method failed. There was an error message in the SAS log that the analysis constituted a Heywood case and could not be completed. The analysis was therefore re-run with the "heywood" option (bolded in the code fragment below for emphasis) inserted into the first line of the PROC FACTOR code, and the results shown in Table 4.11 were obtained.

TABLE 4.11. Factor Analysis Summary Tables Comparing Maximum Likelihood and Alpha Factoring Methods for Locurto et al. Data

	Maximum Likelihood Method (with Heywood Option)						Alpha Factoring Method					
	Factor Loadings		Communalities			Uniqueness	Factor Loadings		Communalities			Uniqueness
	Factor 1	Factor 2	Factor 1	Factor 2	Total		Factor 1	Factor 2	Factor 1	Factor 2	Total	
Olfactory	0.270	−0.278	0.073	0.077	0.150	0.850	−0.293	0.460	0.086	0.212	0.298	0.702
Winshift	0.080	0.834	0.006	0.696	0.702	0.298	0.726	−0.077	0.527	0.006	0.533	0.467
Detour	1.000	0.000	1.000	0.000	1.000	0.000	0.406	0.736	0.165	0.541	0.706	0.294
Fear	−0.280	−0.212	0.078	0.045	0.123	0.877	−0.228	−0.248	0.052	0.061	0.114	0.886
Operant	0.030	0.071	0.001	0.005	0.006	0.994	0.076	0.033	0.006	0.001	0.007	0.993
Eigenvalues =			1.159	0.823	1.982	3.018			0.836	0.821	1.658	3.342
Percents of eigenvalues =			23.2%	16.5%	39.6%	60.4%			16.7%	16.4%	33.2%	66.8%

209

```
proc factor data=Locurto method=ml heywood nfact=2;
```

In examining the factor analysis summary table for the maximum likelihood analysis, it can be observed that the communality of 1.000 responsible for the Heywood case is found in the detour task, the third variable in the table. If the analysis had produced an "ultra Heywood case," a communality value greater than 1.000, it would still be possible to run it in SAS by specifying "ultraheywood" rather than "heywood" in the first line of the PROC FACTOR code. However, this sometimes causes convergence problems (see p. 1563 of SAS 2009).

From the foregoing, it seems that in adopting PCA, Locurto et al. (2006) chose the right factoring method for their study, particularly since accounting for the variance seems more important for their purposes than reproducing the correlations. In conclusion, we will turn now to the question of the effects of the individual reliabilities of items on the performance characteristics of various models of factor analysis.

4.10   DATA RELIABILITY AND FACTOR ANALYSIS

The focus of this final section of the chapter is to demonstrate the adverse effects on the fidelity of factor analysis results of having low-item reliabilities in the data set to be factored. The demonstration is based upon simulated data from a study by Lauritzen et al. (2007), much in the spirit of Thurstone's (1947) "boxes" demonstration and Cattell and Sullivan's (1962) "demonstration by cups of coffee." Both involve using data with known properties to reveal the operating characteristics of the factor analytic model. At the beginning of the Cattell and Sullivan paper, they speak of the need for performance evaluations of this kind:

> The psychologists who use factor analysis have found such exciting work to do, in exploring and structuring new data by the model, that practically no one has stopped to check the nature of the model itself; by, for example, applying it to data of already *known* structure (p. 184).

To create the simulation data, we begin with 64 data points in a two-dimensional space. The 64 data points are organized into eight groups of eight points each within the space of two latent variables, as shown in Figure 4.12, organized into a configuration that looks like eight "Olympic rings." The intent of this recognizable and systematic shape is to make perturbations in the structure more visually obvious when one examines the results of various levels of reliability in combination with various factor analytic models.

Now that the 64×2 matrix $\mathbf{Y}$ of factor scores has been created, the next step is to specify the relationship between 12 observed variables and the two

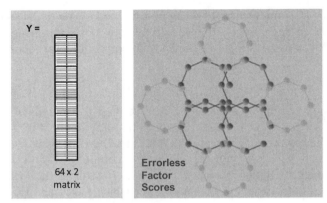

Figure 4.12 Matrix **Y** of 64 factor scores (in eight groups of eight) within a two factor space, and a scatter plot showing their configuration. (Image courtesy of Metrika (metrika.org), © Appzilla, LLC.)

F =	factor 1	factor 2
Maintenance 1	0.996	−0.087
Maintenance 2	0.999	−0.052
Maintenance 3	1.000	−0.017
Maintenance 4	1.000	0.017
Maintenance 5	0.999	0.052
Maintenance 6	0.996	0.087
Task 1	0.087	0.996
Task 2	0.052	0.999
Task 3	0.017	1.000
Task 4	−0.017	1.000
Task 5	−0.052	0.999
Task 6	−0.087	0.996

12 x 2 matrix

Errorless
Clustered
Factor Pattern

Figure 4.13 Hypothetical factor loadings matrix for 12 variables, in two clusters, completely contained within a two-factor space. (Image courtesy of Metrika (metrika.org), © Appzilla, LLC.)

latent variables. This is done by creating a matrix **F**, which is shown in Figure 4.13. The 12 × 2 **F** matrix of factor loadings numerically specifies the relationships. The vector plot on the right shows graphically that we have created a *clustered* factor pattern, with six of the variables clustering around the first factor and six clustering around the second factor.

To relate this simulation data set to a possible research example, suppose that these 12 variables are the same 12 described at the beginning of Section 4.7, a management profiling instrument with the first six variables related to effectiveness in working with people (maintenance) and the last six related to task effectiveness, as they are labeled in Figure 4.13. The eight "rings" in the factor score plot of Figure 4.12 could be eight types of managers, with eight managers in each group.

Equation 4.9 from Section 4.5.2 can now be used to create the standardized scores of the 64 persons on the 12 variables by postmultiplying the 64×2 matrix of **Y** scores by the 2×12 transposed matrix of factor loadings, **F′**.

$$\mathbf{Z} = \mathbf{YF'}. \tag{4.9}$$

When the resulting 64×12 standard score matrix of raw data, **Z,** is factor analyzed, it of course yields factor loadings that are identical to the idealized **F** matrix which was used to create the raw data. The two eigenvalues are each equal to 6.000, with 100% of the variance accounted for with a two-factor solution. Since the data are errorless and a perfect fit to the factor analytic model, the results are equivalent no matter which factor analysis method is used.

We now create simulated data at each of six levels of reliability by adding a random error component of various sizes. To create data with a reliability of 0.90, one part error variance (using a random number generator) is added to nine parts of structure variance. A 1:4 error-to-structure ratio creates a reliability of 0.80, a 1:2 ratio creates a reliability of 0.67, 1:1 creates a reliability of 0.50, 2:1 creates a reliability of 0.33, and 3:1 creates a reliability of 0.25. The six simulation conditions are referred to as r90, r80, r67, r50, r33, and r25, respectively.

The factor analysis summary table for the r90 condition is shown in Table 4.12 for a maximum likelihood factor analysis. With this high individual item reliability, the factor pattern is very similar to the one for errorless data given in Figure 4.13, as seen in the "variance accounted for" row of the summary table of Table 4.12. Although these results are for a maximum likelihood factor analysis, they would be almost identically the same for any of the other four analyses for which the simulation was done—principal component, principal factor, iterated principal factor, or alpha. None of them differ from this analysis in any of their factor loadings by more than 0.001. Harman (1976), Nunnaly (1978), Kline (1993), Rencher (2002), and others have pointed out that the results of the common factors methods differ little from principal components when the data matrices have a large number of variables and strong psychometric properties. For the data reported here, even with the r67 simulation (item reliabilities of 0.67), no loading for a common factor method deviated more than 0.03 from the principal component loadings. However, for all three of the lower reliability simulations, r50, r33, and r25, the pattern of loadings does differ markedly among the methods.

Figure 4.14 displays the vector plots for the factor patterns at each of the six levels of simulated reliability. Clearly, the integrity of the factor pattern has deteriorated substantially in all three of the lower reliability level simulations. The watershed between an acceptable level of structural integrity in the factor loadings seems visually to be somewhere between $r = 0.67$ and $r = 0.50$.

This is also consistent with the results in terms of percent of variance accounted for by a two-factor solution. In the three highest reliability levels,

TABLE 4.12. Factor Analysis Summary Table for Maximum Likelihood Analysis of Clustered Pattern Simulated Data with Item Reliability Set at $r = 0.90$

	Loadings		Communalities			Uniqueness
	Factor 1	Factor 2	Factor 1	Factor 2	Total	
Maint 1	0.991	−0.081	0.981	0.007	0.988	0.012
Maint 2	0.993	−0.059	0.985	0.003	0.989	0.011
Maint 3	0.994	0.000	0.989	0.000	0.989	0.011
Maint 4	0.993	0.027	0.987	0.001	0.988	0.012
Maint 5	0.993	0.063	0.986	0.004	0.990	0.010
Maint 6	0.992	0.074	0.983	0.005	0.989	0.011
Task 1	0.081	0.990	0.007	0.979	0.986	0.014
Task 2	0.080	0.992	0.006	0.984	0.990	0.010
Task 3	0.005	0.994	0.000	0.989	0.989	0.011
Task 4	0.014	0.995	0.000	0.990	0.990	0.010
Task 5	−0.055	0.992	0.003	0.983	0.986	0.014
Task 6	−0.103	0.990	0.011	0.980	0.991	0.009
Variance accounted for:			5.939	5.926	11.864	0.136
Percents to total variance:			49.5%	49.4%	98.9%	1.1%

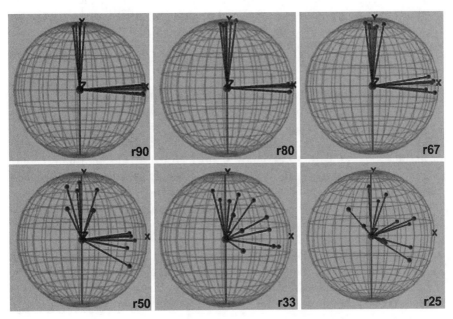

Figure 4.14 Vector plots for simulations at six levels of reliability, based upon the clustered factor pattern of Figure 4.13. (Image courtesy of Metrika (metrika.org), © Appzilla, LLC.)

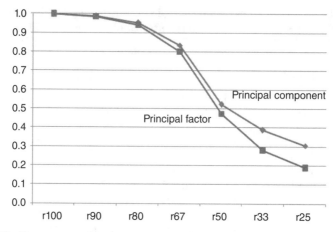

Figure 4.15   Percentage of variance accounted for at each of the six reliability levels and for errorless data, shown for principal component analysis and for principal factor analysis.

r90, r80, and r67, two factors accounts for 99.0, 95.3, and 83.7% of the variance, respectively, using a PCA. On the other hand, only 52.3, 38.9, and 30.9% of the variance is accounted for in the three lowest reliability levels, also using a PCA. Figure 4.15 displays these results both for PCA (the highest of the five methods in variance accounted for) and also for principal factor analysis (the lowest of the five methods in variance accounted for). Notice that the two methods are very similar in values at the highest levels of reliability, but diverge at the lower levels of reliability. The profiles for the other three common factor methods are found somewhere between these two, but very close to the lower profile, the one for principal factor analysis.

Reliability concerns are obviously central to the question of when to stop factoring, that is, how many factors should be extracted. From these simulations, it becomes apparent that when the reliability of individual variables drops below 0.67, the question of how many factors are present usually cannot be correctly answered. Each of these data sets was created from two latent variables and should therefore yield a two-factor solution, as they do at the three highest reliability levels. However, using SAS's default criterion (which differs by method) for selecting the number of factors, the r50 data simulation erroneously led to three factors on two of the five methods (principal factor and maximum likelihood), but was correct in selecting two factors on three of the methods (principal component, iterated principal factor, and alpha).[18] The r33 data set resulted in all five methods returning the wrong number of factors,

[18] It should be noted that the number of factors returned on each of the methods reflects the particular factor selection criterion used by SAS for each method as well as the inherent virtues of the mathematics of a particular method.

with three factors on two of the methods, and four factors on the other three. The r25 data set also resulted in the wrong number of factors for all five methods, with three factors on three of the methods[19] and five factors on two. This is in accord with Rencher's (2002, 444) statement that many of the problems encountered with factor analysis are because correlation matrices and covariance matrices contain both structure and error, and factor analysis cannot separate the two. Factor analysis conflates error with structure, and at lower levels of reliability can be expected to give incorrect assessments of the number of factors. It seems the only answer is to do everything possible to increase the reliability of individual items above the 0.67 level before factoring. It is surprising how many factor analytic studies in the published literature have item reliabilities far below this level. Rencher suggests averaging as a way of increasing item reliabilities when there are replications in the data set. When there are not, it might prove cost effective to include replications to improve the integrity of the factor analysis. Item reliability is probably more important to the fidelity of factor analysis results than the much debated concerns about the number of observations that might be necessary to provide a proper factor analysis.[20] Clearly, when one is operating with item reliabilities in the range of 0.20–0.50, which is often the case, assertions such as "four factors emerged" can be called into question.

Most of the data sets in this chapter, both small ones and also larger ones, have been based upon a clustered covariance structure. That is, the vectors have tended to cluster around the factors. The reason for this is pedagogical— it is easier to understand the functioning of factor analysis with simple data, and most of the data sets have had few variables and few factors. One exception to the clustered patterns was the Locurto et al. data in which the vectors, although few, were fairly well spread out from one another within the two-factor space. Consider the factor pattern shown in Figure 4.16. This factor pattern is an example of a *Toeplitz* covariance structure, in contrast to the clustered pattern of Figure 4.13. More precisely, it is a particular kind of

---

[19] One of the methods that returned three factors in the r25 condition is alpha factoring. However, something strange occurred in the alpha factoring simulations for both the r25 and the r33 conditions. As was indicated in footnote 14 and the accompanying text, the factor score reliability estimates provided by alpha factoring were accurate and helpful in identifying spurious factors given the relatively high-item reliability levels in the Locurto et al. data. However, in these simulations, at the r25 and r33 low item-reliability levels, these estimates performed very badly. The third factors in these cases (which are obviously spurious since the underlying structure is two-factor) had factor score reliabilities of 0.738 for r25 and 0.345 for r33. This is in contrast to the helpful performance at higher item-reliability levels (r90, r80, r67), where spurious factors were marked by impossible negative values. Evidently, there is something about the low fidelity data that "fools" the algorithm in the alpha factoring method for calculating factor score reliabilities.

[20] Kline (1993) reviews several of the positions on the issue of how many observations are needed to construct a good factor analysis. He summarizes with the comment that "the leading authorities vary considerably in their advice. On one point there is complete agreement. For reasons of matrix algebra it is essential that there are more subjects than variables" (p. 121). The issue has not been resolved in the intervening 18 years. Authorities still vary considerably in their advice.

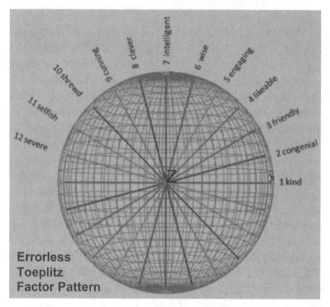

Figure 4.16   Vector plot of the factor pattern for 12 variables in an evenly-spaced Toeplitz configuration. (Image courtesy of Metrika (metrika.org), © Appzilla, LLC.)

Toeplitz covariance structure referred to as autoregressive Toeplitz. The correlation matrix corresponding to such a pattern has approximately equal correlation values in stripes around the diagonal (such as stripes of values of 0.90 on each side of the diagonal, stripes of 0.80 in the next locations out on each side). This section will close with a demonstration of the performance characteristics of factor analysis methods for a Toeplitz covariance structure at various reliability levels, to compare with the results shown above for clustered covariance structures.

This factor pattern, like the Locurto et al., one has the variables spread out, but there are over three times as many variables here, and they are evenly spaced throughout the two factor space.[21] Suppose that these variables correspond to trait ratings in a person-perception study. The horizontal axis (which corresponds to factor one) is defined by kind, congenial, and friendly

---

[21] Notice that the vectors for each of the 12 variables in this Toeplitz structure can be extended through the center of the figure (the origin) to the other side of the circle to show both the positive and also the negative end of each. In other words, when we consider both the positive end of each vector (variable), that is, values above the mean, and also the negative end of each, values below the mean, this Toeplitz covariance structure of vectors has equally spaced vectors throughout the 360° degrees of the circle. The 15° angles between all pairs of the 12 vectors, including their 12 negative opposites, together sum up to 360°. For simplicity, we have only plotted the positive end of each vector when creating vector plots throughout the chapter, and we will continue to do so in Figure 4.17, where the effects of various reliability levels on this pattern are considered, but it should be remembered that the other side of each vector is also implied.

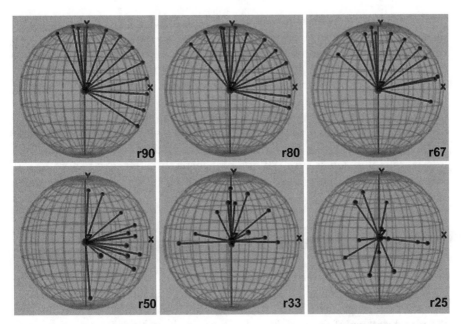

Figure 4.17 Vector plots for simulations at six levels of reliability, based upon the underlying Toeplitz factor pattern of Figure 4.16. (Image courtesy of Metrika (metrika. org), © Appzilla, LLC.)

on the right side, with a polar opposite of selfish and severe (and implied unkind) on the other. The underlying factor seems to be expressive of a generally positive versus a generally negative interpersonal stance toward others. The vertical factor seems to be a general personal competence factor, with the positive end defined by intelligent, clever, and wise, and by implication, the negative end at the bottom of the figure defined by unintelligent, anything but clever, and unwise. We have tried to create the variables in between the two factors to be believable mixes of each factor, with likeable, for example, having some elements of positive interpersonal stance and also elements of general competence, and with shrewd combining general competence with a somewhat negative interpersonal stance. These conceptual relationships among the variables are intended to help us understand what happens in the factor analyses at lower reliability levels when too many factors are extracted, when structure is conflated with error.

The resulting factor patterns from a principal component method factor analysis of the Toeplitz pattern simulations at the six varying reliability levels are given in Figure 4.17. Again, we see that the pattern of the underlying covariance structure holds up visually quite well down to a reliability of about 0.67. Below that, it becomes quite difficult to visually identify the original Toeplitz structure. Indeed, the r33 and r25 simulations here have more resemblance to the r33 and r25 simulations for clustered data than they do to the Toeplitz simulations at higher reliability levels.

The results of the variance-accounted-for statistics are virtually identical to those shown in Figure 4.15 for clustered data: they hold up quite well down to a reliability of 0.67, and then they drop off precipitously below that. All five factor analysis methods give virtually identical results for the r90 simulation and very similar results for r80 and r67, but they diverge markedly at the lower three reliability levels. As expected, the PCA results have substantially higher variance accounted for than do any of the four common factor models, which are all at about the same level as one another.

Using the default SAS settings, the five-factor analysis methods at the three highest reliability levels (r67, r80, and r90) correctly return two factors. At r50, three of the methods correctly return two factors, but two erroneously return three factors. At r33, all five methods extract too many factors, with three methods returning three factors and two methods returning four factors. In the r25 condition, all five methods return too many factors, with three methods returning three factors and two methods returning five factors.

Table 4.13 displays the factor analysis summary table for one of the r33 condition analyses that erroneously returned four factors, in this case a principal component method analysis. When only two or three factors emerge, it is possible to graph them with a vector plot and then identify the factors from the visual pattern of the variables that load on them. With four or more variables, a common practice is to mark all of the loadings in the table that exceed a given value (either positive or negative), and then to name the factors according to the identity of the high-loading variables. We have followed that procedure in Table 4.13, circling all of the loadings that exceed 0.50 in absolute value.

Factor 2 in this pattern brings together three variables (kind, severe, and selfish) that seem contradictory, and were quite at odds with one another in the original factor pattern of the errorless data of Figure 4.16 (with a −0.966 correlation in the errorless correlation matrix between kind and severe, and a correlation of −0.866 between kind and selfish). The coming together of these three variables into one common factor is, of course, the result of error—a two-to-one ratio of error to structure in the r33 condition. It is of interest to note that others of the factors are quite interpretable even though they combine variables from both sides of the original interpersonal stance axis. Factor 1 combines, for example, friendly, intelligent, cunning, and shrewd. This factor is somewhat believable, and would not require too much imagination to identify it as a "sales aptitude" factor. Perhaps there is a lesson here that psychological meaningfulness may not be a good guide to factorial fidelity. The human ability to find meaning in combining oppositional trait descriptions should not be underestimated. There are structural contradictions created by factoring data matrices of low reliability, but the fluid properties of psychological traits may occlude them. Perhaps Thurstone's (1947) example of the factoring of the physical properties of boxes replicated with strong error components would reveal the contradictions.

TABLE 4.13. Factor Analysis Summary Table for Principal Component Method Analysis of Toeplitz Pattern Simulated Data with Item Reliability Set at $r = 0.33$

	Loadings				Communalities					Uniqueness
	Factor 1	Factor 2	Factor 3	Factor 4	Factor 1	Factor 2	Factor 3	Factor 4	Total	
1 Kind	0.237	0.735	-0.137	0.117	0.056	0.540	0.019	0.014	0.629	0.371
2 Congenial	0.137	0.190	-0.617	0.392	0.019	0.036	0.381	0.154	0.590	0.410
3 Friendly	0.618	0.058	-0.186	0.379	0.382	0.003	0.035	0.143	0.563	0.437
4 Likeable	-0.035	0.017	0.001	0.833	0.001	0.000	0.000	0.694	0.696	0.304
5 Engaging	0.222	-0.263	-0.593	0.334	0.049	0.069	0.352	0.112	0.583	0.417
6 Wise	0.197	0.202	-0.068	0.738	0.039	0.041	0.005	0.544	0.629	0.371
7 Intelligent	0.724	0.315	0.019	0.079	0.524	0.099	0.000	0.006	0.630	0.370
8 Clever	0.464	0.145	0.583	0.135	0.215	0.021	0.339	0.018	0.594	0.406
9 Cunning	0.748	0.144	0.122	-0.067	0.559	0.021	0.015	0.004	0.599	0.401
10 Shrewd	0.509	0.009	0.681	0.199	0.259	0.000	0.464	0.040	0.763	0.237
11 Selfish	-0.032	0.672	0.466	0.165	0.001	0.451	0.217	0.027	0.696	0.304
12 Severe	0.240	0.765	0.072	0.024	0.058	0.585	0.005	0.001	0.648	0.352
Variance accounted for:	2.162	1.867	1.832	1.758	2.162	1.867	1.832	1.758	7.619	4.381
Percents of total variance:	18.0%	15.6%	15.3%	14.6%	18.0%	15.6%	15.3%	14.6%	63.5%	36.5%

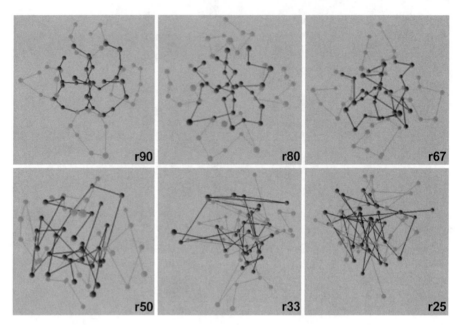

Figure 4.18 Factor score plots of eight groups of eight managers (hypothetical), for simulations at six levels of reliability. (Image courtesy of Metrika (metrika.org), © Appzilla, LLC.)

In Figure 4.18, we examine the effects of the six levels of item reliability upon factor scores. These are the same 64 factor scores that were constructed and displayed in their errorless form in Figure 4.12. The first thing that can be observed is that the effects of error variance on the geometric configuration of factor scores seems to be more stark and noticeable than its effects on the vector plots (shown above in Figs 4.14 and 4.17). Whereas the two highest reliability levels, r90 and r80, produced vector plots similar to errorless data, the same is not true of the factor score plots in Figure 4.18.

In the r90 condition (with nine parts structure and one part error variance), the eight "rings" (groups) of eight managers are still clearly recognizable, as they are in the r80 condition, but the geometric arrangement of the points is somewhat distorted in each of these. In the r67 condition, the rings for the eight groups start to run together, but the four dark inner rings still keep their basic topological location in relation to the four light outer rings. In the three low reliability conditions, the eight groups run together, and their original locations are for the most part visually obscured. This can be quantified with a multivariate analysis of variance (chapter eight), which indicates that whereas in the original errorless pattern shown in Figure 4.12, the eight groups account for 95.7% of the variance in location within the two-factor space, in the r25 condition, that figure drops to 39.2%. From both the vector plots and also the factor scores of the simulations, it is clear that to obtain factor loadings and factor scores that accurately reflect the true state of affairs, one should strive to have individual item reliabilities of 0.67 or higher in preparation for a factor analysis.

4.11 SUMMARY

Mulivariate data share variance. For instance, biomedical and social scientists may collect data on several components of a skill observed during the study, such as multiple cognitive abilities. Data from those several measures of cognitive ability should clearly be related to one another. Although a correlation matrix is one way to ascertain whether the observed data approximate theoretical expectations, this chapter has shown that factor analysis and principle component analysis can identify underlying latent variables that parsimoniously account for multivariate data. These methods parse correlation and covariance matrices into factors or components fewer in number than the observed variables. The matrix of factor loadings clarifies the relationship between the latent variables and the observed manifest variables, and can reveal geometric patterns in the data. Hence, the shared variance characteristic of multivariate data can be used to a researcher's advantage when examined through the methods described in this chapter.

Using factor analysis and generating factor scores also reduces the adverse effects of measurement error that typically beset social science instruments. Factors often explain greater proportions of variance in the data than individual measures and have greater reliability. Moreover, extraction of factors makes for parsimonious data interpretation. Data have underlying characteristics that are often revealed by factoring, but would otherwise remain obscured. Factors may represent some kinds of data more accurately than the individual manifest variables from which the factors are constructed.

There is a clear difference between principal components as a descriptive data simplification method, and common factor analysis as a way of going beyond the surface manifestations of observed variables to the deep structure of the latent variables underlying them. The philosophical foundation of the common factor model is logically compelling. It seems reasonable that a particular latent variable, such as verbal intelligence, is more fundamental and more real than any particular way of measuring it. However, when one examines the *performance characteristics* of the common factor methods in comparison with those of principal components, there is little basis to prefer common factors. There is substantial agreement among both proponents and opponents of the common factor model that with data that would be considered to be most appropriate for factoring (large matrices, many variables, stable item reliabilities), the differences in results among the factor analytic methods (including principal components) are negligible. On the other hand, when the data are less optimal for factoring (small matrices with few variables), or less appropriate for factoring (low item reliabilities), there is little evidence from the performance evaluation data in Sections 4.9 and 4.10 that would recommend the common factor methods. They do well in reconstructing the correlation matrix, but quite badly in accounting for variance in observed variables. We are left to wonder with Cattell (1978) whether perhaps the complexity and sophistication of methods, such as the maximum likelihood common factor model, are overkill for "appropriate" factor analytic data sets, and not equal to the task of remedying the flaws in weak data.

As Rencher (2002) has also pointed out, PCA as a method "requires essentially no assumptions" (p. 448), whereas the common factor methods tend to be quite encumbered, both by assumptions and also by a multitude of recommendations about required number of observations, appropriate types of data, best methods for estimating factor scores, and so forth. Given its unassuming simplicity, there is much to recommend the principal component method of factor analysis, particularly when it is functioning in the role of an "exploratory factor analysis," a method for exploring data structure inductively in preparation for a careful evaluation of a theoretical model using confirmatory factor analysis. It is perhaps in confirmatory model testing where the sophistication of maximum likelihood estimation could be put to better use.

Over a century after its inception, factor analysis in its many varieties remains a powerful tool. It reveals patterns of relationships among variables. It helps to uncover the structural properties of data. It enables researchers to evaluate how well the observed data account for the phenomena of interest. It allows for the evaluation of particular measures relative to other measures. Its value as a tool could be compared with spectrum analysis in chemistry, whereby observed properties help to identify underlying structures. Its exploration and discovery have come to us from the mathematics of matrix decomposition, eigenvectors/eigenvalues, and related methods, which also characterize many of the multivariate methods covered in remaining chapters. The principles learned in this chapter will therefore have much application in the remaining chapters, one of the ways in which factor analysis is the quintessential multivariate method.

STUDY QUESTIONS

A. Essay Questions

1. Define the following factor analysis terms: factor loadings, communalities, factor scores, eigenvalues, eigenvectors, transformation matrix, matrix of first factor residuals, reproduced correlation matrix, trial eigenvector, uniqueness, and percents of eigenvalues.

2. Describe the logic underlying the decision to stop factoring when using the Kaiser (1960) criterion of an eigenvalue greater than one and the scree plot (Cattell, 1966). That is, why should factoring decisions be made based on the proportion of explained variance?

3. Explain why factor loadings are the coordinates for creating a graphical representation of the factor structure. That is, explain why loadings are the coordinates of each variable within the orthogonal factor space, given that a correlation coefficient is the cosine of the angle between two variables as demonstrated in Chapter 2.

4. Given that $\mathbf{\Lambda}^{1/2}\mathbf{\Lambda}^{1/2} = \mathbf{\Lambda}$ and that $\mathbf{F} = \mathbf{K}\mathbf{\Lambda}^{1/2}$, explain how the reconstruction of $\mathbf{R}$ from the fundamental factor theorem $(\mathbf{R}^{+} = \mathbf{F}\mathbf{F}')$ relates to Equation 4.12, the reconstruction of $\mathbf{R}$ by the method of quadratic and bilinear forms $(\mathbf{R}^{+} = \mathbf{K}\mathbf{\Lambda}\mathbf{K}')$.

5. Describe the steps by which you obtain the loadings on the first factor by the "method of successive squaring" or the "unit vectors method" of factor analysis. Make your outline so precise that one could actually accomplish the computations from it.

6. Compare and contrast factor analysis with principal components analysis.

7. Why are rotated factor loadings not orthogonal? You may wish to examine Section 4.6 on PCA and Section 4.7 on rotation.

8. Explain how data reliability impacts the results of factor analysis and PCA. See Section 4.10.

B. Calculation Questions

1. Using Excel or another spreadsheet, calculate the loadings and the eigenvalue for factor 1 for the 5×5 correlation matrix from the Locurto et al. (2006) study, Tables 4.1 and 4.2, and compare your results with those reported in the paper. You may refer to the calculations shown in Section 4.3 or those in Section 4.4 for guidance.

2. Using Excel or another spreadsheet, calculate the loadings and the eigenvalue for factor 2 as a follow-up to the calculation of factor 1 in question 1.

3. For the following simplest case data set, use Excel or another spreadsheet to do a complete factor analysis by the principal components method. Create a factor analysis summary table that includes: factor loadings, communalities, eigenvalues, percents of eigenvalues, and factor scores.

$X =$	10	9	19
	12	7	19
	11	6	17
	7	6	13
	8	5	13
	6	3	9

C. Data Analysis Questions

1. Enter the 5×5 correlation matrix from the Locurto et al. (2006) study, Table 4.1, and obtain the factor analysis results using SAS, Stata, or SPSS by each of the following methods: PC method of factor analysis, principal factors method, iterated principal factors method, maximum likelihood method. Compare the results obtained across these methods.

2. Do the analysis of question 1 above using the PC method, but with a varimax rotation, and then for comparison, with an oblimin rotation. Compare the results to one another and then to the results in question 1.

3. For the following data set, obtain factor analysis results using SAS, Stata, or SPSS by each of the following methods: PC method of factor analysis, principal factors method, iterated principal factors method, maximum likelihood method. Compare the results.

$X =$	32	64	65	67
	61	37	62	65
	59	40	45	43
	36	62	34	35
	62	46	43	40
	32	64	65	67

4. Do the analysis of question 4 above but with a varimax rotation, and then for comparison, with an oblimin rotation. Compare the results to one another and then to the results in question 4.

5. Using SAS or SPSS or Stata, analyze the simple data from Section 4.5.3 in three ways: (1) PCA of raw data, (2) PCA of standardized data, (3) factor analysis by the principal components method. Compare the results using the summary table method used in the chapter.

6. Using SAS, Stata, or SPSS obtain factor scores for each of the analyses of questions 1 through 5 above.

7. Conduct a principal components analysis of the data for questions 1 and 3 above (such as by the PRINCOMP procedure of SAS). How do the results differ from the factor analytic results obtained in questions 1 and 3?

REFERENCES

Burt, C. 1909. Experimental tests of general intelligence. *British Journal of Psychology*, 3, 94–177.

Burt, C. 1941. *The Factors of the Mind: An Introduction to Factor Analysis in Psychology*. New York: MacMillan.

Cattell, R. B. 1966. The scree test for the number of factors. *Multivariate Behavioral Research*, *1*, 141–161.

Cattell, R. B. 1978. *The Scientific Use of Factor Analysis*. New York: Plenum.

Cattell, R. B., and Sullivan, W. 1962. The scientific nature of factors: A demonstration by cups of coffee. *Behavioral Science*, *7*(2), 184–193.

Coombs, C. H. 1964. *A Theory of Data*. New York: John Wiley and Sons.

Coombs, C. H., Dawes, R. M., and Tversky, A. 1970. *Mathematical Psychology: An Elementary Introduction*. Englewood Cliffs, NJ: Prentice-Hall.

Cronbach, L. J. 1951. Coefficient alpha and the internal structure of tests. *Psychometrika*, *16*, 297–334.

Cudeck, R., and MacCallum, R. C. 2007. *Factor Analysis at 100: Historical Developments and Future Directions*. Mahwah, NJ: Erlbaum.

Gutman, L. 1956. Best possible systematic estimates of communalities. *Psychometrika*, *21*, 273–285.

Harman, H. H. 1976. *Modern Factor Analysis*. Chicago, IL: The University of Chicago Press.

Heywood, H. B. 1931. On finite sequences of real numbers. *Proceedings of the Royal Society, Series A*, *134*, 486–501.

Holzinger, K. J. 1930. *Statistical Resume of the Spearman Two-factor Theory*. Chicago, IL: University of Chicago Press.

Holzinger, K. J. 1942. Why do people factor? *Psychology*, *7*, 147–156.

Jolliffe, I. T. 2002. *Principal Component Analysis, Second Edition*. New York: Springer.

Kaiser, H. F. 1958. The varimax criterion for analytic rotation in factor analysis. *Psychometrika*, *23*, 187–200.

Kaiser, H. F. 1960. The application of electronic computers to factor analysis. *Educational and Psychological Measurement*, *20*, 141–151.

Kaiser, H. F., and Caffery, J. 1965. Alpha factor analysis. *Psychometrika*, *30*, 1–14.

Kaiser, H. F., and Derflinger, G. 1990. Some contrasts between maximum likelihood factor analysis and alpha factor analysis. *Applied Psychological Measurement*, *14*(1), 29–32.

Kline, P. 1993. *The Handbook of Psychological Testing*. New York: Routledge.

Lauritzen, M., Hunsaker, N., Poppleton, L., Harris, M., Bubb, R. R., and Brown, B. L. 2007. Measurement error in factor analysis: The question of structural validity. American Statistical Association 2007 Proceedings of the Section on Statistical Education (2256-2259). Alexandria, Virginia: American Statistical Association.

Lawley, D. N. 1940. The estimation of factor loadings by the method of maximum likelihood. *Proceedings of the Royal Society of Edinburgh*, *60*, 64–82.

Lawley, D. N., and Maxwell, A. E. 1971. *Factor Analysis as a Statistical Method*. New York: Macmillan.

Locurto, C., Benoit, A., Crowley, C., and Miele, A. 2006. The structure of individual differences in batteries of rapid acquisition tasks in mice. *Journal of Comparative Psychology*, *120*(4), 378–388.

Lunneborg, C. E., and Abbott, R. D. 1983. *Elementary Multivariate Analysis for the Behavioral Sciences*. New York: North-Holland.

MacCorquodale, K., and Meehl, P. E. 1948. On a distinction between hypothetical constructs and intervening variables. *Psychological Review*, *55*, 95–107.

Nunnaly, J. O. 1978. *Psychometric Theory*. New York: McGraw-Hill.

Parlett, B. N. 1998. *The Symmetric Eigenvalue Problem, Second Edition, Classics in Applied Mathematics, Vol. 20*. Philadelphia: Society for Industrial and Applied Mathematics (SIAM).

Pearson, K. 1901. On lines and planes of closest fit to systems of points in space. *Philosophical Magazine Series 6*, *2*, 559–572.

Pearson, K., and Moul, M. 1927. The mathematics of intelligence. I. The sampling errors in the theory of a generalized factor. *Biometrika*, *19*, 246–292.

Rencher, A. C. 2002. *Methods of Multivariate Analysis, Second Edtion*. New York: Wiley.

SAS Institute Inc. 2009. *Chapter 32: The FACTOR Procedure. SAS/STAT ® 9.2 User's Guide, Second Edition*. Cary, NC: SAS Institute Inc.

Spearman, C. 1904. General intelligence, objectively determined and measured. *The American Journal of Psychology*, *15*, 201–293.

Spearman, C. 1934. The factor theory and its troubles. V. Adequacy of proof. *Journal of Experimental Psychology*, *25*, 310–319.

Thomson, G. H. 1951. *The Factorial Analysis of Human Ability*. London: London University Press.

Thurstone, L. L. 1931. Multiple factor analysis. *Psychological Review*, *38*, 406–427.

Thurstone, L. L. 1935. *The Vectors of the Mind*. Chicago, IL: The University of Chicago Press.

Thurstone, L. L. 1947. *Multiple-factor Analysis: A Development and Expansion of the Vectors of the Mind*. Chicago, IL: The University of Chicago Press.

CHAPTER FIVE

MULTIVARIATE GRAPHICS

5.1   INTRODUCTION

Graphs are everywhere. Perhaps because of their familiarity and ubiquity, and our habitual use of them, we fail to fully appreciate what an amazing human artifact they really are. They have enormous informative, persuasive, and transformative power.

To be convinced of the incredible illuminating power of graphics, you only need to look through a column of daily Dow Jones Industrial Average closing prices, and then compare what you have learned from the table with the amazing clarity found in even a simple line graph of the closing price data.

One of the most powerful graphs is the bivariate scatter plot in its myriad forms. Indeed, a major percentage of this chapter is based upon one or another variant of the bivariate scatter plot. To witness firsthand the power of scatter plots in instructing, informing, edifying, and even *entertaining*, we suggest that you take out a few minutes right now, before reading this chapter, and view Hans Rosling's Internet scatter plot show.[1] It will set your mind reeling with

---

[1] Given the Hericlitean nature of the Internet, it is always a little dangerous to suggest a website in a published book, but our guess is that Rosling's show will be around for awhile, and if the URL given here doesn't work, a Google search on his name probably will bring up multiple possibilities. Our favorite show is one of the earliest, perhaps even the first, of his online movies, currently found at this web location: http://www.ted.com/talks/hans_rosling_reveals_new_insights_ on_poverty.html.

Multivariate Analysis for the Biobehavioral and Social Sciences: A Graphical Approach,
First Edition. Bruce L. Brown, Suzanne B. Hendrix, Dawson W. Hedges, Timothy B. Smith.
© 2012 John Wiley & Sons, Inc. Published 2012 by John Wiley & Sons, Inc.

possibilities. As you will see in the next few dozen pages, this is truly the multimedia chapter of the book, and graphics are great story builders.

Rosling's real-time method could perhaps be called a "moving picture time-series scatter plot." It begins with a screen full of what could best be called *balloons*, varying in size to represent the relative size of the nations of the world. The two axes of the graph are fertility rate (*x*-axis) and life expectancy (*y*-axis). The story begins with the year 1962, and then it starts to move. Successive years appear in large but unobtrusive numerals in the background as you witness the balloons shifting (left and right, up and down, but mostly to the left and up), indicating how our world has been transformed over the past 50 years. There are also some historical data flashbacks, some as far back as the 19th century, and a great deal of energy and enthusiasm, and much informative dialogue. It would not be exaggerating to say that the presentation is gripping.

The Rosling show impresses us with the power of graphics to tell a story, but we need to be reminded that graphs can also be used for sinister purposes as well as good. They can be used to obfuscate and deceive. The all-time best-selling book in statistics is Darrel Huff's (1954) *How to Lie with Statistics* (Steele, 2005). A large share of Huff's book is devoted to demonstrating how graphs can be highly misleading and even deceitful, either intentionally or unintentionally so. Even though the tone is light, it is clear that Huff is aware that he is dealing with an important ethical issue (Brown and Hedges, 2009). Interestingly, Huff was not a statistician, but rather a journalist who saw a need, or rather was alerted to a major social problem that needed correcting. With humor and wit, he calls into question the veracity of much of what passes for statistical presentation. The book has sold over half a million copies in English and has been now translated into a number of languages. Obviously, Huff hit a nerve.

Another major player in the business of raising the graphics bar and setting standards for graphicity is Edward Tufte (1997, 2001), referred to in the New York Times as "the Leonardo da Vinci of data." Tufte *is* a statistician. His focus, however, has been primarily positive. He has gathered many examples of outstanding graphs—a best practices strategy rather than social commentary. His books are art books, every bit as engrossing as some that are found on coffee tables and in doctor's offices. Figure 5.1, for example, is one of Tufte's most notable finds—a visual account of Napoleon's Russian campaign.

This graphic departs substantially from the sterile bar charts and ordinary line graphs we are accustomed to. Minard could have presented the information in the most pedestrian of ways, with a handful of separate graphs and charts and a map of the path from Paris to Moscow and back. But instead, he gives a holistic picture by combining the elements all into one graph. The use of the path's width to indicate the current size of the army, and the accompanying temperature line at the bottom create a dramatic effect. These many kinds of combined information could make his graphical story tedious, but Minard's artistic talent makes it at once transformative, entertaining, highly

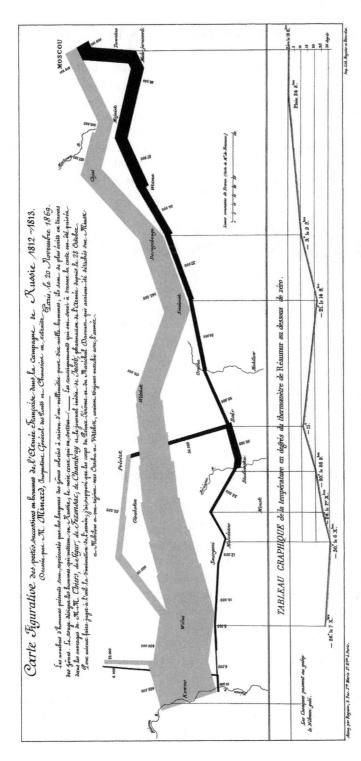

Figure 5.1 Minard's 1869 drawing of Napoleon's Russian campaign. Tufte (2001) comments that it seems to "defy the pen of the historian with its brutal eloquence."

informative, and heuristic. This graph grabs us by the lapels and literally throws in our face the immense human suffering of Napoleon's tragic quest.

It must be mentioned that Tufte also occasionally adopts the social critic role, such as his criticisms of "chartjunk" (graphical embellishments that add nothing to the information in a chart) and his chiding lessons on misleading rectangular information in histograms or barcharts. He combines in an effective way his talent for esthetics and his considerable expertise in statistical methods. He is, however, also known for his acerbic wit, as demonstrated in his lively little paper "Powerpoint is evil" (Tufte, 2003), in which he takes to task the ever-present and inane intrusions of "Powerpoint shows" into the otherwise rational world of scientific dialogue at our national academic meetings.

Given Tufte's (1997) incisive critiques, and similar warnings from Huff (1954), it would seem advisable to begin this chapter with at least some discussion of the question of standards and expectations in the creation of multivariate graphics. Not that we would be so presumptuous as to propose rules. The creation of graphs is, after all, a creative undertaking. However, we might venture to identify the seven guidelines we have used in selecting and in now presenting the multivariate graphical methods of this chapter. We will leave it to the reader to judge how well we follow our own advice.

DESIRABLE CHARACTERISTICS IN MULTIVARIATE GRAPHICS

1. Do no harm. In concert with Huff, we propose that multivariate graphs should not be in any way misleading.
2. Graphs should be esthetically pleasing, in keeping with the high standards set by Tufte's work.
3. Multivariate graphs should take a Gestalt approach to *variables*, revealing the holistic pattern of multivariate sets of variables.
4. Graphs should also take a Gestalt approach to *individual observations*, revealing their holistic pattern.
5. The information in graphs should be convergent and consonant with the information in any accompanying tables and significance tests. They should be mutually informative and clarify one another.
6. Multivariate graphs should also reveal internal *relationships among variables* within the holistic pattern.
7. Graphs should clarify and reveal *relationships among individual observations* in the multivariate space.
8. Graphics should be realistic, in the sense of relating to life and helping us to understand the world better (as does Minard's drawing of the Russian campaign.)

5.2   LATOUR'S GRAPHICITY THESIS

There have been many approaches to the philosophy of science and many prescriptions for the proper conduct of science. None has been more surprising and refreshing than Bruno Latour's *graphicity thesis*. Latour holds that, Popper (1959) and Kuhn (1962) to the contrary, graphicity is at the bottom of the history of science and the real demarcation criterion for its successful practice.

Popper's (1959) demarcation criterion of falsifiability has been highly influential, as has Kuhn's (1962) well-known sociological analysis of the salutary effects of revolutionary stages upon the progress of science. But perhaps one of the most surprising and refreshing approaches to the evaluation of scientific method is Bruno Latour's (1983, 1990) *graphicity thesis*. Latour holds that graphicity is the real story, the real demarcation criterion for the successful practice of science. The refreshing thing about Latour's approach is that, right or wrong, it breaks out of logical armchair philosophy into an anthropological performance evaluation, an empirical analysis of what actually can be demonstrated to be foundational in the immense progress of science.

Essentially, Latour is saying that although the work of Popper, Kuhn, Feyerabend, and others is logically compelling, it has little to do with the actual determinants of success in the scientific enterprise. In actual practice, he argues, the effective use of graphs is the primary determinant of good science. In 2002, Laurence D. Smith[2] and his colleagues (Smith et al., 2002) published a paper in *American Psychologist* entitled "Constructing knowledge: The role of graphs and tables in hard and soft psychology."[3] The authors carefully construct the case that graphics are the essential element in establishing and communicating knowledge claims and in convincing other scientists:

> Especially important for the construction and negotiation of knowledge claims are the representational techniques that Latour and Woolgar (1986) called "inscription devices"—such as graphs, tables, and diagrams—which scientists use in recruiting allies to their viewpoint and persuading members of competing camps (p. 750).

They demonstrate the use of *fractional graph area* (FGA) to measure and quantify graphicity within the journals of any particular discipline. FGA is defined as the fraction of the total page area in an article that is devoted to graphs. They then draw upon a number of studies to show that the "hard" sciences have notably high FGA values (0.14 overall for chemistry, physics, biology, and medicine), with much lower values among the "soft" sciences (0.03

---

[2] Dr. Laurence D. Smith is Emeritus Professor of Psychology at the University of Maine. This stimulating paper is representative of a number of studies he and his colleagues have conducted on the effects of various scientific practices on the effectiveness of research within psychology.

[3] This is in fact a follow-up to an earlier study by Smith et al. (2000).

overall for sociology, psychology, and economics). From the studies they review, they find that the correlation coefficient between rated "hardness" of each of the sciences and FGA is amazingly high (.97). Smith et al. apply this same measure to the many and varied disciplines within psychology and demonstrate comparable results. The FGAs range from a high of 0.13 for *Behavioral Neuroscience* to a low of 0.01 for *Journal of Counseling Psychology*.[4] Consistent with the findings of others from the scientific world at large, Smith et al. find the correlation within the disciplines of psychology between journal "hardness" ratings and graphicity is also large (0.93).

One might argue that this is in fact a reflection of the extent to which each discipline employs quantitative methods or perhaps even a reflection to a commitment to empiricism, but the authors convincingly demonstrate that such is not the case. They argue that although tables are just as quantitative as graphs, they are much less informative and much less effective as "inscription devices." They are less powerful in their ability to persuade and convince. Nor do they correlate with measures of hardness—with Smith et al. actually finding a negative correlation between the use of tables and hardness.

A commitment to quantification is also clearly not the issue. As the authors argue, "economics is among the most quantified of all scientific fields . . . yet it consistently ranks low both in hardness and its use of graphs" (p. 756). Cleveland (1984) and Porter (1995) have both observed that the social sciences, though ranking low in scientific hardness, are inundated with huge amounts of quantitative data. They are typified by much quantification but few graphs. The extensive use of statistical significance tests seems to actually be a marker of softness and effort to establish credibility. Smith et al. evaluated the relationship between the use of *statistical significance tests* and hardness, and found an inverse relationship ($r = -0.77$), which they relate to a quote from Wainer and Thissen (1981) that "the use of graphics by psychologists has lagged far behind their use of elegant statistical procedures."

They are proposing that an empirical evaluation of the effectiveness of practices will be more helpful to the progress of science than logic-based theories of how science should proceed. And the verdict that comes from their empirical analysis of practices is that graphs are essential to good science. In his 1990 work "Drawing Things Together," Latour claims that the mindset of hard scientists is one of intense "obsession" with graphism—the primary factor responsible for their success and for the preeminence of science.

---

[4] In our writing team, we have some experience with both ends of this spectrum. Dawson Hedges is the former Director of our Neuroscience Program at BYU, and Tim Smith currently serves as Chair of the BYU Department of Counseling Psychology (which was ranked seventh nationally in faculty productivity in the *Chronicle of Higher Education*). Suzanne Hendrix, with a PhD in mathematical statistics, is also more toward the hard science side, working as a freelance researcher in the clinical trials industry. Bruce Brown helps Smith to balance the soft side, working on the application of various methods to the psychology of language and to applied problems insocial psychology.

In Chapter 4, we saw that there is a strong preference among many factor analysts for common-factor based models over the principal-component model, even though the performance characteristics of the models would hardly justify such a preference. The arguments for the common-factor approach are logically compelling, even as the prescriptions and arguments of philosophers of science are compelling. There is a major difference in focus in the two approaches to factor analysis that parallels the empirical versus philosophical approaches to evaluating science. The common factor approach emphasizes theory while the advocates of the principal component approach are seeking simplified data description. We will see in this chapter that multivariate graphics are for the most part based upon the principal component approach, precisely because of the emphasis on description (often graphical) rather than model testing.

John Tukey (1977, 1980), credited by many as being the founding father of exploratory data analysis (EDA), contrasted it with confirmatory data analysis (the testing of hypotheses) and saw each as having its place, much like descriptive and inferential statistics. He referred to EDA as a primary reliance on description and visual display, and at root a fundamental difference in focus and attitude. It is inherently graphical. Tukey is responsible for much of the development of EDA in the 20th century, with his highly creative methods, such as the stem-and-leaf display and the box-and-whiskers plot. These, together with a host of other graphical display methods (including Pareto charts, dot plots, residual plots, jiggle plots, Multi-Var charts, and a wide variety of scatter plot methods), have dominated the field of EDA. With rapid advances in computer graphics 30–40 years ago, Tukey became particularly interested in the value of using rotating three-dimensional scatter plots as a way of uncovering data structure. Three-dimensional scatter plots (like Metrika) have now become commonplace. They are consonant with and a part of several of the graphical demonstrations of this chapter.

5.3 NINETEENTH-CENTURY MALE NAMES: THE CONSTRUCTION OF CONVERGENT MULTIVARIATE GRAPHS

Before discussing some of the many varieties of multivariate graphs, we will first present a demonstration of how three graphical methods can be brought together—cluster analysis, principal component plots, and line graphs—to give a rather complete picture of the relational structure to be found within a data set. It is our intent to use this data set example in a pedagogical way, to present some of the principles of reading and also constructing combined and interacting sets of multivariate graphs. We have chosen a data set that is particularly simple and clear in demonstrating how convergent multivariate graphical information can bring closure. The source of the data is U.S. Census records from the 19th century, from which we have selected the 100 most frequent male names in America over the 10 decades of that century.

The data set has 100 rows (names) and 10 columns (decades). The entries reflect the frequency of occurrence of each name within each decade.[5] One advantage of the data set is that it a recording of actual historical facts. There is no error term. It is simple, descriptive, demographic data, and yet it can be expected to contain interesting structural patterns reflective of sociological processes. Our primary interest here is in a *descriptive* graphical application of the multivariate mathematical equations and formulas (Brown, 2010).

Figure 5.2 is the dendrogram produced by a cluster analysis (using the statistical package *R*) of the 100 × 10 matrix of logarithms of frequencies of occurrence of the 100 male names in each decade of the century. A dendrogram is essentially a tree structure that shows the hierarchical groupings of data points according to their distance from one another. From the dendrogram, we see six clear clusters of the 100 names. The first cluster on the left, for example, consists of eight highly popular names (John, William, James, George, Charles, Joseph, Thomas, Henry). As we shall see, these eight names maintain their popularity well throughout the century.

The same 100 × 10 matrix of logarithms of frequencies is now submitted to a principal component method factor analysis, and the vector plot of Figure 5.3 is obtained. The cumulative percent of eigenvalues (Chapter 4, Section 4.3) for this factoring is 89.5%, so this is a good capture of the essential two-dimensional properties in the 10 decade vectors. This kind of vector plot is particularly illuminating because the vectors constitute a time series. The vectors for the first two decades of the 19th century point upward, loading high on factor 2, and then move incrementally in a clockwise direction until the vector for the last decade coincides with factor 1, the *x*-axis of the graph. To summarize, the vectors covering the entire century spread across a 90° arc.

Notice that the vectors do not progress in equal increments. The first five decades are quite tightly grouped together in the vertical dimension of factor 2, and then the later decades begin to spread out. This indicates that the patterns of naming did not change much at the beginning of the century, but began to change more in mid-century. In some ways, these data, because of their simplicity, can be seen as a paradigm case to guide the application of PCP time-series vector plots to more crucial and substantive scientific questions.

The semantic space defined by this configuration of vectors has a clear interpretation. At the top of the latent variable space will be the data points for names that were popular in the first decades of the century, and at the bottom are those that were rare in the early decades. On the right are those that were popular in the closing decades of the century, and on the left are those that were rare in the closing decades. A data point in the upper right of

---

[5] Actually, the entries are *logarithms* of frequencies of occurrence rather than actual frequencies. Since distributions of names are known to be highly skewed, with a few names having high frequencies of occurrence, and many names having low frequencies of occurrence, cluster analysis will not work well on the data unless logarithms are used; that is, the data are log transformed.

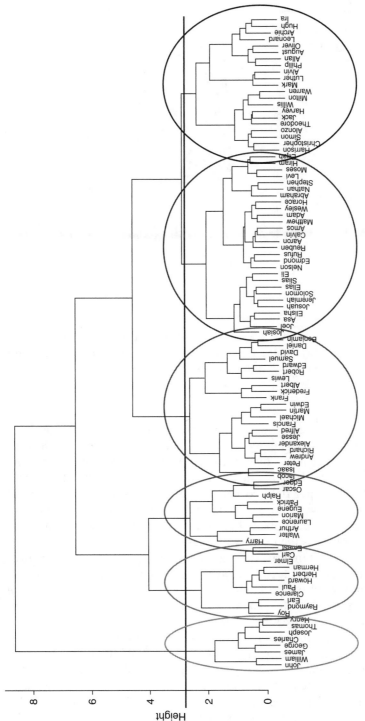

Figure 5.2 Dendrogram from complete-linkage hierarchical cluster analysis of the most frequent male given names in the 19th-century United States within the space of the 10 decades. We have placed a horizontal line across the dendrogram which we raise and lower to select what appears to be the "best" number of clusters. If the line were near the top, we would have two clusters; farther down, we would have three, and so forth.

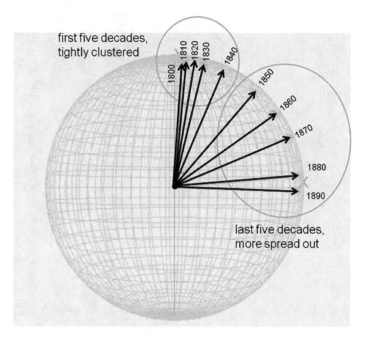

Figure 5.3 Vector plot of the 10 decades of the 19th century from a principal component method factor analysis of the frequencies for each of 10 decades of the most frequent male names. (Image courtesy of Metrika (metrika.org), © Appzilla, LLC.)

this space therefore represents a name that was high in frequency throughout the century, both at the beginning and also at the end. On the other hand, a data point at the lower left is one that was low in frequency throughout the century. An imaginary diagonal line can be drawn from the lower left corner of this space to the upper right corner. Any data points above this diagonal represent names that were more popular at the beginning of the century and then declined, and any data points below this diagonal represent names that become more popular as the century progresses.

Figure 5.4, the next multivariate graph in this process, is a principal-component scatter plot displaying the location of the means of the six clusters within the latent variable space defined in Figure 5.7. This is somewhat unusual in that the points in the scatter plot are not observations, but means of groups of observations. It could be called a *principal-components means plot*.

Notice in this figure that we have three cluster means that fall almost perfectly on the imaginary diagonal line from lower left of the figure to upper right, indicating that name popularity for these will be about equal across the century. The top right one of these is the cluster of eight names (John, William, James, etc.) that are highly popular across the century. The other two are also consistent across the century, but less popular. Above the diagonal, we see the mean of a cluster of names that declined in popularity across the century.

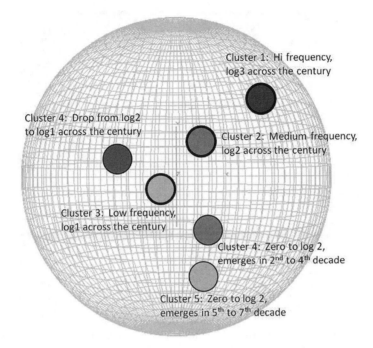

Figure 5.4 Scatter plot of mean locations of each of the six cluster groups of male 19th-century names, plotted within the latent variable space defined by the 10 decade vectors shown in Figure 5.3. (Image courtesy of Metrika (metrika.org), © Appzilla, LLC.)

Below the diagonal are the means of two clusters that gained in popularity across the century.

The next step in this progression of multivariate graphs is a principal-components star plot that connects the 100 data points for the individual names to their means for each of the six clusters. This is given in Figure 5.5. This scatter plot indicates how well the clustering worked, that is, how well it separated the names into clearly differentiated groups. We see that it did very well indeed. There is hardly any overlap among the groups within the latent variable space. In fact, when we run a one-way multivariate analysis of variance (MANOVA) with the six clusters as groups and the ten decade vectors as dependent variables, the Wilks' lambda value is 0.0056, which indicates that the multiple R-square between cluster groups and dependent variables is 0.9944 (99.4% variance accounted for). Truly, this cluster analysis does "cut nature at the joints." A graph like this has a great deal of topological precision, both because a high percentage of variance is accounted for in the two-dimensional representation, and also because the clustering worked.

In Section 5.4.7, we will see that with other names data sets a three-dimensional representation will be necessary. With some data sets, particularly those with a higher amount of error, even a three-dimensional representation

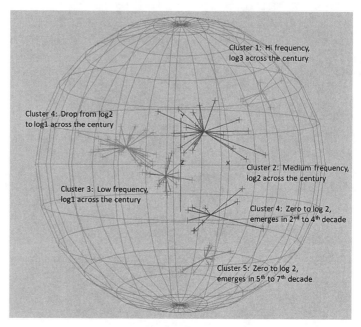

Figure 5.5 Star plot of the 100 19th-century male names organized into six cluster groups. The data point for each name is connected to its cluster mean, plotted within the latent variable space defined by the 10 decade vectors shown in Figure 5.3. (Image courtesy of Metrika (metrika.org), © Appzilla, LLC.)

will not account for enough variance to give this kind of precision. These kinds of graphs become less useful when the amount of variance accounted for in the factor analysis drops below about 60%.

In the next step of this progression, Figure 5.6, we superimpose a set of three univariate graphs upon the scatter plot of means. These are a graphical verification of the accuracy of the entire process, a check that indeed the clusters are homogeneous. These graphs are the least challenging to interpret. They are simply line graphs. They have the 10 successive decades of the nineteenth century along the x-axis, and frequency of occurrence in logarithms as the y-axis.

Let's examine the line graph for the upper right cluster as an example. It consists of eight line plots, one for the name John, one for the name William, one for the name James, and so on, all grouped quite closely together and all staying quite steady with a log frequency of occurrence of about three throughout the 19th century. The next line graph down is also fairly well grouped and fairly steady. It is, however, a log value of about 2. This indicates that these names are about one-tenth as frequent as those in the first cluster. The third line plot, the one in the lower left, hovers around a log value of 1, and occasionally one of the names in one of the decades drops down to a zero frequency of occurrence.

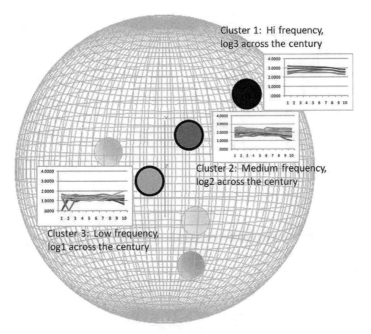

Figure 5.6   Mean locations of each of the six cluster groups of male 19th-century names, with ordered profile plots superimposed for each of the three clusters that fall on the upper right to lower left diagonal of the space. (Image courtesy of Metrika (metrika.org), © Appzilla, LLC.)

The final graph in this progression, Figure 5.7, is the same as Figure 5.6, but displaying instead the three off-diagonal clusters with superimposed line graphs. The line graph for the upper left cluster indicates a decline in popularity for these 28 names. They start out at nearly a log of 2 and decline to a log of less than 1 by the end of the century. These are names like Josiah, Joel, Asa, Elisha, Jeremiah, Solomon, and so on. They are mostly Biblical names, which may be an informative reflection of social processes. It is especially noteworthy that the decline is quite steady, particularly in contrast to the other two clusters, the ones below the diagonal.

Both line graphs for clusters below the diagonal rise rather quickly. In the upper graph of the two, the names start at zero frequency and arise rapidly in the first three decades to reach a log value of nearly 2.5 by the end of the century. These are names like Harry, Walter, Marion, and Eugene, clearly non-Biblical. The lower group of names are seen to arise a little later, about in mid-century. They also arise quickly from zero to a log value of about 2. This cluster consists of "modern" names like Roy, Raymond, Earl, and Clarence.

Each successive graph illustrates different relationships in the data set with a complete story by the end of the progression. This particular data set was chosen because of its simplicity and its clarity and therefore its pedagogical

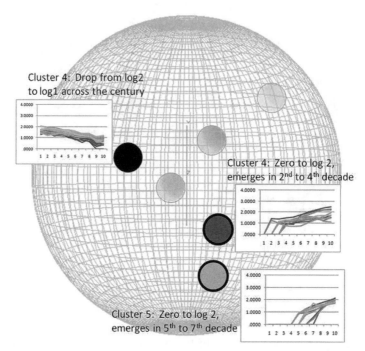

Figure 5.7 Mean locations of each of the six cluster groups of male 19th-century names, with ordered profile plots superimposed for the cluster that falls above the diagonal, and the two clusters that fall below the diagonal in the space of the 10 decades of the 19th century. (Image courtesy of Metrika (metrika.org), © Appzilla, LLC.)

value. Obviously, two-dimensional representations are better for pedagogical purposes than three-dimensional ones. The very simple structure of male names in the 19th century enables a high percentage of variance to be accounted for by the principal-component plots making the relationships precise and somewhat mechanical. All of the graphs and the descriptive statistics (percents variance) converge very nicely to tell a common story, as recommended in the "Desirable Characteristics in Multivariate Graphics" of Section 5.1.

5.4 VARIETIES OF MULTIVARIATE GRAPHS

There have been a number of useful books summarizing the state of the art in multivariate graphical methods (Everitt, 1978, 1994; Chambers and Kleiner, 1982; Cleveland and McGill, 1983; Chambers et al., 1987; Cleveland, 1993, 1994). There is a wide variety of methods from which to choose, and many of them are based upon factoring methods. In this chapter, we will focus primarily upon several varieties of principal-component plot—vector plots, various types of PC scatter plots—and also on other scatter plot-based methods.

5.4.1 Principal-component Plots

One of the earliest applications of principal-components plots appeared in Brown's dissertation in 1969. The topic was evaluative reactions to various dialects of spoken Canadian French. Recorded speech samples were rated on 20 paired-opposite adjectives. Figure 5.8 shows the linked factor analysis vector plot of the 20 adjectives and the factor analysis scatter plot of the 24 speaker means. One unusual characteristic of this particular application of factor analysis is that the factoring was not done on the individual observations, but rather *on the means*. With principal components, as we shall see in the discussion of the Gabriel biplot, the factoring of a data matrix of means is a defensible procedure. However, in this dissertation study, the principal-factor method was used, which is not recommended for small data sets, such as 24 means.[6] The results, however, are highly consistent with what is obtained when one reanalyzes with principal component analysis, as embodied in the Stata implementation of the Gabriel biplot to which we now turn.

5.4.2 Ruben Gabriel's Biplot

The most popular of the multivariate graphing tools in common use is probably the biplot. Thanks to Stata, it has become very broadly accessible and is simple to use. The biplot was first published 40 years ago by Ruben Gabriel (1971),[7] but it has stimulated much interest and several good books since that time, explaining its rationale and use (Gabriel, 1995, 2002; Yan and Kang, 2003; Gower and Hand, 2011; Gower et al., 2011).

The two figures in Figure 5.9 appeared in the original 1971 paper that introduced the biplot method. The mathematical basis for the biplot is principal component analysis. If you go back to the principal component mathematical demonstration in Chapter 4, Section 4.5, you will find two figures—one that displays the vectors (Fig. 4.8, with vector length indicating variance size), and one that displays the scatter plot of the data points (Fig. 4.6). If you were to superimpose these two graphs, you have essentially constructed a biplot.

Some have thought that the "bi" in biplot is because they are plotted in a two-dimensional space, but such is not the case. There are also three-dimensional bi-plots. The "bi" indicates that two kinds of information are superimposed, the vector plot of variables and the scatter plot of data points.

---

[6] This point is somewhat controversial. It is of interest to note that at least one of the examples that Rencher (2002) gives of clear and useful applications of factor analysis (p. 443) is from one of the early studies in which we employed factor analysis of means (Brown et al., 1973). It can be argued that particularly when one is using principal components descriptively, perhaps even for graphing purposes, there are some advantages in using means in that they are more stable and representative of population values.

[7] When K. Ruben Gabriel first published the biplot, he was a young professor of statistics at the Hebrew University of Jerusalem. Gabriel has made a number of important contributions to statistics in addition to his biplot method. He taught in the Department of Mathematics at the University of Rochester until his death in 2003.

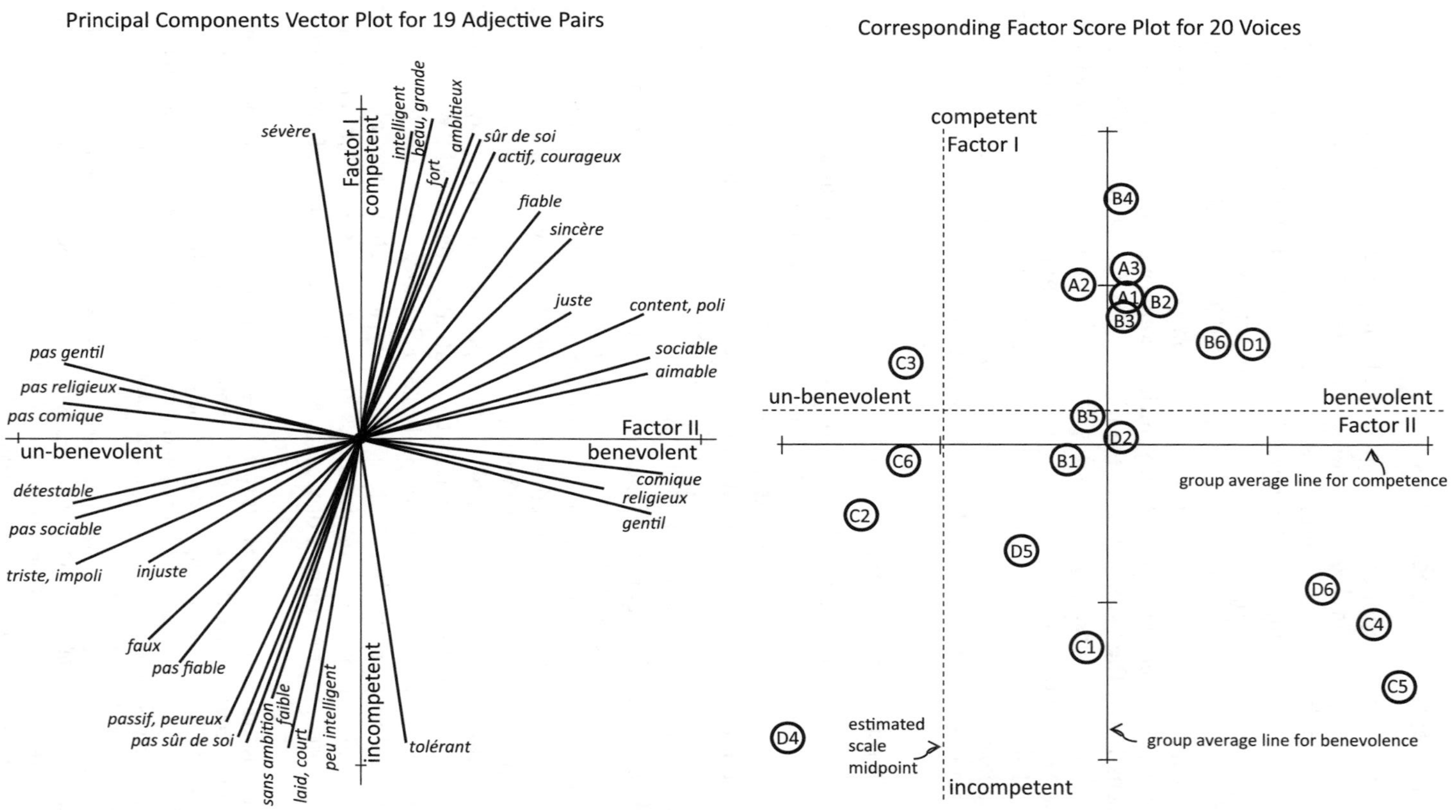

242

Principal Components Vector Plot for 19 Adjective Pairs

Corresponding Factor Score Plot for 20 Voices

Figure 5.8 An early approach to the use of principal components to plot both the vectors for the variables and also the data points plotted within the latent variable space defined by those vectors (Brown, 1969). (Reprinted courtesy of the author and McGill University.)

The panel on the left is artificial data that Gabriel uses to demonstrate some of the properties of the biplot. The first thing to notice is that the vectors for the four variables have very different lengths. That is because this is principal components of a covariance matrix (rather than a correlation matrix), and the length therefore indicates the standard deviation[8] of each variable. The vector pointing left and the vector pointing right in this left panel graph are seen to have four or five times the standard deviation of the other two variables. In contrast to the biplot, the other principal-component plots we present in this chapter are *standardized principal-component plots*, which makes the vectors all of unit length. When a vector appears short in a standardized plot, it is not. It is still a vector of unit length (a length of 1.0), but it is inclined toward the viewer or back into the page (and thus we are not seeing its full length), and the short appearance of the vector indicates that all of the variance is not accounted for in the two-dimensional plane of the graph.

The right panel of Figure 5.9 is real data, and gives some insights about the land of Israel in 1967, with modern quarters on the left, poorer quarters on the right, rural areas at the top, and the vector for electricity pointing to the bottom (showing that rural areas didn't have electricity). Gabriel comments:

> The modern quarters appear to have a particularly high prevalence of baths, water inside the dwelling and refrigerators (all three vectors pointing to the left), whereas the poorer quarters have relatively high prevalence of toilets and electricity. Evidently the last two items were pretty generally available in all urban sections and thus are not indicative of better living conditions, whereas the former three items were much more available in better off homes (Gabriel, 1971, 45).

As we shall see in the next section, both the covariance matrix approach to principal-component plots and also the standardized approach to them have their own advantages. Clearly, the covariance approach used in the classic form of the biplot gives more information in the vector plot in that you can not only see how correlated variables are (the angle between them) but also their relative dispersion values (their standard deviations). This, of course, presupposes that the scales of the variables are *commensurate*, such that comparisons of their standard deviations are meaningful.

The real power in biplots is that they combine the information in the scatter plot and the information in the vector plot such that the reader can easily see which data points are highest on which variables. In that sense, it adds one more desirable multivariate-graph characteristic to our list of eight—it enables

---

[8] In matrix algebra, the dot product of a vector of data and its transpose is a sum of squares (SS) of that vector, and of course this is converted to a variance by centering the vector and dividing the SS by its degrees of freedom. The square root of the dot product is the "magnitude" of the vector, which corresponds to the standard deviation of the variable.

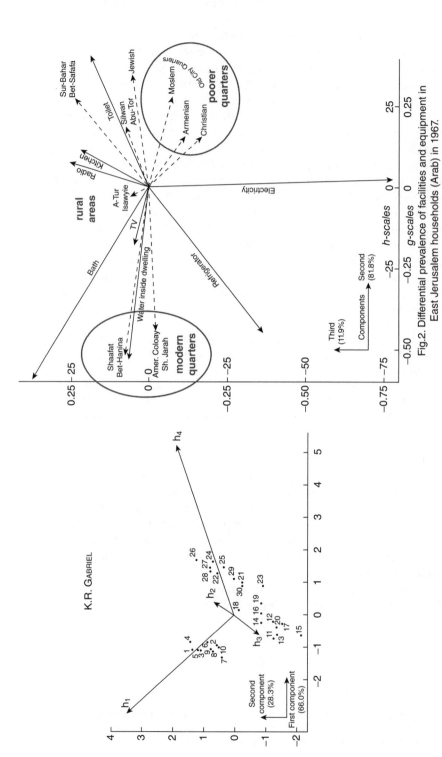

Fig.2. Differential prevalence of facilities and equipment in East Jerusalem households (Arab) in 1967.

Figure 5.9 These two biplot displays are from Ruben Gabriel's (1971) original publication of the biplot. The one on the left, his figure 3 in the paper, is artificial data to illustrate the concept. The one on the right, his figure 2, is from actual household "facilities and equipment" data gathered in 1967 in East Jerusalem. (From Gabriel, K. R., 1971. The biplot graphic display of matrices with application to principal component analysis. *Biometrika*, 58(3), 453–467, by permission of Oxford University Press and the Biometrika Trust. *Biometrika*

one to see convergent properties between the location of variables and the location of data points.[9]

The biplot is one of the most useful tools in any data analyst's toolbox, particularly in the highly useful embodiment of Stata's *biplot* procedure. Good instructions for its use are given in the Stata manual, in the section entitled "Base Reference," under the listing "biplot."

The process is not a difficult one. The Stata instructions shown below will create the data matrix

```
input str7 persons aleph bet gimel
al 8 10 18
bill 6 8 14
charlie 5 12 17
dan 5 16 21
ed 4 4 4
frank 2 10 12
end
```

shown at the bottom of the left panel of Figure 5.10, and the additional Stata instruction of

biplot aleph bet gimel, rowlabel(persons)

will produce the graphic shown at the top of the left panel in Figure 5.10. Notice in this figure that the variable labeled "gimel" has a longer vector (a larger standard deviation) than either of the other two variables. That is because, as you can see in the data matrix below the figure, we have created variable gimel to be the sum of the other two variables.[10] If, instead, we create gimel to be the average of aleph and bet, we obtain a vector for gimel that is midway in length between aleph and bet (as shown in the middle panel). In the far right panel, gimel is created as double the sum of aleph and bet, giving a vector for gimel that is very long indeed.

Hopefully, the point is not lost on the reader that this graphics tool is amazingly enabling, with myriad possible uses in one's research. Rather than just reading a paper of interest, it is possible for you to immediately examine both the spatial organization of the variables and also a scatter plot of the observations in a linked space. The graphical information is not only helpful, but actually a crucial part of understanding the other data analysis results, as demonstrated in Chapters 4, 5, and 6.

---

[9] Actually, this is somewhat along the lines of our fifth desirable characteristic, "information in graphs should be convergent and consonant. . . ."

[10] Hopefully, the reader is aware that data usually are what they are. You don't usually go around creating one variable to be the sum of two others, but here we are creating artificial data to explore some of the operating characteristics of the biplot graphical toolbox.

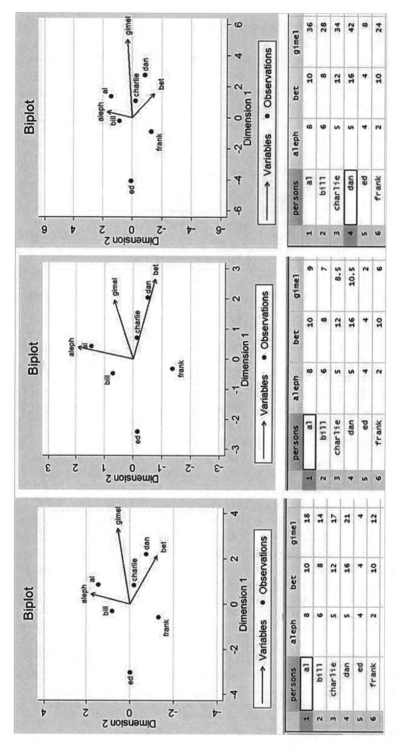

Figure 5.10 Three demonstrations of the biplot with simplest case data, showing the input data beneath each and the biplot result above. In the left panel, the third variable, gimel, is the sum of the other two. In the second panel, gimel is the mean of the other two, and in the third panel, gimel is twice the sum of the other two. (StataCorp, 2009. Stata: Release 11. Statistical Software. College Station, TX: StataCorp LP.)

It is surprising how many papers in the published literature actually do have tables of means, proportions, frequencies, percentages, or other summary statistics that lend themselves to a biplot analysis. Just to try out that proposition, we looked through the March 2011 issue of the *General Archives of Psychiatry*, one of the top two or three journals by impact factor in the field of psychiatry to find possible tables for graphical/spatial analysis. Nearly every paper had graphs, some of them highly informative and engaging, and about half of the papers had a table that could be analyzed graphically with Stata's biplot. One of them is shown in Figure 5.11. This is from a paper by Lee et al. (2011). The study deals with how social environment affects cognitive functioning in the elderly however with the additional question of whether the apolipoprotein E (APOE) ε allele modifies this association.

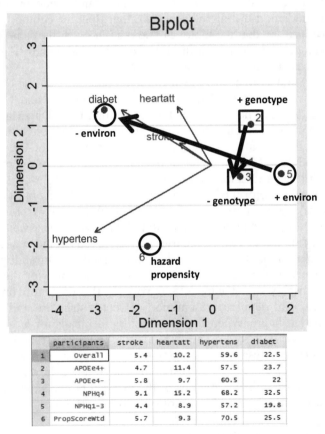

	participants	stroke	heartatt	hypertens	diabet
1	Overall	5.4	10.2	59.6	22.5
2	APOEe4+	4.7	11.4	57.5	23.7
3	APOEe4-	5.8	9.7	60.5	22
4	NPHq4	9.1	15.2	68.2	32.5
5	NPHq1-3	4.4	8.9	57.2	19.8
6	PropScorewtd	5.7	9.3	70.5	25.5

Figure 5.11 Application of the biplot program of Stata to an empirical set of data from Lee et al. (2011). The cell entries in the matrix below are percentages of four diseases found at six locations. The four diseases are named in the column headings across the top, and the locations are listed in the first position for each of the six rows. (StataCorp, 2009. Stata: Release 11. Statistical Software. College Station, TX: StataCorp LP.)

The 6×4 matrix of data for Figure 5.11 are taken from their table 1, which lays out the percentage occurrence of four diseases across five different characteristics of participants, plus the figures for the overall combined sample. The two "bounding vectors" are the one for myocardial infarction (heartatt), which is primarily vertical and to the left, and hypertension, which is primarily horizontal, pointing to the left and somewhat down. The two are at nearly a 90° angle, indicating that these two diseases are quite independent of one another. The other two diseases are in between. Data point 5 is the 881 participants who have a positive psychosocial environment, and data point 4 is the 243 participants who have a negative environment. The direction of change indicated with the large connecting arrow between these two points shows that those in the negative environment are at higher risk for all four diseases, but particularly the vertical ones. Other relationships can be seen, but our interest is primarily in demonstrating the process rather than the content of the findings. Suffice it to say that the biplot can be read in much the same way as the principal-component plots from the 19th century male names (Figs. 5.3–5.7), but with an easier process for creating the figure.

Although Stata's biplot has the advantage in speed and ease of use, there are also relative advantages of the Metrika visualization tool and the DataMax tool discussed in the next section. For one thing, both DataMax and also Metrika are fundamentally three-dimensional visualization tools that use visual rotation to create a three-dimensional experience of the data. Also, Metrika[11] and DataMax both have substantially more visual precision in dealing with large and complex data structure patterns. In Metrika, the user has a high level of control of every aspect of the visualized objects (lines and object shapes), but the process also requires fine-tuned work from the creator of the principal-components plot. The Stata biplot works very well with up to a few dozen data points to give the user a quick and useful idea of data structure, but it is not intended for the kind of fine-tuned structural display in, for example, Figure 5.18. Shown in Figure 5.2 is a Stata biplot for the Hawaiian names data displayed in Figure 5.18. Obviously, this does not compare well with the clear and detailed presentation of vector structure and data point structure of that figure. When the visualization task calls for graphs like the star plot graphs of Figures 5.5 and 5.18 to visualize the fine-tuned structure of hundreds, or even thousands, of data points and perhaps 50 or even 100 vectors, Metrika is a better tool of choice. It should be mentioned, however, that there are many software implementations of the Gabriel biplot, some of which do have the capability of rendering a more fine-tuned structure.

5.4.3 Isoquant Projection Plots

In the 1980s, Brown and several colleagues developed a data visualization system that, like Gabriel's biplot, superimposes the vector plot over the top

---

[11] Metrika, like the R system, is an open-source analysis system, available to be downloaded free from the Internet, at the location *metrika.org*.

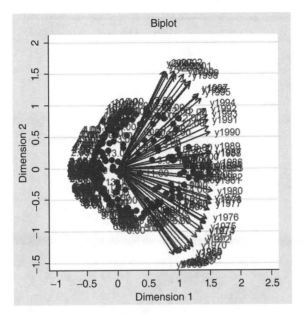

Figure 5.12 Application of the biplot program of Stata to the time-series vector plot data of Figure 5.19 for the 48 vectors of consecutive years superimposed on the scatter plot of the 320 Hawaiian names. Absent from the biplot is three-dimensional rotation capability, the star plot grouping capability, and the time-series connections on the vectors that make Figure 5.19 so systematic, clear, and illuminating. (StataCorp, 2009. Stata: Release 11. Statistical Software. College Station, TX: StataCorp LP.)

of the scatter plot of individual observations. The method is called an *isoquant projection plot*. The approach is different than Gabriel's in that it always uses standardized variables, such that one can (when a reasonably high percentage of variance is accounted for in the factoring) reconstruct the metric information in the original variables from the combined holistic plot, hence the name isoquant[12] projection plot. The software embodiment of isoquant projection plots is DataMax. It is typically used with industrial data that deals with close tolerances and a small error component, such as the Cray data described in Section 5.4.6. Because it is an industrial tool and has not been used extensively in academics, there are only two published papers using isoquant projection plots (Hendrix and Brown, 1990; Hirsh and Brown, 1990), and there are no published works explaining it, other than the DataMax technical manual.

---

[12] Isobars are lines linking locations of equal barometric pressure, and isotherms are for equal temperature. Isoquants are lines constructed perpendicular to a variable at a set of equally spaced metric locations that connect all locations within the space that share that metric value on that variable. This process is explained in Figure 5.13 and the accompanying text.

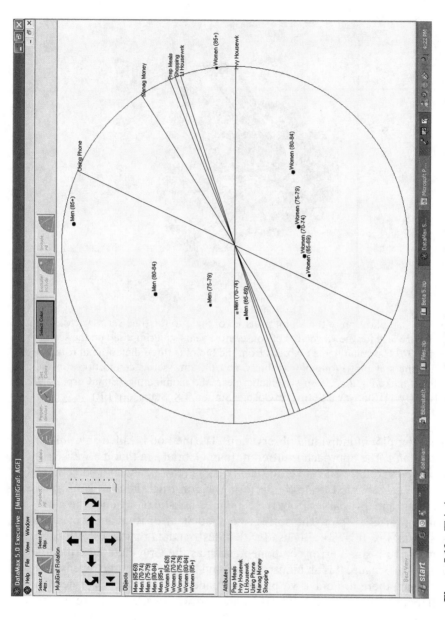

Figure 5.13 This is a screen capture example of the isoquant projection plots approach employed in DataMax Executive. This particular plot is from a principal component factor analysis of simple data from a sociological paper investigating the comparative competence of elderly persons in dealing with typical household tasks. DataMax is an exclusive product of ECHO Solutions, Inc.

The visual analysis screen of DataMax,[13] is shown in Figure 5.13. We will illustrate the visualization information in DataMax with a small set of sociological data.[14] This graph comes from a 10 × 5 matrix showing what percent of elderly persons have trouble with each of five household tasks. There are five data points for men at successively higher ages, and five for women, making a total of 10 rows of data. The five household tasks (variables) are: using the phone (the vertical bounding vector), managing money, preparing meals, shopping, and finally, light housework, as the bounding vector at the right. As we move upward in a vertical direction, it reflects more and more difficulty dealing with social tasks, such as using the phone. The horizontal dimension, on the other hand, seems to reflect difficulties with physical things like housework as one moves to the right. We see a clear pattern in men that as they get older, they have more and more trouble with social things (with successive age points moving upward), while women have more and more trouble with physical things (with successive age points moving to the right). The last age point for women (85+ years), however, begins to also move up, indicating they also begin to have trouble with social things at this advanced age.

In the upper left corner of the DataMax screen are several three-dimensional rotational buttons that enable fine-tuned control of graph movement to a very specific visual location (within one degree). DataMax also has the capability to project from the data points to the vector for each variable (i.e., to drop a perpendicular line to each vector), and thereby to compare the holistic reconstructed metric for each variable to its actual empirical metric. If one accounts for all the variance in the factoring, which is seldom achieved, one can perfectly reconstruct the metric information in the original data.[15] But even when one accounts for only 90%, or even 80%, of the variance with factoring, the reconstruction will be reasonably good. That can often be achieved with everyday data sets through the wise use of replicating and of averaging methods.[16]

Figure 5.14 can be used to make the rationale of isoquant projection plots more clear. This is a Stata biplot, showing the same data displayed in Figure 5.10, but with the "std" option turned on, so that Stata biplot standardizes the data before performing principal components analysis on it. Notice that the six data points for the first variable, "aleph," are 8, 6, 5, 5, 4, and 2. When perpendiculars for each of the data points for the six persons are constructed to

---

[13] DataMax software was in fact used to create the graphics for the Cray supercomputer performance evaluation project described in Section 5.6, and shown in Figure 5.14.

[14] Although the primary use of DataMax is industrial and generally with large data sets, we have used a simple set of data here to more clearly show how it operates.

[15] To understand better how this works, it is recommended that you go back to Chapter 4, Section 4.5.4, and review how factor analytic data transformations are completely reversible when all of the variance is accounted for.

[16] This is not as impossible a task as some who are accustomed to factoring rating-scale data might think. As an example, the four studies of empirical databases of names reported here, and displayed in Figures 5.3, 5.16–5.18, account for 89.5, 95.7, 91.6, and 81.3% of the variance in the variables of their respective semantic spaces.

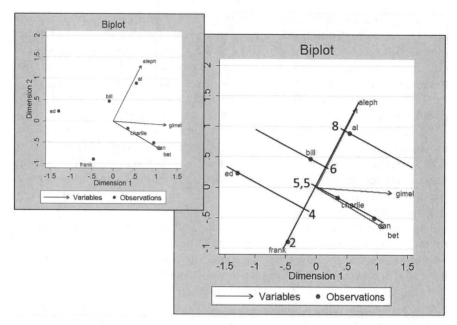

Figure 5.14 The biplot program of Stata is used here to demonstrate how isoquant projection plots are interpreted. The small figure in the upper left shows how the biplot (from the data of Fig. 5.10 with a "standardized" setting) looks when first created. In the lower right, isoquants for the first variable, aleph, are superimposed upon the PCP plot, and we see that they perfectly reconstruct the metric properties of this variable. (StataCorp, 2009. Stata: Release 11. Statistical Software. College Station, TX: StataCorp LP.)

intersect with the vector for this "aleph" variable, we see that it perfectly reproduces the metric information in that variable. Of course, this also is true for variable "bet" and for variable "gimel." The reason it works on all three is because the *rank* of this simple artificial data set is two, that is, all of the variance in the three variables is accounted for in a two-dimensional space, as explained in Chapter 3, Section 3.7.

5.4.4 Cluster Analysis

Cluster analysis is a well-developed art form within graphical data analysis methods, and incorporates a wide variety of approaches. Although we use cluster analysis to good advantage in this chapter, we will not review the wide variety of methods nor give any computational details. The interested reader is referred to Everitt et al. (2011) and Rencher (2002) for discussions of computational methods, and to Arabie et al. (1996) for an edited book touching upon a variety of applications. The SAS, Stata, and SPSS manuals also give considerable helpful information, conceptual as well as methodological.

We have seen in Section 5.4 that cluster analysis helps us to "cut nature at the joints" to find the natural groupings in a multivariate set of data. The primary division in "types" of cluster analysis is between the partition-clustering methods and the hierarchical clustering methods. The hierarchical methods are more common and in many ways more useful. Stata, for example, offers only two partition-clustering methods but seven hierarchical methods. We have found the hierarchical clustering methods to be very effective for the way we are using them in this chapter—to obtain natural groupings of data points to clarify the internal structure in a principal-component plot. All such demonstrations in this chapter will use the hierarchical clustering (single linkage) method.

5.4.5 Cluster Principal-component Plot

The multivariate graphical analysis demonstration of Section 5.2 was in fact an example of a *cluster principal-component plot*. This same method is also used in four other data analyses in this chapter, always including both a vector plot and also a cluster plot for each: the "names" data from three additional locations (Fig. 5.1 for Austrian names, Fig. 5.18 for French names, and Fig. 5.19 for Hawaiian names, all in Section 5.4.7), plus an analysis of the Flourishing Families data set of Section 5.5, with Figure 5.22 giving the vector plot and Figure 5.25 giving the cluster plot.

The essential idea in all of these is to use a cluster analysis to find the natural groupings within the data points that will "cut nature at the joints." The vector plot is then used to define the semantic space on the basis of the relational pattern among the variables. Finally, the cluster plot is created to group the individual data points within the semantic space defined by the vectors. We can then assess the *effectiveness of factoring* in the vector plot (using the cumulative percent of eigenvalues), and we can likewise assess the *effectiveness of clustering* in the cluster plot using a one-way multivariate analysis of variance (Chapter 8, Section 8.3). In a one-way multivariate analysis of variance, Wilks' lambda is equivalent to 1 minus the R^2 in accounting for the grouping variable by what we are here calling the "vector variables" (Chapter 8).

It is instructive to note that both the factoring operation and also the clustering operation are more effective when the data are objective and have a small error term. The four "names" data sets are all examples of this. All involve objective, factual data. In contrast to this, the Flourishing Families data set is a good example of rating scale data, the kind of data on which most of the social sciences move forward. It is a good example because of careful methodology and impressive psychometrics, but it would still be expected to have a larger error term than would objective data.

In evaluating the *percent variance accounted for by the factoring operation*, the four "names" data sets have 89.5, 95.7, 91.6, and 81.3% of the variance accounted for, respectively, while the Flourishing Families psychological-scales data set has 60.4% of the variance accounted for. These statistics are an indication of how well the information in the original variables is accounted for within the three-dimensional[17] holistic plots. With regard to the *quantitative evaluation of the effectiveness of clustering*, the R^2 values (from the Wilks' lambdas) for the four names data sets are $R^2 = 0.9944, R^2 = 0.9565, R^2 = 0.9161$, and $R^2 = 0.8133$, respectively, and the value for the Flourishing Families data set is $R^2 = 0.9928$. The clustering operation is notably successful for all five data sets, Hawaiian names being the ones with the lowest R-squared "goodness of cluster" indicator.

The purpose of a cluster principal-component plot is to group the data points according to empirically derived natural internal structure. On the other hand, a MANOVA-based principal-component plot, explained in the next section, examines the effects of external manipulations (exogenous variables) on the structure of the data points. It is obviously easier to find a coherent internal structure than it is to be successful in experimentally manipulating or quantitatively predicting data point locations on the basis of exogenous independent variables. However, in the next section, we will examine the MANOVA PC plot in an area where we have a high level of prediction, and the MANOVA PC plot is highly effective—in dealing with hard science data.

5.4.6   MANOVA-Based Principal Component Plot

At the American Statistical Association meetings about 20 years ago, Suzanne Hendrix (Hendrix and Brown, 1988)[18] put forth an unusual thesis with respect to ANOVA, MANOVA, and multivariate graphics. She first stated that statisticians seldom analyze an interaction higher than a three-way in actual practice, because four-way and higher interactions are difficult if not impossible to understand. There was much assent from the audience. She then made the somewhat surprising counter assertion that while this is true univariately, it is *not* true multivariately—in a multivariate graphical presentation, the geometry is such that higher-order effects (four-, five-, and six-way means) often create shapes[19] that can be readily discerned. We will see in this section that her statement is indeed correct.

Neil Hirsh, the director of capacity planning at Cray Research (the company responsible for developing and marketing the premier supercomputer in the world at the time) was in the audience, and proposed to Hendrix afterward that perhaps her multivariate graphical method and his complex and expensive Cray supercomputer performance evaluation data set would be a good combination for a paper to be submitted to an upcoming IEEE volume on visualization in scientific computing. Hirsh, Hendrix, and Brown began working

---

[17] In fact, it is a three-dimensional factor space for four of the data sets, but only a two-dimensional factor space for the 19th-century male names data.

[18] See also Hendrix and Brown (1990).

[19] The mantra of this multivariate graphical approach to understanding data is that abstract data has a shape, and that in some ways what we are trying to build is a pattern detection/identification system.

on the project, and it was eventually published in the IEEE volume (Hirsh and Brown, 1990).[20]

Figure 5.15 shows one of the data analyses from the Hirsh and Brown (1990) chapter, a four-way multivariate analysis of variance of (A) operating system, SYST; (B) transfer format, TRFO; (C) blank compression, BLCO; and (D) segment size, SEGS. Figure 5.14 is a principal-component plot showing the four-way means. When you look at a multivariate display of four-way means like this, you are not only seeing a four-way interaction of factors A, B, C, and D. You are seeing the combined information of the single four-way interaction (ABCD), the four three-way interactions (ABC, ABD, and ACD), the six two-way interactions (AB, AC, AD, BC, BD, CD), and the four main effects (A, B, C, D). That is why a MANOVA-based PC plot is so important. If you can examine and understand the pattern in the four-way means, you have simultaneously comprehended the Gestalt of the main effects and all of the interactions.

In tackling the Cray data set, we used graphics and MANOVA-based statistics jointly in every analysis. In the process, we discovered an interesting thing. Even though the two kinds of information are mutually constitutive—that is, the numerical adds helpful precision to the graphical, and the graphical brings clarity and meaning to the numerical—they are far from being on an equal footing. The graphical analysis can stand alone, but the numerical results cannot. In other words, one really cannot tell what is going on when only looking at the numerical. A comparison of Figures 5.15 and 5.16 makes this clear.[21] In looking at Figure 5.16, which gives R-squared values for each of the four factors and interactions, all one can see is that operating system (SYST) has by far the largest effect, especially on factor 2. Transfer format (TRFO) and segment size (SEGS) also have fairly strong effects, but the effects of blank compression (BLCO) and all of the interactions are quite small by comparison.[22]

Knowing from the R^2 values that operating system (SYST) has a large effect is one thing, but understanding the nature of that effect is quite another, and that is only found in a multivariate graphic like Figure 5.15. The graphic makes it immediately obvious that operating system 1 has a profound magnifying effect on the effects of the other three factors, creating a geometric figure of

---

[20] The work reported in the Hirsh and Brown (1990) chapter became the foundation for a spin-off company to graphically deal with capacity planning in supercomputers.

[21] Figure 5.15 is a true multivariate graphic that captures the holistic pattern in the data, but Figure 5.16 is just bar graphs of R^2 values, only a slight improvement over presenting the R-squared value for each factor-combination in a table.

[22] It is instructive to note that all 15 of the effects, all one way, all two way, all three way, and the four way, are highly significant, $P < 0.0001$. This is what happens in a MANOVA when the data are mechanical and error-free. Obviously, significance tests are not particularly informative in this kind of work, and yet the mathematics of MANOVA and the graphs that can be created from it are of great value—a MANOVA without significance tests.

256

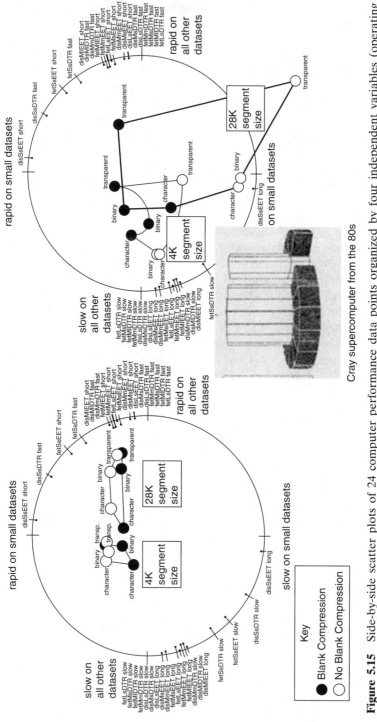

Figure 5.15 Side-by-side scatter plots of 24 computer performance data points organized by four independent variables (operating system, segment size, data transfer format, and blank compression) within the 2D latent variable space of 20 dependent variables. The panel on the left shows the three-way means for operating system 1, and the panel on the right shows the three-way means for operating system 2. (Reprinted from Hirsh and Brown (1990, 200-201). © 1990 IEEE.)

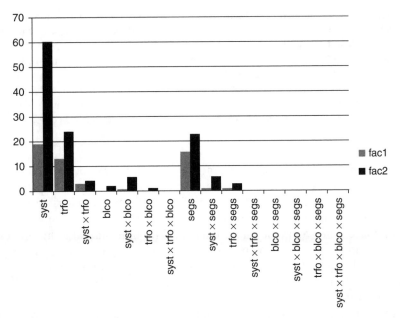

Figure 5.16   Bar graph comparisons of the magnitude of the sum of squares values on factor 1 and factor 2 from a four-way multivariate analysis of variance. There are four main effects (SYST, TRFO, BLCO, and SEGS) and 11 interaction terms.

much two-dimensional spread, while on the other hand, operating system 1 (left panel) has no such effect. This observation requires a graph, and not just any graph, a multivariate one that can show the pattern of everything in the data taken together. Such observations in no way appear in the numerical results until you know from the multivariate graph what is happening, which will then enable you to devise a way to indicate it numerically. The graphics are primary.

After the analysis, we met with Hirsh at Cray headquarters in Mendota, Minnesota in the spring of 1989, and showed him the convergent results from the MANOVA analyses and the MANOVA-based PC plots. He commented that he had learned more about his data in the first half hour of investigating it graphically than in the past dozen years looking at similar data statistically. From our side, it was also a valuable joint experience. It was refreshing and exhilarating to apply multivariate analysis tools to mechanistic data with very small error terms, after decades of wrestling with human judgment data having large error terms, low-reliabilities, and large standard errors. And in the process, we found the graphics to be highly informative at every turn and absolutely indispensable.

In many ways, this experience confirms the vivid account in the Smith et al. (2002) paper of the place of graphics in hard science:

Scientists in action surround themselves with graphical displays that make productive talk possible (Roth and McGinn, 1997). When deprived of those resources, scientists lose their shared rhetorical space and are found to stutter, hesitate, and talk nonsense (Latour, 1990, 22), regaining their powers of articulation only when new inscriptions are scribbled with whatever materials are at hand. In view of the striking dependence of scientists on such inscriptions, Latour (1990) concluded that scientists display an extraordinary "obsession for graphism" (p. 39) and that their prevalent use of graphs is, in fact, what distinguishes science from nonscience.

One of the lessons from applying multivariate statistical tools and multivariate graphing tools to "hard" computer science data is that the question of "what is the appropriate analysis here?" that so often appears in statistics books does not seem relevant. It is more a question of which arrangements of the data are more heuristic and illuminating in understanding the processes we are trying to observe. With a solid, precise, highly reliable, and complex data set like this, there are many possible "inscriptions," many possible arrangements of the data to choose from, and they tend to be illuminating in different ways.

Another observation is that computers are much better behaved than humans. They are more consistent and predictable, and their "behavior" is therefore easier and more illuminating to analyze, and they also help us to understand our quantitative analysis tools better, and in new ways. The mathematics of multivariate analytical methods seems to work particularly well when applied to hard, objective data, something beyond a 5-point scale. That might be part of the reason for the phenomenon observed by Latour (1990) and Smith et al. (2002) that graphics play a much bigger role in hard science than in soft science. As we have just seen with the Cray data, when there is very little error of measurement, significance tests are not particularly informative. In fact, nothing from the numerical MANOVA of the Cray data set rivaled what was learned from the patterns shown in figures like Figure 5.15.

5.4.7 PCP Time Series Vector Plots

Adopting sequential time points, such as years or decades, or milliseconds in an EEG wave, as the variables in a vector plot is often quite illuminating. It is also usually esthetically interesting, as shown in the vector plots of Figure 5.17–5.19. Usually, with such data, the location of each successive vector only changes by a small increment from the preceding vector, which tends to create a clear, simple, and often insight-producing pattern. We will compare the time series vector patterns of three additional "names" data sets, one from Austria, one from France, and one from Hawaii, to see how differential structure manifests itself through this medium of time-series plots.

Principal-component plots can be thought of as having two aspects, the *vector plots of variables* and the *scatter plots of data points*, and both of these

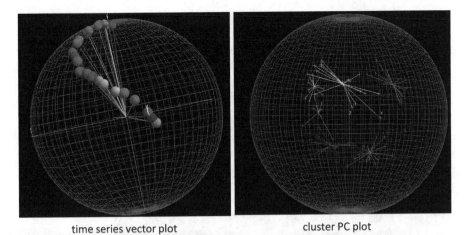

time series vector plot   cluster PC plot

Figure 5.17   Time-series vector plot of 24 consecutive years, 1984–2007, from an analysis of the changing frequencies of male given names for each of those years in Austria. On the right is a scatter plot of the 100 most frequent male names in Austria during those years, grouped by clusters from a cluster analysis dendrogram, plotted within the latent variable space defined by the vector plot on the left. (Image courtesy of Metrika (metrika.org), © Appzilla, LLC.)

have a unique kind of structure. As suggested in the "desirable characteristics" list, items 5 and 6, at the beginning of the chapter, it is important to look for structure both in the pattern of the vectors, and also in the patterns of the scatter plot of data points. We have seen in several previous sections how well *cluster principal-component plots* identify the natural structure in the scatter plot of data points. In this section, we will examine how *time series structure* can help us to understand the corresponding patterns in vector plots.

The interesting thing about years and decades is that they have no direct semantic space meaning for us. In contrast, paired-opposite adjectives, for example, have semantic immediacy. A factor structure based on them can be directly interpreted on the strength of the meaning of the adjective pairs. In the French Canadian dialects graphical display of Figure 5.8, the semantic space is named and understood on the basis of the semantic properties of the 20 paired-opposite adjectives used to define the space. The means for the ratings of 24 dialects are then plotted and understood within that established semantic space.

The principal-components plots for names are different: the years obtain some of their meaning from their relation to the names, and the names obtain some of their meaning from their relation to years. The two mutually constitute one another semantically. Both names and years obtain a kind of relational meaning from one another in their intersecting patterns. All of this becomes quite abstract and not easy to put your mind around. In the 19th-century male names of Section 5.2, we were able to make it all comprehendable by also including line graphs into the mix. Line graphs are simple and easy to

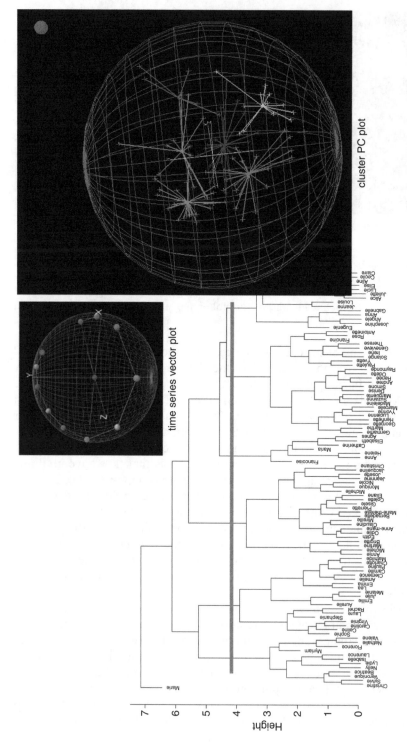

Figure 5.18 Dendrogram, time-series vector plot (in a spiral pattern) and star plot of six clusters, with the single name "Marie" as an outlier, for the 100 most frequent female names in France during 11 decades covering the 20th century. (Image courtesy of Metrika (metrika. org), © Appzilla, LLC.)

260

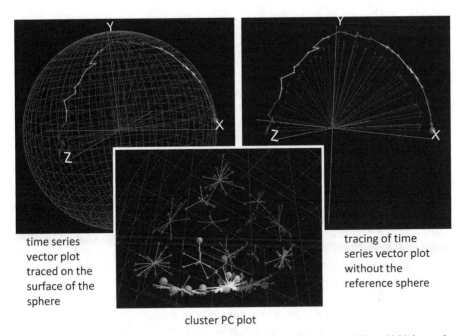

time series vector plot traced on the surface of the sphere

tracing of time series vector plot without the reference sphere

cluster PC plot

Figure 5.19 Time-series vector plot of 49 consecutive years from 1960 to 2000 for male names in Hawaii at the end of the 20th century. The path of the vectors for each decade are traced on the surface of the reference sphere, which is included on the left but left out on the right. Below is a scatter plot of the 320 most frequent male names in Hawaii during those years, grouped into 15 clusters. (Image courtesy of Metrika (metrika.org), © Appzilla, LLC.)

understand—merely the ups and downs of popularity of a particular name or group of names across a time sequence. They have semantic immediacy. And incorporating the line graphs into the scatter plot, as shown in Figures 5.6 and 5.7, helped to give that same kind of concrete meaning to the various locations within the scatter plot space.

We will not go into that kind of detail with these three sets of names, the Austrian, the French, and the Hawaiian. We will simply make a few comments about the implications of their unique and varied vector plot patterns. Notice that in each of these figures, we have included the cluster PC plot to the side of each vector plot, with the expectation that the pattern of name clusters can be drawn upon to help to explain the interpretation of the time series pattern of vectors, given that the two are linked.

We will begin with the Austrian male names vector plot, Figure 5.17. It is an esthetically pleasing vector plot, with a softly arching shape revealing the pattern of 24 consecutive years of male names in Austria. The pattern looks a little like a reversed question mark, with the first time point at the bottom of the shape (near the x-axis) corresponding to the year 1950. The vector for each successive year progresses upward and then off to the left, and then back to the right and ending with 1999 at the top of the y-axis.

This three-dimensional pattern can be understood as a variation on the theme that was found in the 19th century male names data in Figure 5.3 at the beginning of the chapter, the simple 90° arc. The thing that makes this differ from that 19th century pattern is that we could account for a high proportion of the variance in the American names with a two-dimensional space, whereas the Austrian names require a three-dimensional space as shown in the cluster PC plot of the right panel of Figure 5.16. The arch scribed by the vector pattern indicates to us that the clusters of names on the left hand side of figure tended to increase in popularity during the late middle years of the 24 year period, pulling the vectors for those years off to the left of the semantic space.

The mechanics of the data set for the French names are in contrast to the Austrian names in two ways. First, they are female names, whereas the Austrian names are male. Second, the data are organized into decades rather than years, resulting in a time series vector plot with a more gross structure. The pattern of vectors in Figure 5.18 would be more esthetically pleasing were we to insert the many vectors (over 100) for the individual years that make up each decade.

The time series begins in the decade ending in the year 1900, with the first vector pointing to the right of the space in the time series vector panel of Figure 5.18. The vectors then move clockwise around the front-middle of the sphere and around the back to complete a circle. They progress quickly at first, in 45-degree jumps. Then, in the last decades of the century, they progress in smaller increments until they have almost scribed a complete circle. However, on closer look, it is apparent that rather than being a circle, it is a spiral, ending up higher in the space than it began.

The idea of defining the semantic space with a circle of vectors is an intriguing one. Vectors on one side of the circle have a high negative correlation with vectors on the other side, indicating that names that were popular 50 years ago are now unpopular. This suggests profound and sweeping, but slowly moving changes in the fashionability of names. In the cluster PC plot panel of Figure 5.18, we see the six clusters of names that map into this space, and a single outlier name that looks like a moon in the in the far upper right—the name Marie that maintains popularity throughout the century.

How very interesting that the pattern would begin to cycle again like this after approximately one century, implying an unconscious century clock in the subtle social forces[23] underlying this pattern. Naming patterns can be viewed as "projective" cultural artifacts that provide clues and reflections of the particulars of human culture.

Comparing the semantic spaces of names in various cultures is a little like collecting insects in the sense that it is a taxonomic and highly empirical science, probably not easily driven by theory. There seems to be early evidence

---

[23] This is reminiscent of sociological studies indicating (with some controversy) that persons have a statistical tendency to wait until after their birthdays to die (Phillips and Feldman, 1973; Kunz and Summers, 1979–1980).

of a wide variety of patterns over time. However, our interest is not in the content of these studies except for what they can reveal to us about method, and the process by which time series vector plots can be used to explicate structure.

We will close with a consideration of the structure of 321 male names in Hawaii during the 49 years from 1960 to 2000. In some ways, this is the most interesting pattern of all in that it gives clear evidence of three separate "models" of name popularity, each of which exerts its influence for a time. The first model begins with the vectors pointing to the right, corresponding to the *x*-axis. The pattern then moves up through a 90° arc to the top of the figure near the *y*-axis. Up to this point, the pattern is much like that of the 19th-century male names. But then the vectors for years move in a direction 90° away from the first arc. This creates a second arc that is 90° away from the first one and ends up at the *z*-axis for the final vector, the one corresponding to the year 2000.

This time series pattern can be used to explain the name clusters in the cluster PC plot panel of Figure 5.19. The clusters on the far right are those names that were popular in 1960, with those at the top being popular in 1990 and those over on the left of the space being popular at the end of the century. Notice that with more vectors in the plot, we have changed our method of graphing. We keep all the vectors grey in the center and locate the brighter end points of each vector on the surface of the reference sphere, which creates a "trace line." The segments on the trace line correspond to 10-year periods. Time series plots of this kind work well with many kinds of complex and high-resolution waves, including those produced by EEG recordings.

5.4.8 PCP Time-Series Scatter Plots

The organizing properties of a sequential time series can be used to order vectors, with time point considered as a variable, as just demonstrated in Section 5.4.7. Time series organization can also be used to organize a scatter plot of data points. For example, we could create a three-dimensional space defined by three types of financial market indices. Perhaps the vertical factor (*y*-axis) could be defined by several metrics of the Dow Jones Industrial Average. The horizontal factor (*x*-axis) could be defined by several similar metrics of the Nasdaq, and the third factor (the *z*-axis) could be defined by the long term stabilized effects of market growth and volume. The data points could now be 650 trading days over a 3-year period.

This very figure is in fact created as Figure 6.5 of Chapter 6, and it shows the time series tracing of the 650 trading days surrounding the market crash in the spring of 2000. The tracing that is created has an identifiable and memorable shape, like a three-legged stool, that invites interpretation as to the dynamics of the market crash. Rather than reproduce that figure and explain it here, the reader is referred to Figures 6.4 and 6.5 and the accompanying explanation in Chapter 6.

5.4.9   PCP Vector Plots for Linked Multivariate Data Sets

This plot by definition requires a data analysis method that can connect multiple sets of multivariate data to one another, perhaps connecting an X set of variables to a Y set, or perhaps even three or more sets. Canonical correlation deals with the combining of two sets of multivariate variables and explicating their relationship to one another. Structural equations modeling is a more general analytical approach that can handle the relationships among three or more sets.

Figure 6.1 of Chapter 6 shows how to use a *vector plot for linked multivariate* data to examine the relationship between general interpersonal orientation (the X set of variables) and various measures of prejudice (the Y set of variables), in data from a published study.

5.4.10   PCP Scatter Plots for Linked Multivariate Data Sets

The linking between two or more multivariate sets of variables can be examined in the linked scatter plots, as well as in the linked vectors, discussed in the previous section. This type of graph is shown in the combined information from Figure 6.5 and Figure 6.6 of Chapter 6. The linking in this study is between an X set of market indices (Dow Jones Industrial Average, Nasdaq, etc.) and a Y set of performance variables on TIAA-CREF mutual funds. Canonical correlation creates multidimensional spaces for these two sets of variables that have the maximum possible correlation with one another. One can then put the two spaces in juxtaposition to one another to examine convergences and divergences in pattern as discussed in the accompanying text for these two figures.

5.4.11   Generalized Draftsman's Display

The idea of a generalized draftsman's display is simple—in any size correlation matrix, say a 9×9, one could place a bivariate scatter plot in the position of each of the $N(N—1)/2$ correlation coefficients, so that rather than having a matrix of coefficients, you have a matrix of bivariate scatter plots as shown in Figure 5.20. An actual three-view drawing created by a draftsman contains a front view, a side view, and a top view. This is often used to represent an object in three-dimensional space.

This process can be generalized so that there are more variables than three in the display. Stata refers to this as a "graph matrix," and it is one of their most useful tools. Consider the display of Figure 5.20. This is a generalized draftsman's display, or a graph matrix, of the scatter plots of 132 wives' responses to nine scales in the Flourishing Families data set. The command for creating this matrix of scatter plots is about what you might expect it to be, just the term "graph matrix" plus a listing of the variables to be included:

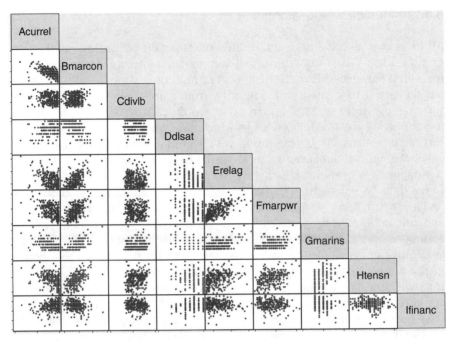

Figure 5.20   Generalized draftsman's display of the scores of 132 wives on nine romantic partner scales of the flourishing families study. This graph is created by the "graph matrix" command of Stata. The nine scales are: current relationship, marital conflict, division of labor, relational aggression, marital power, marital instability, tension, and finances. (StataCorp, 2009. Stata: Release 11. Statistical Software. College Station, TX: StataCorp LP.)

```
graph matrix Acurrel Bmarcon Cdivlb Ddlsat Erelag Fmarpwr
Gmarins Htensn Ifinanc
```

This Stata tool is like the biplot in that it is handy, easy to use, and quickly creates a great deal of useful diagnostic and/or data analytic information. It is a good way to get a fairly complete picture of the Gestalt of all of your data within a many-dimensional space. You can see bivariate normality or departures therefrom quite well with this plot, and you see it for all possible pairs.

Notice in Figure 5.20 that two of the variables in this particular set of nine show gaps and are not quite continuous data. These are the data for the Division of Labor Satisfaction Scale (Ddlsat), which only consists of two items, and the Marital Instability (Gmarins) scale that only consists of three items. This is a clear illustration that it is important to create scale scores that combine enough items to let the central limit theorem do its work (perhaps five or more items). As you summate over more items, the resulting scale scores are more likely to begin to fit the necessary assumptions such as multivariate normality.

5.4.12 Multidimensional Scaling

Multidimensional scaling is quite different from the preceding multivariate graphical tools. It is not a principal-component plot, and it does not need measured variables. Much of the appeal of multidimensional scaling is that it has the power to create a scale implicitly from paired comparison judgments, for example, from preferences. In other words, rather than I as the experimenter giving you the dimensions by which you judge a certain set of "things" (artifacts, persons, etc.), I can, using your paired comparison judgments, discover the implicit dimensions you are using for your judgments.

Suppose, for example, I give you a set of five apples to taste and evaluate. I give them to you in 10 pairs (the number of non-redundant possible pairs) and ask which apple you prefer and how strongly. Given certain assumptions and calculations, your preferences can be analyzed with multidimensional scaling, and we can create a two- or three-dimensional map of your semantic space for the five apples according to their taste. By examining the topology of the map, we can perhaps identify the "natural" dimensions you are using for judgment. For example, we might find that your map has all the sweet apples on the right, tart on the left, and all the soft apples at the bottom, and crisp ones on the top, and we see that your implicit scales of judgment center on two dimensions, tartness versus sweetness, and softness versus crispness.

It's a little like asking persons for distances between cities in the United States and then metrically constructing a map from those distances. In fact, that is one of the classical demonstrations of the original *metric*, or *classical,* multidimensional scaling algorithm, and of course it always produces a map of the United States, although in various rotations, and sometimes flipped left to right.

The map-from-distances problem, however, is a much simpler one to solve than the problem of creating an implicit semantic space from human judgments, since human judgments are seldom truly metric in the way that intercity distances are. Suffice it to say that Shepard (1962) and Kruskal (1964) solved the problem by the development of a measure of "badness of fit" that Kruskal calls STRESS, and there are now many models of *nonmetric* multidimensional scaling that can accomplish maps from human judgments.

Lattin et al. (2003) give the history of the development of both the *classical* model of multidimensional scaling (pp. 211–216) in which Torgerson (1958) played the major role, and also the *nonmetric* method of multidimensional scaling (pp. 219–225) in which Shepard (1962) and Kruskal (1964) were the major players. The Lattin et al. (2003) source is a good one for a terse but accurate overview, the authors themselves having contributed to the MDS literature. Carroll himself was a major part of the development of MDS as the author of INDSCAL, a method for individual differences multidimensional scaling (Carroll and Chang, 1970). For those interested in learning more about multidimensional scaling, much of it is summarized in a two-volume edited set of books (Romney et al., 1972; Shepard et al., 1972).

An example of the use of multidimensional scaling in the published literature is given in Figure 5.20. This is taken from a paper by Segerstrom et al. (2003). Although it is published in the *Journal of Personality and Social Psychology*, it has a decidedly clinical theme—the relation to one another, indeed the semantic structure, of a number of personality measures in concert with a number of negative behaviors (worry, rumination, depressive rumination, repetitive thoughts, etc.) that are disruptive of mental health. They administered a large battery of tests, consisting of six repetitive thoughts (RT) measures and five personality measures. The subjects were 978 undergraduate students at three universities. The personality scales, including the NEO Five-Factor Inventory and several additional personality scales and RT scales, were factored to produce multiple scale scores, producing the eighteen that are shown in Figure 5.21.

The manner in which Segerstrom et al. (2003) apply multidimensional scaling to their data is unusual and creative, and shows the breadth of possible applications for this useful and general multivariate graphing method. Rather than using paired comparison judgments as is often the case, they used multidimensional scaling to combine scale scores and factor scores from their battery of measures into a common two-dimensional graph showing the pattern of similarities and oppositions, and suggesting a super-ordinate structure by which these 18 measures can be organized.

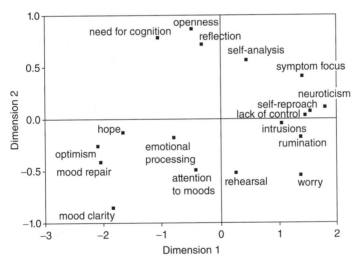

Figure 5.21 Multidimensional scaling two-dimensional spatial representation of 18 repetitive thought and personality measures from figure 1 of the Segerstrom et al. (2003) paper. Mood repair, mood clarity, and attention to moods in the lower left area relate to emotional intelligence; rehearsal and lack of control relate to the rumination scale; and self-analysis, symptom focus, and self-reproach relate to the response style questionnaire.

In a general way, it could be said that the horizontal dimension has negative affect behaviors on the right and positive on the left, and that the vertical dimension has problem-solving (closing off) behaviors at the bottom, and opening up behaviors at the top. Their summary of the vertical scale is that it "appears to reflect searching for new ideas and experiences versus solving problems and improving certainty and predictability." The semantic space produced by this process is interesting and instructive in many ways.

5.5   FLOURISHING FAMILIES: AN ILLUSTRATION OF LINKED GRAPHICS AND STATISTICAL ANALYSES IN DATA EXPLORATION

In the winter of 2005, five School of Family Life professors at Brigham Young University began planning for a major, long-term data gathering process "to assess core family strengths in couples and families" using a multi-method approach (Day et al., 2011). Over 6000 responses are gathered from each family each year, with hundreds of families from two western urban areas (Seattle, Washington, and Provo, Utah) participating in the study. Each family in the initial wave had one or more children between the ages of 7 and 16. The study has now been conducted in four yearly waves of data gathering, from 2007 to 2010. In this section of the chapter, we will draw upon the Provo, Utah, data from Wave II of the study to demonstration the application of multivariate graphics to family science data. A number of studies have been published and are in process based upon this rich database, dealing with such topics as relational aggression in marriage (Carroll et al., 2010), adolescent parenting (Day and Padilla-Walker, 2009), and other current issues in family science.

Our particular use of the Flourishing Families data set in this chapter will be to demonstrate cluster-group principal-components plots. The approach is to first use cluster analysis to identify reasonably homogeneous subgroups of couples and then principal-components plots to display them. The primary data set consists of the joint responses of 132 couples from Provo Utah on 12 Romantic Partner scales from Wave II of the Flourishing Families database. The 12 scales are (1) Current Relationship Quality, (2) Marital Conflict, (3) Division of Labor, (4) Division of Labor Satisfaction, (5) Relational Aggression, (6) Marital Power, (7) Marital Instability, (8) Couple/Parenting Spillover, (9) Couple Financial Communication, (10) Couple Communication (Gottman), (11) Triangulation, and (12) Marital Love.

The data were first cluster-analyzed using the complete-linkage hierarchical clustering method (hclust) of the open-source statistical package R. Figure 5.22 is the dendrogram of the cluster groupings of the 132 couples. The bar across the dendrogram shows the point at which the 132 couples were divided into 13 couple groups.

Both the cluster analysis and also the principal-component method factor analysis were calculated on the same input data matrix. The matrix has 132 rows, one for each couple, and 24 columns—12 columns containing the scale

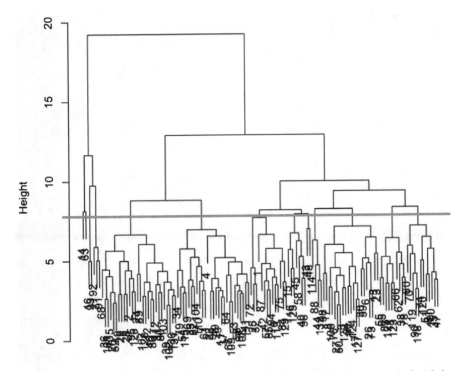

Figure 5.22 Dendrogram from complete linkage hierarchical cluster analysis of the 132 couples from the Flourishing Families data set, wave 2, within the space of the couple scores on the 12 Romantic Partner Scales. (Data courtesy of Day et al., 2011.)

scores of the wife and 12 columns containing the scale scores of the husband. The rationale of this approach is to capture the overall pattern of the dyadic relationships between the couples within the latent variables space.[24] This cluster analysis step has given us 13 groups into which the 132 couples can be organized in preparation for factoring and creating the principal-components plots.

The vector plot for the factor analysis of the 24 scale scores is shown in Figure 5.23. Actually, it is two vector plots, the top two panels giving the front view and the side view of the graph for a varimax (orthogonal) factor rotation, and the bottom two panels giving the front view and the side view of the graph for an oblimin (oblique) factor rotation. In the varimax rotation, the two

---

[24] The data were also analyzed by the more usual approach of having 264 rows of data, one for each person, and 12 columns of Romantic Partner scale scores. This converts it from a 132×24 matrix of 132 dyadic units with both "his" and also "her" scores appearing on each line, to a 264×12 matrix of individual person observations (rows) with only 12 variables (columns). One can then connect the data point of each wife to that for her husband. The two approaches produce different patterns of results. Both approaches are informative, but the dyadic approach fits better our purposes in this chapter.

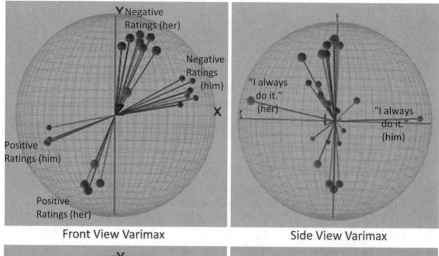

Figure 5.23   Vector plot of the 24 scale scores (12 for wife and 12 for husband) of the romantic partner set of scales from the Flourishing Families wave 2 data. (Data courtesy of Day et al., 2011. Image courtesy of Metrika (metrika.org), © Appzilla, LLC.)

"bundles" of vectors, the vertical one and the horizontal one, are seen to be almost orthogonal, but somewhat correlated with each other. The oblimin rotation places the two factors in the geometric center of the vector bundles, thus giving the factors a more empirically correct interpretation, as representing the bundles of vectors surrounding them.

The two bundles of vectors in the front view of Figure 5.23 are the scale score ratings by the wife (the mostly vertical bundle) and the scale score ratings by the husband (the mostly horizontal bundle). Interestingly, even though the topics of the 12 scales seem quite diverse, they empirically fall quite close together within the vectors for females into a kind of "good marriage" versus "bad marriage" factor for wives, with a similar bundle of vectors for husbands.

Seven of the 12 scales are worded in such a way that a high score indicates a more unhappy marriage. These are the seven vectors at the top of the figure for the bundle of vectors for wife ratings and the bundle of seven vectors at the right for the husband ratings. The four vectors at the bottom (wife) and four at the left (husband) are for the four scales on which high numerical ratings indicate a happy marriage.

It could be stated that the vertical axis in this latent variable space represents "her" evaluation of the marriage, with data points at the top representing a high conflict and low satisfaction marriage, and data points at the bottom representing a low conflict and high satisfaction marriage. Similarly, the horizontal axis represents "his" evaluation of the marriage, with data points on the right representing those couples where the husband sees the marriage as high in conflict and low in satisfaction, and data points on the left representing low conflict and high satisfaction.

The two factors are reminiscent of Bernard's (1972) view of spouses often holding differing and contrasting perceptions of marital life: "her marriage" and "his marriage." For those couples in the upper right quadrant of this latent variable space, the two views of the marriage are consonant—both partners view it as relatively unhappy. Similarly, those in the lower left are consonant. Both view it as relatively happy. Couples that fall in either of the other two quadrants have differing views of the marriage. In the upper left quadrant are those couples where the wife perceives the marriage as relatively unhappy but the husband views it as happy, and in the lower right is the reverse of this, where the husband sees it as relatively unhappy and the wife happy.

We have only accounted for 22 of the 24 scale scores. The others are the two Division of Labor scales, the one for her ratings and the one for his. They actually define the third factor in this latent variable space, as shown in the side view of Figure 5.3. "Her" scale score on the Division of Labor scale is the one extending to the left. It loads highly on the positive end of the third factor, the z-axis. "His" scale score on Division of Labor extends in an opposite direction to the right. It loads highly on the negative end of the third factor. This indicates that data points that are on the positive side of the z-axis represent couples where she says "I always do this" (and he tends to agree), and those on the negative side of the z-axis are for couples where he says "I always do this" (and she tends to agree).

Clearly, there is a high negative correlation between her rating and his rating on the Division of Labor Scale. However, because of the logic of the situation, this means that they have a strong tendency to *agree*. That is, when she has a high rating on "I always do this" he has a low rating on his scale, indicating "She always does this." The converse is also true.

It must be mentioned, however, that it is a much rarer occurrence for both of them to agree that he is always doing the tasks than for them to both agree that she is doing them, as shown in Table 5.1. This table is a *paired dyadic*

TABLE 5.1. Table of Summated Values for the Dyadic Shared Reality Matrices of 143 Couples in the Provo Location, Wave 2, of the Flourishing Families Database (Day et al., 2011)

			Her Ratings			
		1	2	3	4	5
		"He always.."		"Equally"		"I always.."
His ratings	5 "I always.."	77	40	24	2	7
	4	50	85	117	36	7
	3 "Equally"	14	45	639	269	98
	2	1	11	135	293	347
	1 "She always.."	1	3	40	136	347
						Total = 2711

response matrix summed over 143 couples.[25] Notice that of the 2711 joint dyadic responses recorded in this table, 347 of them (12.8%) are indications of both of them agreeing that she always does the tasks. Conversely, there are only 77 summed joint responses of both of them agreeing that he always does the task (2.8%).

Occasionally, the husband and wife will arm-wrestle, with each of them contending "I am the one doing household things." However, as shown in Table 5.1, this is a fairly rare event. When it does happen, the couple tends to have conflict.

We will now examine the factor scores for this analysis to locate the 132 couples within this latent variable space. It will be instructive to compare the factor scores from the varimax analysis with those in the oblimin analysis. Figure 5.24 shows a "graph matrix" plot and a correlation matrix from Stata for the three factors in each of these approaches to the analysis. The correlation matrix and the graph matrix for varimax factor scores (orthogonal) are in the left panel. The correlation matrix and the graph matrix for the oblimin factor scores (oblique) are in the right panel. Notice that the factor scores are uncorrelated in the varimax rotation, but they are correlated in the oblimin rotation. Now that the two have been compared, for pedagogical purposes, we will use the oblimin rotation throughout the remainder of the chapter.

Figure 5.25 is the oblimin principal components factor score plot (with star plot structure) that shows where the means (indicated by spheres) and also the individual couples (indicated by plus signs) are for the 13 clusters of the 132 couples. Only the front view will be shown. There are two outlier couples

---

[25] Not all of the 143 couples who were analyzed with the dyadic shared reality matrices had complete data on the other 11 scales. Those who did have complete data constitute the 132 couples whose data appear in the other analyses of this section (Figs. 5.22–5.27).

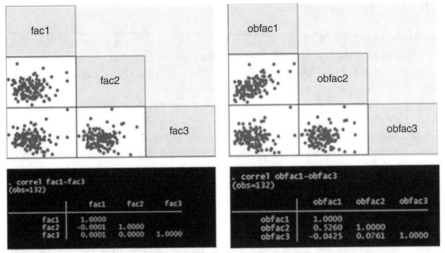

Graph matrix and correlation
matrix for varimax rotation.

Graph matrix and correlation
matrix for oblimin rotation.

Figure 5.24 Correlation matrix of the three factor scores of the *Romantic Partner Scales* from a Varimax-rotated solution compared with an Oblimin-rotated solution. Also, a generalized draftsman's display comparison of the two. (StataCorp, 2009. Stata: Release 11. Statistical Software. College Station, TX: StataCorp LP.)

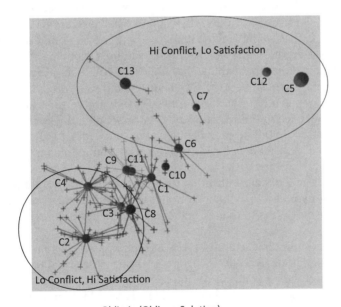

Oblimin (Oblique Solution)

Figure 5.25 Scatter plot in star plot form of each of the 132 couple (shown as + signs) from the flourishing families romantic partner scales data set. Each couple is connected to its individual cluster group mean (shown as a sphere) with a line. The 13 clusters were constituted from the dendrogram in Figure 5.22, and the points are plotted within the latent variable space defined by the 24 vectors of Figure 5.23. (Data courtesy of Day et al., 2011. Image courtesy of Metrika (metrika.org), © Appzilla, LLC.)

on the right side of the figure in the upper right quadrant. They are represented with spheres rather than plus signs because they are clusters consisting of only a single couple each. Their extreme locations in this space indicate they are the two couples least satisfied with their marriage.

If you draw an imaginary diagonal line from the bottom left corner of the figure to the upper right, that line represents equal ratings of their marriage by the husband and the wife. One of these two outlier points is clearly very close to that line, indicating that they are in agreement about their marriage being bad. The other outlier point, however, is below that line, indicating that the husband sees the marriage more negatively than does his wife.

Another important observation is the cluster of three couples in the upper left of the figure. This location indicates that the wife sees the marriage quite negatively, but the husband does not. These internal patterns within the data have much in common with qualitative research. The process can be quite tedious, but it can lead to a deeper and more precise understanding of the results of a study, substantially deeper than just examining the overall correlations among variables. It can lead us to look for meaning in the data.

In wave II, a new scale was added for the children of these couples, a Hope scale adapted from work of Peterson and Seligman (2004). It includes hope and optimism items, such as "I always look on the bright side," "I have a plan for what I want to be doing five years from now," and "If I feel down, I always think about what is good in my life." One of the surprises of the analysis of this data set is that we have not found a strong or compelling statistical relationship between the scales we have just examined (couple satisfaction) and child optimism. One might think that children would be affected negatively by parental conflict and positively by a strong and satisfying marriage.

This question can be asked pictorially by looking up the Hope scores for the children of each of the 132 couples. The scatter plot of Figure 5.25 is reproduced in Figure 5.26 but in three forms, panel *a* indicating (with spheres) those couples who have a child with a hope score above three (on a five-point scale), panel *b* indicating those couples having a child with a Hope score between two and three, and panel *c* indicating those couples having a child with a Hope score below two. The first thing to notice is that there are more couples in either panel *b* or panel *c* than in panel *a*—the higher hope scores are more rare.

The second thing one can observe is a trend in panel *a* consonant with the expected "child hope hypothesis." There is a noticeable visual tendency for the parents of high Hope children to not be found in the high-conflict locations of the latent variable space. There are high Hope children, however, among some of the average couples in the middle of the figure.

There are two times when multivariate graphics are particularly useful for a study. The first is when all of the significance tests are positive. One knows something is going on, but one needs a visual tool to help explain *what* is happening. The second case is when one fails to identify a significant relationship, but there is still something heuristic in the pattern of the data. This is one of

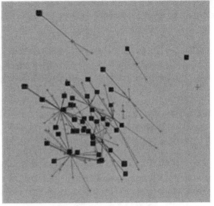

(a) Hope score of 3 or more

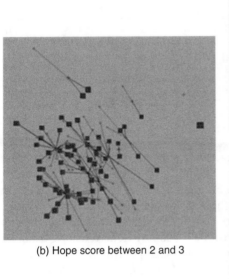

(b) Hope score between 2 and 3

(c) Hope score below 2

Figure 5.26   Three views of the Flourishing Families star plot from Figure 5.25 showing the location of families with a child at a given level on the hope scale. In panel a, those families having a child a high hope score (3 or above) are indicated by spheres, with the remaining families represented with a small + sign. In panel b, the medium hope scores are represented with cubes and in panel c those with low hope scores are represented with cubes. (Data courtesy of Day et al., 2011. Image courtesy of Metrika (metrika.org), © Appzilla, LLC.)

the latter cases. It is of interest to note that relatively high levels of optimism in children (above an average rating of three on a 5-point scale) can occur in a fairly wide range of families, extending from those who rate their marriage as very positive to those who rate it quite average.

Table 5.2 is helpful in giving us a better understanding of the relative distribution of child hope and the relative distribution of positive and negative scores among parents on the Romantic Partner Scales. The columns of this two-way table of frequencies are the three groups of parents identified in Figure 5.25: the five clusters of parents in the upper right of the figure, the "hi conflict, lo

TABLE 5.2. Two-Way Contingency Table of the Relationship between Child Hope and Romantic Partner Scale Dyadic Scores

	Parent Dyadic Scores			
	Most Positive Ratings	Middle Group	Most Negative Ratings	
Score not above three	74	29	13	116
Child hope score above three	12	3	1	16
	86	32	14	132

satisfaction" area of the graph; the four clusters of parents in the lower left, the "lo conflict, hi satisfaction" area of the graph; and the four clusters of parents in the middle. We can see from Table 5.2 that well over half of the couples in the study (65.2% to be exact, 86 of the 132) are in these four most positive clusters of the graph. Truly, they are flourishing families. However, it seems that their children may not be flourishing to quite the same extent. When we examine the distribution of the 132 families according to the hope scores of their children, only 16 children of the 132 have hope scores of three or above on a five-point scale. It must be remembered also, that a three is right dead center in a five point scale. It may be that what these numbers are saying is that adolescence and also preadolescence is a difficult time for many of our youth, and it is not always easy for them to maintain optimism and a heads up attitude.

There is clearly a frequency imbalance in both the rows and also the columns of this table, but it goes different directions for parents than for children. The parents have a substantial positive imbalance in their marriage ratings, but the children have a negative imbalance in their ratings of hope and optimism. We analyzed this contingency table with a loglinear analysis in order to see which of the three aspects of the table are most salient: rows, columns, or interaction. One of the useful properties of the likelihood ratios in log-linear analysis is that they are absolutely additive. That is, the likelihood ratio statistic G^2 for rows, the G^2 for columns, and the G^2 for row-by-column interaction, will all sum to the G^2 for the total matrix (a measure of departure of all cell frequencies from an expected value of all of them being equal). When we do this, we find that the G^2 for rows (child hope) is 57.3% of the total value, while the G^2 for columns (parent RPS scores) is 42.1%, and the G^2 for the interaction between the two is 0.6%. In other words, this contingency table does not have a strong predictability of child hope from parent marriage satisfaction on these scales. We will return to the log-linear analysis of this Flourishing Families data in Chapter 9, Section 9.6.

We also analyzed the contingency between parent RPS scores and child Hope scores using logistic regression. The strength of logistic regression in a data configuration like this is that with odds ratios, we can answer the question "How much more likely is a Hope/Optimism score over three for those in cluster A than for those in cluster B?" In other words, we first identify the "odds" of an "above three rating" child being found in each of the clusters, and then compare clusters in terms of their ratios of odds. Interestingly, of the 13 clusters, seven of them (primarily in the upper right quadrant) have no child with an above-3 Hope score. The least of those that do have one is cluster C8, which has only one. Since there are 10 couples in cluster C8, one having a high Hope child and nine not having one, the odds are 1 to 9 of having a high Hope child in that cluster.

These results are given in Figure 5.27, which could perhaps be called an "odds ratio clustered principal components plot." The appropriate odds ratio is written next to each cluster. Notice first the location of those seven clusters

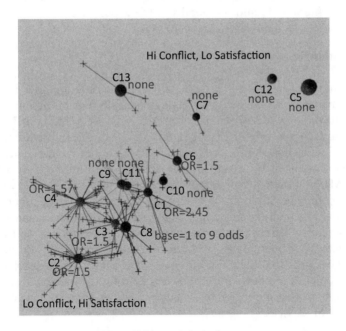

Oblimin (Oblique Solution)

Figure 5.27 Scatter plot in star plot form showing all 132 families in their 13 cluster groups with the odds ratio of each written in red. Seven of the thirteen groups had no children with a hope score above 3 on the 5-point scale, and they are marked by "none." Of the other six, cluster C8 has the lowest odds of a score above three, at "1–9." This cluster is taken as the base and the odds ratios of the remaining five groups who have at least some "hopeful" children are calculated from that. For example, several groups have odds ratios of 1.5, which means they have odds of "3 in 18." (Data courtesy of Day et al., 2011. Image courtesy of Metrika (metrika.org), © Appzilla, LLC.)

that do not have a child with a high Hope score. It includes four of the five clusters in the upper right quadrant, all but cluster C6. Of those clusters that do have at least one, the lowest odds are in cluster C8, as discussed above, so C8 becomes the comparison cluster, the base cluster, with odds of 1 to 9. There are three clusters (C3, C2, and C6) that have odds 1.5 times as great as C8 (odds of 3 out of 18), so their Odds Ratio scores are each 1.5. Cluster C4 has an Odds ratio of 1.57, and cluster C1 has an Odds Ratio of 2.45. That means that the odds of having a high hope child in cluster C1 are almost two-and-one-half times as great as they are in the comparison cluster, C8.

This OR cluster graph begins moving us into the deeper look of qualitative research, but combined with quantitative-based tools. What kind of narratives, for example, would we obtain from the three high-Hope children in cluster C3? What kinds of narratives would we have from low Hope children who are in a family with high relationship satisfaction scores? In this human research enterprise, we are, after all, not just correlating variables; we are trying to understand human life and the depths of life experience.

5.6 SUMMARY

The advantage of multivariate methods over univariate methods is that multivariate methods can deal with the data holistically. Given the importance of graphics to the progress of science, it makes sense to create holistic graphing methods to accompany multivariate significance tests, and multivariate descriptive statistics (such as multivariate R^2 values). Quantitative biobehavioral and social scientists have been active in developing such methods over the past century. Many of them are principal components based (the many varieties of principal components plots), but there are also a number of others, such as cluster analysis, multidimensional scaling, generalized draftsman's displays, Andrews plots, Chernoff faces, surface plots, trellis graphics, and a host of others. Some have been presented here, but a large number of additional ones have not. We hope that the point has been made that the data analyst has only completed half of the task, or perhaps less than that, if the conclusions of the analysis are not also supported by one or more strong and convincing "graphical inscription devices."

STUDY QUESTIONS

A. Essay Questions

1. Explain how the number of graphs, tables and P-values are associated with each other and with the "hardness" of the science being presented in journal articles.

2. Describe each step of the following list of progressive graphical analysis processes and specify what is achieved by the addition of graphical information at each step. Cluster Analysis/Dendrogram $\rightarrow$ Vector Plot $\rightarrow$ Means Plot $\rightarrow$ Star Plot $\rightarrow$ Profile Plots

3. In what ways is it helpful to factor a data set based on means instead of on individual values? In what ways can it be inappropriate or not helpful?

4. Describe Gabriel's biplot, specifically mentioning what the "bi" stands for.

5. In what cases is it helpful to use standardized scores for PCP plots, such as the biplot, rather than raw scores, and when are raw scores more appropriate?

6. How is the R^2 value relevant for deciding whether to look at a plot in two or three dimensions? What kind of R^2 value would indicate that a plot of the data may not capture the important aspects of the data?

7. Describe what multidimensional scaling accomplishes and what type of data it utilizes.

B. Computational Questions

1. Use Excel to calculate a principal component analysis on the Lee et al. data of Figure 5.11. Create your own biplot and compare it to the Stata biplot in that figure.

2. Use Excel to create a bivariate scatterplot of the first two variables (alpha and beta) in the left panel data of Figure 5.10. Compare it to the biplot of that figure.

C. Data Analysis Question

1. Use Stata to recreate the biplot of the Lee et al. data of Figure 5.11. Also create a graph matrix for these data and discuss the convergent information.

REFERENCES

Arabie, P., Hubert, L. J., and De Soete, G. (Eds.) 1996. *Clustering and Classification*. River Edge, NJ: World Scientific Publishing.

Bernard, J. 1972. *The Future of Marriage*. New York: Bantam Books.

Brown, B. L. 1969. The social psychology of variations in French Canadian speech styles. (Unpublished doctoral dissertation). McGill University.

Brown, B. L. 2010. Descriptive statistics. In Salkind, N. J., Dougherty, D. W., and Frey, B. (Eds.), *Encyclopedia of Research Design*. Thousand Oaks, CA: Sage.

Brown, B. L., and Hedges, D. W. 2009. Use and misuse of quantitative methods: Calculation and presentation of results. In Mertens, D., and Ginsberg, P. (Eds.), *Handbook of Social Science Research Ethics.* Thousand Oaks, CA: Sage.

Brown, B. L., Strong, W. J., and Rencher, A. C. 1973. Perceptions of personality from speech: Effects of manipulations of acoustical parameters. *The Journal of the Acoustical Society of America, 54,* 29–35.

Carroll, J. D., and Chang, J.-J. 1970. Analysis of individual differences in multidimensional scaling via an N-way generalization of "Eckart-Young" decomposition. *Psychometrika, 35,* 283–319.

Carroll, J. S., Nelson, D., Yorgason, J. B., Harper, J. M., Hagmann, R., and Jensen, A. 2010. Relational aggression in marriage. *Aggressive Behavior, 36,* 315–329.

Chambers, J. M., and Kleiner, B. 1982. Graphical techniques for multivariate data and clustering. In Krishnaiah, P. R., and Kanal, L. N. (Eds.), *Handbook of Statistics, Vol. 2: Classification, Pattern Recognition and Reduction of Dimensionality.* New York: North-Holland.

Chambers, J. M., Cleveland, W. S., Kleiner, B., and Tukey, P. A. 1987. *Graphical Methods for Data Analysis.* Belmont, CA: Wadsworth.

Cleveland, W. S. 1984. Graphs in scientific publications. *The American Statistician, 38,* 261–269.

Cleveland, W. S. 1993. *Visualizing Data.* Murray Hill, NJ: AT&T Bell Laboratories.

Cleveland, W. S. 1994. *The Elements of Graphing Data,* revised edition. Murray Hill, NJ: AT&T Bell Laboratories.

Cleveland, W. S., and McGill, M. E. 1983. *Dynamic Graphics for Statistics.* Belmont, CA: Wadsworth.

Day, R. D., and Padilla-Walker, L. M. 2009. Mother and father connectedness and involvement during early adolescence. *Journal of Family Psychology, 23,* 900–904.

Day, R. D., Bean, R. A., Coyne, S., Harper, J., Miller, R., Walker, L., Yorgason, J., and Dyer, J. (project authors), and Blickfeldt, S., Voisin, J., and Dickson, D. H. (report writers). 2011. *The Flourishing Families Project Progress Report, January 2011.* Provo, UT: Brigham Young University School of Family Life.

Everitt, B. S. 1978. *Graphical Techniques for Multivariate Data.* Amsterdam: North-Holland.

Everitt, B. S. 1994. Exploring multivariate data graphically: A brief review with examples. *Journal of Applied Statistics, 21*(3), 63–94.

Everitt, B. S., Landau, S., Loess, M., and Stahl, D. 2011. *Cluster Analysis, Fifth Edition.* New York: Wiley.

Gabriel, K. R. 1971. The biplot graphic display of matrices with application to principal component analysis. *Biometrika, 58*(3), 453–467.

Gabriel, K. R. 1995. MANOVA biplots for two-way contingency tables. In Krzanowski, W. J. (Ed.), *Recent Advances in Descriptive Multivariate Analysis.* Oxford: Clarendon.

Gabriel, K. R. 2002. Goodness of fit of biplots and correspondence analysis. *Biometrika, 89*(2), 423–436.

Gower, J. C., and Hand, D. J. 2011. *Biplots.* London: Chapman & Hall.

Gower, J. C., Lubbe, S., and leRoux, N. 2011. *Understanding Biplots.* New York: Wiley.

Hendrix, S. B., and Brown, B. L. 1988. Graphical manova: A global hyperspace analysis method. Paper presented at the 148th Annual Meeting of the American Statistical Association, New Orleans, LA.

Hendrix, S. B., and Brown, B. L. 1990. The surface of ordered profiles: A multivariate graphical data analysis method. American Statistical Association 1990 Proceedings of the Section on Statistical Graphics. Alexandria, Virginia: American Statistical Association.

Hirsh, N., and Brown, B. L. 1990. A holistic method for visualizing computer performance measurements. In Nielson, G. M., Shriver, B., and Rosenblum, L. J. (Eds.), *Visualization in Scientific Computing*. Los Alamitos, CA: IEEE Computer Society Press.

Huff, D. 1954. *How to Lie with Statistics*. New York: Norton.

Kruskal, J. B. 1964. Multidimensional scaling by optimizing goodness of fit to a nonmetric hypothesis. *Psychometrika, 29*, 1–27.

Kuhn, T. 1962. *The Structure of Scientific Revolutions*. Chicago, IL: University of Chicago Press.

Kunz, P. R., and Summers, J. A. (1979–1980). A time to die: A study of the relationship of birthdays and time of death. *Omega: Journal of Death and Dying, 10*(4), 281–289.

Latour, B. 1983. Give me a laboratory and I will raise the world. In Knorr-Cetina, K. D., and Mulkay, M. (Eds.), *Science Observed*. London: Sage; pp. 141–170.

Latour, B. 1990. Drawing things together. In Lynch, M., and Woolgar, S. (Eds.), *Representation in Scientific Practice*. Cambridge, MA: MIT Press; pp. 19–68.

Latour, B., and Woolgar, S. 1986. *Laboratory Life: The Construction of Scientific Facts* (Rev. ed.). Princeton, NJ: Princeton University Press.

Lattin, J., Carroll, J. D., and Green, P. E. 2003. *Analyzing Multivariate Data*. Belmont, CA: Brooks/Cole.

Lee, B. K., Glass, T. A., James, B. D., Bandeen-Roche, K., and Schwartz, B. S. 2011. Neighborhood psychosocial environment, apolipoprotein E genotype, and cognitive function in older adults. *Archives of General Psychiatry, 68*(3), 314–321.

Peterson, C., and Seligman, M. E. P. 2004. *Character Strengths and Virtues: A Handbook and Classification*. Washington, DC: Oxford University Press.

Phillips, D. P., and Feldman, K. A. 1973. A dip in deaths before ceremonial occasions: Some new relationships between social integration and mortality. *American Sociological Review, 38*, 678–696.

Popper, K. R. 1959. *The Logic of Scientific Discovery*. New York: Harper.

Porter, T. M. 1995. *Trust in Numbers*. Princeton, NJ: Princeton University Press.

Rencher, A. C. 2002. *Methods of Multivariate Analysis, Second Edition*. New York: Wiley.

Romney, A. K., Shepard, R. N., and Nerlove, S. B. (Eds.) 1972. *Multidimensional Scaling: Theory and Applications in the Behavioral Sciences, Volume 2, Applications*. New York: Seminar Press.

Roth, W.-M., and McGinn, M. K. 1997. Graphing: Cognitive ability or practice? *Science Education, 81*, 91–106.

Segerstrom, S. C., Stanton, A. L., Alden, L. E., and Shortridge, B. E. 2003. A multidimensional structure for repetitive thought: What's on your mind, and how, and how much? *Journal of Personality and Social Psychology, 85*(5), 909–921.

Shepard, R. N. 1962. The analysis of proximities: Multidimensional scaling with an unknown distance function. *Psychometrika*, *27*, 125–140; 219–246.

Shepard, R. N., Romney, A. K., and Nerlove, S. B. (Eds.) 1972. *Multidimensional Scaling: Theory and Applications in the Behavioral Sciences, Volume 1, Theory*. New York: Seminar Press.

Smith, L. D., Best, L. A., Stubbs, D. A., Johnston, J., and Archibald, A. B. 2000. Scientific graphs and the hierarchy of the sciences: A Latourian survey of inscription practices. *Social Studies of Science*, *30*, 73–94.

Smith, L. D., Best, L. A., Stubbs, D. A., Archibald, A. B., and Roberson-Nay, R. 2002. Constructing knowledge: The role of graphs and tables in hard and soft psychology. *American Psychologist*, *57*(10), 749–761.

Steele, J. M. 2005. Darrel Huff and fifty years of *How to Lie with Statistics*. *Statistical Science*, *20*, 205–209.

Torgerson, W. S. 1958. *Theory and Methods of Scaling*. New York: John Wiley & Sons, Inc.

Tufte, E. R. 1997. *Visual Explanations: Images and Quantities, Evidence and Narrative*. Cheshire, CT: Graphics Press.

Tufte, E. R. 2001. *The Visual Display of Quantitative Information, Second Edition*. Cheshire, CT: Graphics Press.

Tufte, E. R. 2003. Powerpoint is evil. *Wired*, *11*, 1059–1028.

Tukey, J. W. 1977. *Exploratory Data Analysis*. Reading, MA: Addison-Wesley.

Tukey, J. W. 1980. We need both exploratory and confirmatory. *American Statistician*, *34*(1), 23–25.

Wainer, H., and Thissen, D. 1981. Grahical data analysis. *Annual Review of Psychology*, *32*, 191–241.

Yan, W., and Kang, M. S. 2003. *GGE Biplot Analysis*. Boca Raton, FL: CRC Press.

CHAPTER SIX

CANONICAL CORRELATION: THE UNDERUSED METHOD

6.1 INTRODUCTION

Nearly 80 years ago, canonical correlation analysis was introduced by Harold Hotelling (1935, 1936). Essentially, it is a method for identifying the holistic relationship between two multivariate sets of variables, an obvious next step after factor analysis and principal component analysis made their appearance in the first decades of the twentieth century. Canonical correlation deals with a single sample on which two sets of multivariate measurements are made, an X set of measures and a Y set of measures. It is similar to multiple regression, and the multiple correlation index on which multiple regression is based, except that rather than predicting one Y variable from a set of X variables, it deals with the prediction of an entire set of Y variables from an entire set of X variables.

It can also be compared with factor analysis. At least in some applications of canonical correlation, it can be considered to be like running two factor analyses in parallel, one on an X set of variables and one on the Y set of variables, but with a "linking function" that enables one to uncover not only the internal structure of each set, but the relationship between the corresponding latent variables in the two sets. This "double factor analysis" way of considering canonical correlation also helps one to understand the mathematical and conceptual structure of the method. Like factor analysis and PCA, canonical

Multivariate Analysis for the Biobehavioral and Social Sciences: A Graphical Approach,
First Edition. Bruce L. Brown, Suzanne B. Hendrix, Dawson W. Hedges, Timothy B. Smith.
© 2012 John Wiley & Sons, Inc. Published 2012 by John Wiley & Sons, Inc.

correlation analysis also finds latent variables that are linear combinations of the observed/manifest variables. However, the criterion by which we obtain the linear combinations differs. Rather than seeking a factor or component that has the least amount of squared distance from the internal set of variables, canonical correlation seeks two linear combinations simultaneously, one on the X set of variable and one on the Y set of variables, subject to the criterion that the correlation coefficient between the two paired latent variables is *maximized*.

The correlation coefficient between the two is the *first canonical correlation*. To extract the second canonical correlation, all of the variance and covariance accounted for by the first is removed, and then the second is extracted by applying the same computations to the residual matrix. In other words, like PCA and some models of factor analysis, canonical correlation obtains latent variables that are orthogonal to one another. However, in canonical correlation analysis, the latent variables come in pairs, one in the pair coming from the X set and the corresponding one from the Y set, and there is a canonical correlation index indicating the strength of relationship between the two. The process will become clearer as we demonstrate the underlying calculations in Section 6.3.

It would seem that this would be a highly informative kind of analysis, particularly in such areas as cognitive neuroscience (where the linking function between cognitive variables and neurological ones is of paramount importance), or finance (where linking functions between such things as market indices and mutual fund performance could be highly informative). However, paradoxically, there are few examples of canonical correlation analysis in the published literature. Tabachnick and Fidell (2007) identify the problem of interpretability as one of the major reason for the paucity of published canonical correlation analysis examples. In this chapter, we introduce three tools to help with the interpretation of results from a canonical correlation analysis— canonical correlation summary tables, canonical correlation linked vector plots, and linked scatter plots of the structure of the individual datapoints.

It is our intent to demonstrate in this chapter that when combined with some of the conceptual tools introduced in the preceding chapters, including summary tables and graphs, canonical correlation can indeed be a highly informative and powerful multivariate method. For that reason, we have entitled the chapter *Canonical Correlation: The Underused Method*. We could just as well have entitled it *Canonical Correlation: The Misunderstood Method*, for truly the strengths of this method have been profoundly underestimated. It has been undersold in the textbooks on the subject, and few of the handful of studies in which it has been applied have exploited its full potential.

In preparation for what follows, we will identify seven types of information that can be obtained from canonical correlation analysis. These are unique aspects of the mathematics of canonical correlation analysis that make it a strong data analytic strategy. The essential strength of canonical correlation analysis is that it can be used to uncover the same kind of structural information as is found in exploratory factor analysis, revealing both the covariance structure of the variables (analogous to factor loadings) and also the internal structure of individual observations (analogous to factor scores). But the unique feature of canonical correlation is that it does latent variable structural analysis in pairs, simultaneously, with a linking function between the X set of variables and the Y set of variables that reveals strength-of-prediction information of the kind identified by multiple regression.

SEVEN TYPES OF INFORMATION FROM CANONICAL CORRELATION ANALYSIS

1. *Summary Table for the* X *Set of Variables.* The canonical variate loadings are analogous to factor loadings in exploratory factor analysis, and can be summarized in a canonical correlation summary table (see Table 6.2 for an example). Like the factor analysis summary table, this gives a considerable amount of rich descriptive information about the covariance structure.

2. *Linked Summary Table for the* Y *Set of Variables.* The same kind of summary table is constructed for the Y set of variables. It is placed near the table for the X set, as shown in Table 6.1, to facilitate comparison between the two sets in such things as amount of variance accounted for, which variables have more common or more unique variance, and so forth.

3. *Vector Plot for the* X *Set of Variables.* The structure defined by canonical loadings is displayed in a vector plot as exemplified by Figure 6.1. As demonstrated in Chapters 4 and 5, vector plots are particularly effective in revealing covariance structure.

4. *Vector Plot for the Y Set of Variables.* A similar vector plot for the Y set of variables is placed next to or superimposed upon the plot for the X set of variables, to illustrate how the variables from the two sets connect to one another through the canonical latent variables (see Fig. 6.1).

5. *Canonical Correlation Coefficients as Indices of the Strength of the Connection.* The canonical correlation coefficients are also reported. They are an index of how correlated each pair of latent variables is across the two sets of observed variables. These canonical correlation coefficients are the primary result from the analysis, and sometimes are the *only* results reported from a canonical correlation analysis. However, they have much more meaning in the context of the other six types of information.

6. *Scatterplot of Individual Observations within the Space of the* X *Set of Variables.* Just as factor scores are sometimes created from an exploratory factor analysis, canonical latent variates can be constructed from a canonical correlation analysis. Two- or three-dimensional scatterplots of canonical variate scores are highly useful in interpreting the internal structure of the X data set.

> 7. *Correponding Scatterplot of Individual Observations for the Space of the Y Set of Variables.* The real value here is in comparing the *Y* set internal structure to that of the *X* set. Each gives insight into the spatial organization of data within its own realm, but the real advantage of canonical correlation analysis is in comparing the internal structure of the two linked sets.

6.2   APPLIED EXAMPLE OF CANONICAL CORRELATION: PERSONALITY ORIENTATIONS AND PREJUDICE

In the July 1999 issue of the *Journal of Personality and Social Psychology*, Bernard Whitley[1] published a paper entitled "Right-wing authoritarianism, social dominance orientation, and prejudice." This paper has much to recommend it as an illustration of the usefulness of canonical correlation in social psychological research. Whitley's (1999) paper exemplifies how canonical correlation can be used to examine and test a substantive theory. It has been theorized among social psychologists that social dominance orientation (SDO) and right-wing authoritarianism (RWA) underlie a variety of types of prejudice. The Whitley paper directly tests this hypothesis through the use of canonical correlation analysis.

A major strength of the paper is that contains its own replication. The study was conducted twice, with two separate samples. The "initial analysis sample" consisted of 88 men and 92 women ($n = 181$), and the "replication sample" consisted of 94 men and 88 women ($n = 182$). As we shall see, this replication of the study enables Whitley to effectively deal with a major concern in the use of canonical correlation analysis, that of *overfitting* the model. That is, as Lattin et al. (2003) have rightly pointed out, canonical correlation, like other regression-based models, is susceptible to overfitting. As they state, it is often the case that regression models "explain not only the systematic variation in the dependent variable that we would observe in the population at large but also 'noise' or other error variance, thus artificially inflating the apparent goodness of fit of the regression model" (p. 337). We have already seen in Chapter 4 (Section 4.10) that high error variance, as indicated by low input variable reliabilities, can have devastating effects upon exploratory factor analysis results. Similarly, canonical correlation can have misleading results when the input data set does not have sufficient reliability.

The obvious way to guard against overfitting is to begin with highly reliable measures in the primary data. Whitley's study does well on this count also. Two of his three predictor variables in the *X* set, Social Dominance Orientation (the SDO scale) and Right-Wing Authoritarianism (the RWA Scale), have reliabilities of 0.88 and 0.87 respectively, and the third predictor, gender of respondent, obviously does not have a reliability problem. Likewise, his seven

---

[1] Dr. Bernard Whitley currently serves as chair of the department of psychological science at Ball State University. He conducts research on academic dishonesty and student reactions to it, and also attitudes toward homosexuality.

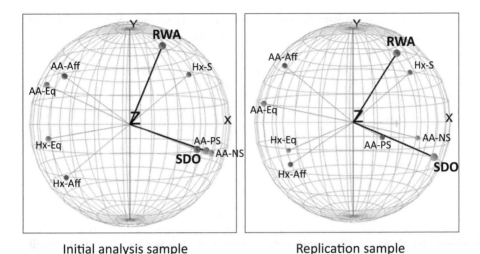

Initial analysis sample   Replication sample

Figure 6.1   Comparison of the canonical correlation vector plot from the initial analysis sample to the canonical correlation vector plot from the replication sample. The dark vectors are for two of the three variables in the X set of predictor variables (RWA and SDO) and the light vectors are for the seven prejudice variables. The two vector plots are based upon canonical correlation analyses of the two correlation matrices given in table 1 of the Whitley (1999) paper. (Image courtesy of Metrika (metrika, org), © Appzilla, LLC.)

measures of prejudice in the Y set, consisting of several stereotyping measures, two affective response measures, and two measures of attitude toward equality, all have reliabilities ranging from 0.81 to 0.95. Tabachnick and Fidell (2007) suggest having 10 observations for each variable in the canonical correlation analysis, and Whitley also meets and exceeds this advice with 181 and 182 subjects in the two samples, to investigate 10 variables. It is probably because of the careful design of this study that Whitley's two replications give essentially equivalent results (as shown in Fig. 6.1).

There is a lesson to be learned here with regard to data preparation for a canonical correlation analysis. Scale scores that are summated or otherwise combined over a number of items are more reliable than single item and will lead to substantially better canonical correlation analysis results. Most of the methods in this book assume a multivariate normal distribution. There are a number of ways to assess a data set for multivariate normality (see Johnson and Wichern, 2007, chapter 4, and Rencher, 2002, chapter 4), but the most important thing one can do to avoid skewed or kurtotic data in the first place is to use stable summed, averaged, or otherwise combined measures, such as scale scores with high Cronbach's alpha indices. As mentioned in Chapter 5, the distributions for summated or averaged data, such as scale scores, can be expected to approach a Gaussian distributional shape as the number of items summed-over increases, consistent with the central limit theorem.

One additional laudable feature of Whitley's (1999) paper is the inclusion in table 1 of the correlation matrices for both the initial analysis sample and also the replication sample, thus enabling the reader to perform the same analyses that Whitley reports in the paper. The paper can thus be made to function as a tutorial for the effective application of canonical correlation to a substantive problem within the discipline of social psychology.

The following SAS code calculates the canonical correlation analysis on Whitley's second data set, the replication sample, directly from the covariance matrix. Notice how the SAS data step in the first paragraph of the SAS code enters the correlation matrix directly. Of course, the usual procedure is to enter raw data and calculate the canonical correlation analysis from that, but this method works well also. However, since SAS does not have access to the raw data in this case, it is necessary to indicate the number of subjects from which the correlation matrix was constructed. Notice the option "EDF = 181" at the end of the first line of the PROC CANCOR code. This informs the CANCOR procedure that the sample size for this correlation matrix in Whitley's replication sample is $n = 182$, so that the significance tests on the canonical correlations can be accurately calculated.

```
***************************************************************************
***      Program: CCwhitley2.sas
***      Canonical correlation of Whitley (1999), dataset of the initial analysis
***************************************************************************/;
data Prejudice2 (type=CORR); _TYPE_='CORR';
  input _NAME_ $ x2RWA x1SDO x3Gen y1AAaff y5HXaff y2AAps y3AAns y6HXs y4AAeq y7HXeq;
  datalines;
x2RWA    1.00    .20    -.13    -.09    -.46    .07    .26    .46    -.32    -.43
x1SDO     .20   1.00     .17    -.65    -.44    .26    .59    .37    -.68    -.52
x3Gen    -.13    .17    1.00    -.19    -.22    .09    .15    .15    -.15    -.31
y1AAaff  -.09   -.65    -.19    1.00     .35   -.18   -.61   -.17    .64     .34
y5HXaff  -.46   -.44    -.22     .35    1.00   -.08   -.30   -.65    .43     .79
y2AAps    .07    .26     .09    -.18    -.08   1.00    .59    .41   -.39    -.19
y3AAns    .26    .59     .15    -.61    -.30    .59   1.00    .48   -.64    -.34
y6HXs     .46    .37     .15    -.17    -.65    .41    .48   1.00   -.41    -.64
y4AAeq   -.32   -.68    -.15     .64     .43   -.39   -.64   -.41   1.00     .53
y7HXeq   -.43   -.52    -.31     .34     .79   -.19   -.34   -.64    .53    1.00
;
run;

proc print data=Prejudice2;
run;

proc cancorr data=Prejudice2 all corr EDF=181
   vprefix=DV vname='Independent Variables'
   wprefix=IV wname='Dependent Variables';
   var x2RWA x1SDO x3Gen;
   with y1AAaff y5HXaff y2AAps y3AAns y6HXs y4AAeq y7HXeq;
   title 'Canonical correlation from Whitley 1999, replication sample';
run;
```

In table 2 of Whitley (1999) the canonical correlations are listed as 0.82, 0.47, and 0.23 for the analysis of these data, the replication sample. The first two of these canonical correlation coefficients are indicated in Whitley's table to be statistically significant. The same results are obtained with regard to the three canonical correlations and their statistical significance, as shown in the following snippet from the SAS output that results from this program.

```
Canonical correlation from Whitley 1999, replication sample              3
                                                      12:56 Friday, March 25, 2011

                           The CANCORR Procedure

                      Canonical Correlation Analysis

                        Adjusted      Approximate        Squared
            Canonical    Canonical     Standard        Canonical
           Correlation  Correlation      Error        Correlation

        1   0.819174     0.810644       0.024451       0.671046
        2   0.458129     0.429167       0.058729       0.209882
        3   0.238597     0.202631       0.070098       0.056928

                                      Test of H0: The canonical correlations in
           Eigenvalues of Inv(E)*H    the current row and all that follow are zero
             = CanRsq/(1-CanRsq)
                                             Likelihood Approximate
    Eigenvalue Difference Proportion Cumulative   Ratio     F Value Num DF Den DF Pr > F

 1    2.0399    1.7743     0.8622     0.8622 0.24511578    14.87    21  494.44 <.0001
 2    0.2656    0.2053     0.1123     0.9745 0.74513806     4.57    12   346 <.0001
 3    0.0604               0.0255     1.0000 0.94307159     2.10     5   174 0.0675


                    Multivariate Statistics and F Approximations

                        S=3      M=1.5     N=85

    Statistic                       Value    F Value   Num DF   Den DF   Pr > F

    Wilks' Lambda                 0.24511578   14.87     21     494.44   <.0001
    Pillai's Trace                0.93785666   11.30     21     522      <.0001
    Hotelling-Lawley Trace        2.36594147   19.26     21     353.05   <.0001
    Roy's Greatest Root           2.03994322   50.71      7     174      <.0001

        NOTE: F Statistic for Roy's Greatest Root is an upper bound.
```

The first of the three tables in this snippet (taken from page three of the output) displays the canonical correlation values, the adjusted canonical correlation values, their standard error, and the squared canonical correlation, the latter figure indicating the strength of the linking function for the three matched latent variables for the X set and the Y set. The connection is fairly strong for the first latent variable pair (a value of 0.671 of squared canonical correlation indicating 67.1% common variance), but the connection for the second latent variable pair is substantially less strong (21.0%), and the third pair only has about 5.7% common variance. The second of the three tables in the example SAS output uses a likelihood ratio test to evaluate each of the three canonical correlations for significance. The first two canonical correlations are significant, with $P < 0.0001$, but the third does not quite reach significance ($P = 0.0675$). The last of the three tables in this SAS output snippet is an omnibus test of all three canonical correlations, employing all four of the multivariate test statistics (the Wilks', the Pillai, the Hotelling–Lawley, and

Roy's) that will be introduced in Chapter 8 in connection with MANOVA. These four tests evaluate the overall null hypothesis that there are no significant canonical correlations in the paired multivariate set, which null hypothesis is of course found here to be rejected.

Table 6.1 is a comparison of the canonical variate loadings from four sources: (1) the loadings from the "initial analysis sample" given in Whitley's table 2, (2) the "replication sample" loadings from Whitley's table 2 (both shown in the top half of Table 6.1), (3) loadings from SAS reprocessing of the "initial analysis" sample, and (4) the loadings from SAS reprocessing of the "replication sample." The SAS analyses in the two latter subtables are both based upon the correlation matrices given in table 1 of the Whitley paper. The reason we have gone to so much trouble to carefully compare the canonical loadings from these four sources is to demonstrate the agreement between the two replications of Whitley's study. Although the agreement between the two samples is not apparent from the information presented in Whitley's table 2, it is very apparent from the SAS reanalysis of the correlation matrices for the two samples.

Notice in the first three row entries of each of these four subtables of Table 6.1 that the three predictor variables (SDO, RWA, and Gender) each correspond quite well to the three latent variables,[2] LV1, LV2, and LV3, respectively. Each of the large loadings of these three is emphasized with a box. We have used this large loading on each latent variable to equivalence the four tables with regard to polarity in order to facilitate comparison.

The polarity on the canonical correlation latent variables can (and often does) switch from positive to negative in a seemingly random fashion without affecting the essential results. The only negative effect of such polarity shifts is in the spatial orientation for plotting and for comparisons. For that reason, we have "corrected" these arbitrary variations by reversing the polarity on any latent variable in the four sub-tables that didn't have a positive value in the boxed "three large loadings" of each of the three predictor variables. All switches in polarity are shown in the table with the entries being printed in italicized bold type.

On the right of Table 6.1 are calculations for the "absolute values of discrepancy scores." These compare the table in the first three columns (the one for the initial analysis sample) to the table in the next three columns (the one for the replication sample). These discrepancy scores are an indication of how the factor pattern of the loadings changed between the two samples—an indication of sampling variation.

---

[2] These three can be referred to as latent variables, canonical variates, or canonical variate scores. The loadings, like loadings in factor analysis are the correlation coefficients between scores on observed variables and scores on these latent variables. "Latent variables" is the generic term. Factor scores, principal component scores, and canonical variate scores are all particular kinds of latent variables. Latent variables can be meaningful theoretical constructs, as factor scores are usually considered to be, or merely convenient and useful linear combinations as in principal component scores or canonical variate scores.

TABLE 6.1. Comparison of the Canonical Variate Loadings for the Initial Analysis Sample and the Replication Sample, as Presented in Whitley (1999) Table 2 and as Calculated by SAS PROC CANCOR from the Correlation Matrices of Table 1

	LV1	LV2	LV3	LV1	LV2	LV3	LV1	LV2	LV3
	Initial Sample, Whitley Table 2			Replication Sample, Whitley Table 2			Absolute Values of Discrepancy Scores		
Predictor Variables (*X* Set)									
Social dominance orientation	.91	*-.27*	-.32	*.91*	*-.39*	*-.12*	.00	.12	-.20
Right-wing authoritarianism	-.31	*.92*	-.23	*.51*	*.80*	*-.33*	.82	.12	.10
Gender of participants	.66	*.21*	.72	*.32*	*.02*	*.95*	.34	.19	-.23
African American Prejudice (*Y* Set)							*Largest discrepancy*		
Affect	-.66	*.43*	.27	*-.74*	*.58*	*.04*	.08	-.15	.23
Positive stereotype	.55	*-.22*	.63	*.30*	*-.13*	*.05*	.25	-.09	.58
Negative stereotype	.84	*-.26*	.12	*.73*	*-.17*	*-.20*	.11	-.09	.32
Enhance equality	-.82	*.34*	.25	*-.85*	*.17*	*.33*	.03	.17	-.08
Homosexual Prejudice (*Y* Set)									
Affect	-.68	*-.63*	-.20	*-.70*	*-.48*	*-.13*	.02	-.15	-.07
Positive stereotype	.63	*.53*	-.10	*.61*	*.55*	*-.04*	.02	-.02	-.06
Enhance equality	-.82	*-.20*	.05	*-.80*	*-.36*	*-.45*	-.02	.16	.50
	Initial Sample, SAS Re-Analysis			Replication Sample, SAS Re-Analysis			Absolute Values of Discrepancy Scores		
Predictor Variables (*X* Set)									
Social dominance orientation	.90	*-.31*	-.31	*.91*	*-.39*	*-.13*	-.02	.08	.18
Right-wing authoritarianism	.38	*.89*	-.25	*.52*	*.80*	*-.30*	-.14	.09	.05
Gender of participants	.65	*-.25*	.72	*.32*	*-.01*	*.95*	.33	-.24	-.23
African American Prejudice (*Y* Set)									
Affect	-.63	.46	.23	-.73	.60	*.02*	.09	-.14	.22
Positive stereotype	.55	-.20	.64	.31	-.16	*.03*	.24	-.04	.61
Negative stereotype	.82	-.28	.13	.74	-.18	*-.20*	.09	-.10	.33
Enhance equality	-.81	.38	.23	-.85	.18	*.33*	.04	.20	-.10

(*Continued*)

TABLE 6.1. (*Continued*)

Homosexual Prejudice (*Y* Set)									
Affect	−.72	−.59	−.17	−.71	−.47	***−.16***	−.01	−.12	.00
Positive stereotype	.64	.53	−.07	.62	.53	***−.04***	.02	.00	−.03
Enhance equality	−.84	−.14	.11	−.80	−.34	***−.46***	−.04	.20	.56

Maximum discrepancies between SAS analysis and Whitley's
Table 2:

.687	.457	.056	.013	.032	.034

Canonical Correlation Analysis Summary Statistics

	Initial Sample SAS Re-Analysis	Replication Sample SAS Re-Analysis
First Canonical Correlation	.7795	.8192
Second Canonical Correlation	.5206	.4581
Third Canonical Correlation	.2915	.2386

In any case, when we create canonical correlation joint vector plots from the two bottom subtables of Table 6.1, the ones from the SAS analysis, we see a high level of agreement between the "initial analysis sample" (the vector plot on the left of Fig. 6.1) and the "replication sample" (the vector plot on the right of Fig. 6.1), which is a clear indication that the findings of this study are robust.

Not only is the pattern of results highly similar across these two vector plots, the pattern is also clear, informative, and memorable. SDO is primarily horizontal on these vector plots and lines up very well with the African-American prejudice measures. In other words, high SDO lines up with all four measures of prejudice toward African-Americans, with high SDO scores corresponding to high scores on the two African-American stereotyping variables, and also corresponding to low scores on African-American positive affect and equality enhancement attitude. The vector for RWA is primarily vertical (very nearly orthogonal to SDO) and tends to correspond to the homosexual prejudice measures. High RWA is positively related to high homosexual stereotyping, and negatively related to positive affect and equality enhancement for homosexuals. Note that the predictor variables, SDO and RWA, are shown with bold vectors,[3] and the prejudice measures are shown with light vectors.

[3] The bold vector for the third predictor variable, Gender, is left out of these two-dimensional plots to promote clarity, but has been examined three-dimensionally using Metrika (Brown et al., 2011). In the vector plot on the right, the replication sample, the Gender vector comes out of the page toward the reader, slightly to the right of the implied *z*-axis. It is, however, out there by itself in that none of the prejudice variables in this figure extend out into the *z*-dimension. They are almost entirely contained within the two-dimensional plane that can be seen in Figure 6.1. In the vector plot on the left, the Gender vector is located midway between the *x*-axis and the *z*-axis at a 45° angle from each, and the African American positive stereotyping prejudice variable virtually coincides with it. However, given the small amount of common variance (5.7%) from the canonical correlation coefficient of the third latent variable pair, this connection is tenuous.

   Perceptually, these vector plots would benefit from a rotation of about 20° counter clockwise. SDO is about 20° below the x-axis in both the left and the right vector plots. RWA is about 20° to the right of the y-axis in the initial analysis plot and 30° in the replication sample plot. A 20° rotation would essentially make SDO coincide with the x-axis, and RWA coincide with the y-axis for the left vector plot and almost coincide with the y-axis for the right vector plot. However, Rencher (2002) recommends against rotating canonical correlation solutions for the simple reason that rotated canonical latent variables no longer have the two desirable properties of: (1) maximum correlation between the pairs of latent variables and (2) orthogonality of the latent variables with respect to one another (they become correlated).

   As long as we are only dealing with a two- or three-dimensional latent variable space, it could be argued that we actually obtain more information from the vector plots of Figure 6.1 in their unrotated present form, in that we can still see the correspondence between SDO and the African-American prejudice variables, and the correspondence between RWA and homosexual prejudice variables, but we can also see the points of maximum correlation between the canonical variates at the x- and y-axes. In canonical correlation analysis, there is not a great deal of focus on the actual latent variables. Usually, they merely provide reference axes for us to plot the X set vectors and the Y set vectors within a common space. It is in this sense that Johnson and Wichern (2007) maintain that canonical variates (the "factor scores" of canonical correlation analysis) are essentially artificial, that they "have no physical meaning" (p. 545). They are therefore somewhat like principal components, in that we typically do not claim ontological status for them (as we do in some varieties of factor analysis), but view them merely as useful linear combinations of the original variables. In this case, they are matched linear combinations of the original variables that give us a structure for relating the topology of the X set of variables with the topology of the Y set of variables.

   A second comparison can be drawn between canonical correlation analysis and principal component analysis. Just as the amount of variance accounted for by principal components declines substantially after the first component, so too the magnitude of each successive canonical correlation value declines. The first canonical correlation coefficient is often substantially stronger than the following ones, as it is in the Whitley data. The squared canonical correlations indicate the amount of shared variance between the two latent variables in each pair, and for the replication sample of the Whitley data, we saw that they declined from 67.1% in the first to 21.0%, and 5.7% in the second and third latent variable pairs. We read this to indicate that the connection between SDO and the African American prejudice scores is quite strong, but the connection between RWA and homosexual prejudice scores is much less so, and in the third latent variable pair, the connection is so tenuous as to probably not justify much attention.

   Table 6.2 is a canonical correlation summary table, with the top part of the table giving the loadings and other statistics for the dependent variables (the Y

TABLE 6.2. Canonical Correlation Summary Table with the _Y_ Set of Prejudice Variables on Top and the _X_ Set of Predictor Variables on the Bottom, Taken from the Replication Sample Loadings in Table 2 of Whitley (1999)

	Loadings			Squared Loadings				Uniqueness
	LV1	LV2	LV3	LV1	LV2	LV3	Total	
Canonical Variate Loadings for Seven Prejudice Variables (Y set)								
African American Prejudice								
Affect	−.74	.58	.04	.55	.34	.00	.89	.11
Positive stereotype	.30	−.13	.05	.09	.02	.00	.11	.89
Negative stereotype	.73	−.17	−.20	.53	.03	.04	.60	.40
Enhance equality	−.85	.17	.33	.72	.03	.11	.86	.14
Homosexual Prejudice								
Affect	−.70	−.48	−.13	.49	.23	.02	.74	.26
Positive stereotype	.61	.55	−.04	.37	.30	.00	.68	.32
Enhance equality	−.80	−.36	−.45	.64	.13	.20	.97	.03
Sum of squares by columns:				3.40	1.07	.37	4.84	2.16
Percents of sums of squares:				49%	15%	5%	69%	31%
Canonical Variate Loadings for Three Predictor Variables (X set)								
Social dominance orientation	.91	−.39	−.12	.84	.15	.01	1.00	.00
Right-wing authoritarianism	.51	.80	−.33	.26	.64	.11	1.00	.00
Gender of participants	.32	.02	.95	.10	.00	.90	1.00	.00
Sum of squares by columns:				1.19	.79	1.02	3.00	.00
Percents of sums of squares:				40%	26%	34%	100%	0%

	Coefficient	_p_ Value
First Canonical Correlation	.2445	<.0001
Second Canonical Correlation	.0587	<.0002
Third Canonical Correlation	.0701	.0675

set of prejudice measures), and the bottom part of the table giving the loadings and other statistics for the independent variables (the _X_ set of predictors). We have constructed each of the two linked tables, the top one and the bottom one, in the same format as a factor analysis summary table to emphasize the "double factor analysis" approach to interpreting a canonical correlation analysis.

Similar to a factor analysis table, to the right of the loadings in each table, we find an equal-sized matrix of the squared loadings. These are like communalities in that they indicate the amount of common variance between each observed variable (row) and each latent variable (column). They are summed at the right in a fourth column of "total" squared loadings, which indicates the proportion of variance in each observed variable that is accounted for by the three latent variables. These values are subtracted from 1 to obtain the column of uniqueness values on the far right, which indicate how much variance in each observed variable fails to be accounted for by the latent variables.

These indices of variance accounted for in each observed variable are quite informative in the Y set of prejudice variables. The one observed variable that has the least in common with the predictor variables is African American positive stereotyping (with a total sum of squares of 0.11, indicating 11% accounted for). This measure includes such items as "Black people are more hardworking than White people," or more religious, or place more value on family ties, and so on. The uniqueness value of 0.89 indicates that 89% of the variance in this measure is "unique" in the sense of being unrelated to the latent variable space that is linked to the three predictor variables.

On the other hand, two variables in particular have much in common with the predictor variables, the two for "attitudes toward equality enhancement." In other words, these two variables are the prejudice variables that best fit the theory, that are best predicted by SDO and RWA (and to some extent Gender). The stronger of the two is homosexual equality enhancement, having 97% common variance and only 3% uniqueness. African American equality enhancement has 89% common and 11% uniqueness.

There is also something to be learned from the summary statistics at the bottom of each matrix of squared loadings. The first row under the matrix of squared loadings in each table, the "sums of squares by columns" row, is in the same position as eigenvalues in a factor analysis table. These values in fact function like eigenvalues and are interpreted in a similar manner, but they are not actually eigenvalues. They are sums of squares of loadings across all m of the variables (where m_1 is 7 for the Y set and m_2 is 3 for the X set). The maximum possible value of these indices is m, 7.00 in the upper table, and 3.00 in the lower table. The percents of "sums of squares" row beneath each is obtained by dividing the sum of squares by its m.

In the upper table, we see that 69% of the variance in the seven prejudice variables is accounted for by the latent variable space (the "percents of sums of squares" entry in the "total" column). And 49 percentage units of those 69% are due to the first latent variable (the one associated with SDO), with only 15 percentage units in the second latent variable (RWA), and 5% in the third. Nearly a third of the variance in the seven prejudice measures is unrelated to the space defined by the three latent variables in the Y set.

There is not nearly so much information in the percents accounted for in the X set of predictor variables. Since there are only three predictor variables, three latent variables will of necessity account for all of the variance in them, regardless of the geometric configuration.[4] Perhaps there are only two notable observations to be made from the X set table. The first is that the three predictor variables correspond quite closely to the three latent variables (SDO with the first, RWA with the second, and Gender with the third), and this would be even more obvious with the rotation discussed above, in which they could be rotated to be very nearly orthogonal. The second observation is that the three "percents of sums of squares" are 40, 26, and 34% for the first, second, and third latent variables, respectively. In other words, it is clearly the case that the percent variance accounted for statistics are not necessarily in descending order across latent variables in canonical correlation analysis as they are for eigenvalues in principal component analysis.

Let us now recapitulate what can be learned from the example of the Whitley paper about the use of tables and figures to interpret a canonical correlation analysis. Of the seven types of information identified at the beginning of the chapter, five have been used here. The canonical correlation summary tables for both the X set of prejudice variables and also the Y set of predictor variables were shown in Table 6.2. Given that the X set had the same number of predictor variables as latent variables, it was not as informative as it could be, but the summary table for the Y set of prejudice gave good comparative information concerning which prejudice variables are highly related to the linked latent variable space and which are not. The synergistic information that is often present in a comparison of these two tables was not present in this study because of the low number of variables in the X set, but it will be shown in Section 6.4.

The third and fourth kinds of information, a vector plot for the X set of variables and a vector plot for the Y set of variables, were presented for these data by superimposing the two upon one another (Fig. 6.1), and then having side-by-side comparisons of the two replications. The strong correspondence in results between the initial analysis sample and the replication sample was difficult to see in the tables of canonical variate loadings (both here and in the Whitley paper), but very compelling in this side-by-side superimposed vector plot form. It was in this figure that the comparative results of the two studies became most clear. The fifth kind of information, the strength of relationship information in the canonical correlation coefficients themselves, was also useful in interpreting the results of the analysis and the patterns in the vector plots. The somewhat weak connections between the second and third canonical latent variable pairs are a good reminder that by far, the most robust finding in the study is the connection between SDO and the measures of prejudice toward African Americans.

---

[4] If we had decided to include only the first two canonical correlation coefficients and their associated latent variables, this table would in one sense be more informative. We would see that in linked two-dimensional spaces of predictors and prejudice variables, that each set of latent variables accounts for about the same amount of variance in its observed variables, about 64% for the Y set and about 66% for the X set. What we will find in this kind of analysis is in some ways dependent on what we ask.

The sixth and seventh kinds of information, the linked scatter plots of individual data, require access to the raw data in order to examine the internal structure of all the observations within each multivariate set, X and Y. Since we do not have access to original data, these two kinds of scatter plot information could not be created, but will be demonstrated in the next two sections.

In the Whitley study, these final two kinds of individual observation scatter plot information would probably be quite informative. For example, it could be illuminating to examine a scatter plot of all 182 respondents within a scatter plot of the X set of predictors, and next to it a scatter plot of the Y set. Using grouping and clustering (of the kind demonstrated in Fig. 5.1 of Chapter 5), we could examine the spatial locations and topological patterning of such things as: all female respondents, all male respondents, the 5% of respondents with highest (or lowest) SDO scores, the 5% of male (or female) respondents with highest SDO scores, and so forth. This could be a useful data exploration technique for identifying attitude patterns with topological precision.

6.3 MATHEMATICAL DEMONSTRATION OF A COMPLETE CANONICAL CORRELATION ANALYSIS

Canonical correlation analysis is probably the analytical method that draws upon the largest number of matrix algebra tools, including such things as Cholesky decomposition, matrix inversion, eigenvalues and eigenvectors, singular value decomposition, and others. From that standpoint, it is a good review of Chapter 3, matrix algebra, and also an illustration of the practical usefulness of these matrix principles. We will outline the process in seventeen steps and also give command code for canonical correlation in Stata and SAS.

COMPUTATIONAL STEPS IN CANONICAL CORRELATION ANALYSIS

Step 1: Combine the **X** data matrix and the **Y** data matrix to create the **X|Y** adjoined data matrix, and standardize it to get the $\mathbf{Z_x|Z_y}$ adjoined data matrix.

Step 2: Calculate the adjoined **R** matrix, which is partitioned into four submatrices ($\mathbf{R_{xx}}$, $\mathbf{R_{xy}}$, $\mathbf{R_{yx}}$, and $\mathbf{R_{yy}}$), by premultiplying the $\mathbf{Z_x|Z_y}$ matrix by its transpose and dividing by the degrees of freedom.

Step 3: From the $\mathbf{R_{adj}}$ matrix, select the $\mathbf{R_{xx}}$ partition and use the Cholesky algorithm to obtain an $\mathbf{R_{xx}'}^{-1/2}$ matrix from it.

Step 4: Obtain the $\mathbf{R_{yy}^{-1}}$ matrix by taking the inverse of the $\mathbf{R_{yy}}$ partition selected from the $\mathbf{R_{adj}}$ matrix.

Step 5: Create the product matrix by the formula $\mathbf{P} = \mathbf{R}'^{-1/2}_{xx}\, \mathbf{R}_{xy}\, \mathbf{R}^{-1}_{yy}\, \mathbf{R}_{yx}\, \mathbf{R}^{-1/2}_{xx}$.

Step 6: Factor the product matrix by the method of successive squaring (Chapter 4, Section 4.3) to obtain the first eigenvector, $\mathbf{k}_1$, of the product matrix $\mathbf{P}$.

Step 7: Obtain the corresponding eigenvalue λ_1 for the first eigenvector $\mathbf{k}_1$, and take the square root of it to obtain the first canonical correlation coefficient r_1.

Step 8: Create four additional eigenvectors from $\mathbf{k}_1$: the $\mathbf{q}_1$ eigenvector (normalized to 1), the $\mathbf{f}_1$ eigenvector (normalized to the eigenvalue), the $\mathbf{c}_1$ eigenvector (standardized coefficients for the X set), and the $\mathbf{d}_1$ eigenvector (standardized coefficients for the Y set).

Step 9: Calculate the first-factor product matrix $\mathbf{G}_1$ by post multiplying $\mathbf{f}$ by its transpose $\mathbf{G}_1 = \mathbf{ff}'$. Subtract this from the product matrix of step 5 to obtain the matrix of first factor residuals, $\mathbf{P}_1 = \mathbf{P} - \mathbf{G}_1$.

Step 10: Factor the first factor residual matrix $\mathbf{P}_1$ by the method of successive squaring to obtain the second eigenvector, $\mathbf{k}_2$.

Step 11: Obtain the corresponding eigenvalue λ_2 for the second eigenvector $\mathbf{k}_2$, and also obtain the second canonical correlation coefficient r_2 by taking the square root of λ_2.

Step 12: Create three additional eigenvectors from $\mathbf{k}_2$: the $\mathbf{q}_2$ eigenvector (normalized to 1), the $\mathbf{c}_2$ eigenvector (standardized coefficients for the X set), and the $\mathbf{d}_2$ eigenvector (standardized coefficients for the Y set).

Step 13: Adjoin the individual $\mathbf{c}$ and $\mathbf{d}$ eigenvectors to create matrix $\mathbf{C}$, the matrix of standardized coefficients for the X set, and matrix $\mathbf{D}$, the matrix of standardized coefficients for the Y set.

Step 14: Obtain the matrix of canonical variates χ (chi), the latent variables for the X set, by post multiplying $\mathbf{Z}_x$ by its matrix of standardized coefficients $\mathbf{C}$.

Step 15: Obtain the matrix of canonical variates η (eta), the latent variables for the Y set, by post multiplying $\mathbf{Z}_y$ by its matrix of standardized coefficients $\mathbf{D}$.

Step 16: Concatenate the matrices of observed variables, $\mathbf{Z}_x$ and $\mathbf{Z}_y$, with the matrices of latent variables, chi and eta, to create an adjoined total matrix of Z-scores, $\mathbf{Z}_t = \mathbf{Z}_x|\chi|\mathbf{Z}_y|\eta$. Calculate an adjoined total matrix of correlation coefficients $\mathbf{R}_t$ from this, by $\mathbf{R}_t = (1/\mathrm{df})\mathbf{Z}'_t\mathbf{Z}_t$.

Step 17: From the correlation coefficients in the $\mathbf{R}_{x\chi}$ partition of the total $\mathbf{R}_t$ matrix, create the canonical correlation summary table for the X set of variables. From the correlation coefficients in the $\mathbf{R}_{y\eta}$ partition of the $\mathbf{R}_t$ matrix, create the canonical correlation summary table for the Y set of variables.

Step 1: Combine the X data matrix and the Y data matrix to create the X|Y adjoined data matrix, and standardize it to get the Zx|Zy adjoined data matrix.

Suppose that we have 13 traumatic brain injury (TBI) patients, and that each has been given four neurological measurements (X1, X2, X3, and X4), and five cognitive performance tests (Y1, Y2, Y3, Y4, Y5). We wish to explore the holistic pattern of how the neurological measurements relate to the cognitive performance tests. The **X** matrix of neurological measurements is adjoined to the **Y** matrix of cognitive performance tests to obtain the **X|Y** matrix below (read as "**X** adjoin **Y** matrix").

$$
\mathbf{X|Y} = \begin{pmatrix}
7.0 & 3.9 & 9.4 & 4.1 & 42 & 24 & 18 & 14 & 49 \\
5.0 & 9.3 & 7.6 & 3.8 & 36 & 23 & 24 & 9 & 44 \\
1.5 & 11.5 & 4.3 & 5.9 & 29 & 33 & 37 & 3 & 37 \\
5.2 & 11.3 & 10.2 & 5.4 & 43 & 38 & 28 & 18 & 50 \\
8.6 & 5.6 & 10.9 & 2.8 & 51 & 25 & 24 & 25 & 56 \\
2.8 & 6.5 & 2.9 & 7.4 & 39 & 31 & 15 & 12 & 44 \\
4.7 & 10.6 & 7.1 & 5.3 & 45 & 39 & 24 & 19 & 53 \\
3.9 & 11.3 & 2.6 & 6.4 & 37 & 41 & 32 & 12 & 47 \\
4.9 & 2.9 & 7.0 & 4.4 & 50 & 26 & 18 & 25 & 54 \\
6.6 & 8.4 & 7.8 & 4.9 & 52 & 27 & 9 & 26 & 55 \\
3.1 & 13.0 & 6.4 & 15.3 & 35 & 32 & 23 & 11 & 39 \\
4.6 & 7.9 & 9.0 & 2.9 & 48 & 30 & 4 & 20 & 52 \\
8.4 & 4.4 & 11.0 & 6.8 & 39 & 34 & 30 & 14 & 44
\end{pmatrix}.
$$

We also calculate the means and the standard deviations for each of the nine variables and enter them in the $\mathbf{m_x|m_y}$ adjoined vector of means and the $\mathbf{s_x|s_y}$ adjoined vector of standard deviations.

$$\mathbf{m_x | m_y} = (5.1 \quad 8.2 \quad 7.4 \quad 5.8 \,|\, 42 \quad 31 \quad 22 \quad 16 \quad 48)$$

$$\mathbf{s_x | s_y} = (2.105 \quad 3.303 \quad 2.798 \quad 3.184 \,|\, 7.000 \quad 5.902 \quad 9.183 \quad 6.916 \quad 6.096).$$

Each X and Y variable is converted to a standard score by subtracting its mean from it and dividing the resultant deviation score by its corresponding standard deviation. The result is the $\mathbf{Z_x|Z_y}$ matrix shown below.

$$
\mathbf{Z_x} \mid \mathbf{Z_y} = \begin{pmatrix}
.90 & -1.30 & 0.71 & -0.53 & 0.00 & -1.19 & -0.44 & -0.29 & 0.16 \\
-.05 & 0.33 & 0.07 & -0.63 & -0.86 & -1.36 & 0.22 & -1.01 & -0.66 \\
-1.71 & 1.00 & -1.11 & 0.03 & -1.86 & 0.34 & 1.63 & -1.88 & -1.80 \\
0.05 & 0.94 & 1.00 & -0.13 & 0.14 & 1.19 & 0.65 & 0.29 & 0.33 \\
1.66 & -0.79 & 1.25 & -0.94 & 1.29 & -1.02 & 0.22 & 1.30 & 1.31 \\
-1.09 & -0.51 & -1.61 & 0.50 & -0.43 & 0.00 & -0.76 & -0.58 & -0.66 \\
-0.19 & 0.73 & -0.11 & -0.16 & 0.43 & 1.36 & 0.22 & 0.43 & 0.82 \\
-0.57 & 0.94 & -1.72 & 0.19 & -0.71 & 1.69 & 1.09 & -0.58 & -0.16 \\
-0.10 & -1.60 & -0.14 & -0.44 & 1.14 & -0.85 & -0.44 & 1.30 & 0.98 \\
0.71 & 0.06 & 0.14 & -0.28 & 1.43 & -0.68 & -1.42 & 1.45 & 1.15 \\
-0.95 & 1.45 & -0.36 & 2.98 & -1.00 & 0.17 & 0.11 & -0.72 & -1.48 \\
-0.24 & -0.09 & 0.57 & -0.91 & 0.86 & -0.17 & -1.96 & 0.58 & 0.66 \\
1.57 & -1.15 & 1.29 & 0.31 & -0.43 & 0.51 & 0.87 & -2.9 & -0.66
\end{pmatrix}.
$$

Step 2: Calculate the adjoined R matrix, which is partitioned into four sub-matrices ($\mathbf{R_{xx}}$, $\mathbf{R_{xy}}$, $\mathbf{R_{yx}}$, and $\mathbf{R_{yy}}$), by premultiplying the $\mathbf{Z_x}\mid\mathbf{Z_y}$ matrix by its transpose and dividing by the degrees of freedom,

Given that a Pearson product moment correlation coefficient is the sum of products of $\mathbf{Z_x}$ scores and $\mathbf{Z_y}$ scores divided by degrees of freedom,

$$
r = \frac{\sum \mathbf{Z_x Z_x}}{N-1},
$$

the matrix formula given in Equation 6.1, is used to obtain the 9×9 correlation matrix from the adjoined $\mathbf{Z}$ score matrix.

$$
\mathbf{R} = \left(\frac{1}{N-1} \right) \mathbf{Z_x} \mid \mathbf{Z_y' Z_x} \mid \mathbf{Z_y}. \tag{6.1}
$$

Notice that the 9×9 correlation matrix calculated from Equation 6.1 has been partitioned into four submatrices, $\mathbf{R_{xx}}$, which is a 4×4 matrix; $\mathbf{R_{xy}}$, which is a 4×5 matrix; $\mathbf{R_{yx}}$, which is a 5×4 matrix; and $\mathbf{R_{yy}}$, which is a 5×5 matrix. All four of these matrix partitions are used in the calculations that follow.

$$R = \left(\frac{1}{N-1}\right) Z_x |Z_y' Z_x| Z_y =$$

$$
\begin{bmatrix}
1.0000 & -0.6035 & 0.8130 & -0.4159 & 0.5950 & -0.3475 & -0.1561 & 0.5885 & 0.5884 \\
-0.6035 & 1.0000 & -0.3944 & 0.4742 & -0.4952 & 0.5762 & 0.3528 & -0.4410 & -0.4283 \\
0.8130 & -0.3944 & 1.0000 & -0.3367 & 0.4961 & -0.3280 & -0.1686 & 0.4861 & 0.4475 \\
-0.4159 & 0.4742 & -0.3367 & 1.0000 & -0.4976 & 0.3162 & 0.2186 & -0.3962 & -0.6221 \\
0.5950 & -0.4952 & 0.4961 & -0.4976 & 1.0000 & -0.3086 & -0.6572 & 0.9846 & 0.9568 \\
-0.3475 & 0.5762 & -0.3280 & 0.3162 & -0.3086 & 1.0000 & 0.4659 & -0.2287 & -0.1945 \\
-0.1561 & 0.3528 & -0.1686 & 0.2186 & -0.6572 & 0.4659 & 1.0000 & -0.5721 & -0.5195 \\
0.5885 & -0.4410 & 0.4861 & -0.3962 & 0.9846 & -0.2287 & -0.5721 & 1.0000 & 0.9348 \\
0.5884 & -0.4283 & 0.4475 & -0.6221 & 0.9568 & -0.1945 & -0.5195 & 0.9348 & 1.0000
\end{bmatrix}
$$

Step 3: From the R_{adj} matrix, select the R_{xx} partition and use the Cholesky algorithm to obtain an $R_{xx}'^{-1/2}$ matrix from it.

To obtain the Cholesky triangular square root inverse matrix for Rxx (symbolized as $R_{xx}'^{-1/2}$, the matrix in the bottom right of the adjoined calculation structure below), we follow the process outlined in Chapter 3, Sections 3.5.2 and 3.7.2, where the Cholesky process is explained. Essentially, the process is to adjoin an identity matrix to the right of the R_{xx} matrix, creating a 4x8 matrix A from these two. We then operate on matrix A according to the equations of Chapter 3 (and the accompanying instructions, to simultaneously create the upper triangular matrix $R_{xx}^{1/2}$, i.e., the triangular square root of matrix R_{xx}, and also the lower triangular matrix $R_{xx}'^{-1/2}$, i.e., the transpose of the triangular inverse square root of R_{xx}). The latter matrix is the one we use, in step 5, as one of the components of the primary product matrix for canonical correlation analysis

$$
\begin{bmatrix}
R_{xx} & I \\
R_{xx}^{1/2} & R_{xx}'^{-1/2}
\end{bmatrix}
=
\begin{bmatrix}
A \\
T
\end{bmatrix}
=
$$

$$
\begin{bmatrix}
1.0000 & -0.6035 & 0.8130 & -0.4159 & 1.0000 & 0 & 0 & 0 \\
-0.6035 & 1.0000 & -0.3944 & 0.4742 & 0 & 1.0000 & 0 & 0 \\
0.8130 & -0.3944 & 1.0000 & -0.3367 & 0 & 0 & 1.0000 & 0 \\
-0.4159 & 0.4742 & -.3367 & 1.0000 & 0 & 0 & 0 & 1.0000 \\
1.0000 & -0.6035 & .8130 & -0.4159 & 1.0000 & 0 & 0 & 0 \\
0 & 0.7973 & 0.1208 & 0.2799 & 0.7569 & 1.2542 & 0 & 0 \\
0 & 0 & 0.5696 & -0.0568 & -1.5580 & -0.2660 & 1.7557 & 0 \\
0 & 0 & 0 & 0.8634 & 0.1318 & -0.4241 & 0.1155 & 1.1583
\end{bmatrix}.
$$

Step 4: Obtain the R_{yy}^{-1} matrix by taking the inverse of the R_{yy} partition selected from the R_{adj} matrix.

Spreadsheets do not generally have functions for calculating Cholesky triangular square root matrices, but the inverse of R_{yy} can be done in Excel with a single instruction. Suppose that the $\mathbf{R}_{yy}$ matrix is found on the spreadsheet in location *R16:V20* (which is a 5×5 matrix location). You highlight cells in the 5×5 location where you wish to have the inverse matrix and enter the instruction "=*MINVERSE(R16:V20)*" and press control-shift-enter (which is Excel's array processing function key). The following inverse matrix for $\mathbf{R}_{yy}$ will appear in this new location.

$$\mathbf{R}_{yy}^{-1} = \begin{bmatrix} 280.371 & 14.460 & 33.963 & -171.693 & 87.309 \\ 14.460 & 2.028 & 1.116 & -8.815 & -4.621 \\ 33.963 & 1.116 & 5.920 & -19.796 & -10.699 \\ -171.693 & -8.815 & -19.796 & 113.775 & 45.923 \\ -87.309 & -4.621 & -10.699 & 45.923 & 35.153 \end{bmatrix}.$$

Step 5: Create the product matrix by the formula $P = R_{xx}'^{-1/2} R_{xy} R_{yy}^{-1} R_{yx} R_{xx}^{-1/2}$.

There is a fairly simple equation for the product matrix that is predictably similar to the fundamental equation for multiple regression analysis (Chapter 9). The multiple regression equation is $R^2 = \mathbf{r}_{yx} \mathbf{R}_{xx}^{-1} \mathbf{r}_{xy}$, and the "computational engine" is the inverse of the $\mathbf{R}_{xx}$ matrix of predictors. Canonical correlation has two sets of "predictors," so to speak: one on the X side and one on the Y side. The equation for the product matrix that underlies canonical correlation analysis is given as:

$$P = \mathbf{R}_{yy}^{-1} \mathbf{R}_{yx} \mathbf{R}_{xx}^{-1} \mathbf{R}_{xy}$$

This matrix will have the dimensions of matrix $\mathbf{Y}$, 5×5 in our case, as can be seen from the subscripts. There is a corresponding product matrix that will have the dimensions of matrix $\mathbf{X}$, 4×4 in our case, which is given as:

$$P = \mathbf{R}_{yy}^{-1} \mathbf{R}_{yx} \mathbf{R}_{xx}^{-1} \mathbf{R}_{xy}$$

These two product matrices will give exactly equivalent results in all of the canonical correlation analysis statistics, but obviously the larger of the two is a little more tedious to calculate.

Actually, both of these formulas are somewhat problematic and will not be used here. Both of them produce a nonsymmetric matrix, which cannot be factored by the ordinary factoring methods we are using in this book. They can be factored by the *singular value decomposition* method introduced in Chapter 3, Section 3.12.3, but since it is a fairly tedious process, we look for another method. Equation 6.2, taken from Van de Geer (1971), does in fact produce a symmetric product matrix that can be factored by our method.[5]

$$P = \mathbf{R}_{xx}^{-\frac{1}{2}} \mathbf{R}_{xy} \mathbf{R}_{yy}^{-1} \mathbf{R}_{yx} \mathbf{R}_{xx}^{-\frac{1}{2}} \tag{6.2}$$

[5] The derivation of the canonical correlation formulas is given on pages 66–69 and 157–161 of Van de Geer (1971).

This equation uses the Cholesky triangular square root matrix for $\mathbf{R}_{xx}$ on both the outer left side and also on the outer right side, but in transposed form on the left. As can be seen from the subscripts, this equation produces a product matrix with the same dimensions as $\mathbf{R}_{xx}$, 4×4 for our data. There is also a similar equation that produces a matrix with the dimensions of $\mathbf{R}_{yy}$. We will use Equation 6.2 to create the 4×4 product matrix.

$$\mathbf{P} = \mathbf{R}_{xx}'^{-\frac{1}{2}} \, \mathbf{R}_{xy} \, R_{yy}^{-1} \, \mathbf{R}_{yx} \, \mathbf{R}_{xx}^{-\frac{1}{2}}$$

$$= \begin{bmatrix} 1.0000 & 0 & 0 & 0 \\ 0.7569 & 1.2542 & 0 & 0 \\ -1.5580 & -0.2660 & 1.7557 & 0 \\ 0.1318 & -0.4241 & 0.1155 & 1.1583 \end{bmatrix}$$

$$\begin{bmatrix} 0.595 & -0.347 & -0.156 & 0.588 & 0.588 \\ -0.495 & 0.576 & 0.353 & -0.441 & -0.428 \\ 0.496 & -0.328 & -1.69 & 0.486 & 0.447 \\ -0.498 & 0.316 & 0.219 & -0.396 & -0.622 \end{bmatrix} \cdot$$

$$\begin{bmatrix} 280.371 & 14.460 & 33.963 & -171.693 & 87.309 \\ 14.460 & 2.028 & 1.116 & -8.815 & -4.621 \\ 33.963 & 1.116 & 5.920 & -19.796 & -10.699 \\ -171.693 & -8.815 & -19.796 & 113.775 & 45.923 \\ -87.309 & -4.621 & -10.699 & 45.923 & 35.153 \end{bmatrix}$$

$$\begin{bmatrix} 0.595 & -0.495 & 0.496 & -0.498 \\ -0.347 & 0.576 & -0.328 & 0.316 \\ -0.156 & 0.353 & -0.169 & 0.219 \\ 0.588 & -0.441 & 0.486 & -0.396 \\ 0.588 & -0.428 & 0.447 & -0.622 \end{bmatrix} \begin{bmatrix} 1.0000 & 0.7569 & -1.5580 & 0.1318 \\ 0 & 1.2542 & -0.2660 & -0.4241 \\ 0 & 0 & 1.7557 & 0.1155 \\ 0 & 0 & 0 & 1.1583 \end{bmatrix}$$

$$= \begin{bmatrix} 0.584 & -0.165 & 0.093 & -0.213 \\ -0.165 & 0.281 & -0.167 & 0.051 \\ 0.093 & -0.167 & 0.184 & 0.090 \\ -0.213 & 0.051 & 0.090 & 0.667 \end{bmatrix}$$

Step 6: Factor the product matrix by the method of successive squaring (Chapter 4, Section 4.4) to obtain the first eigenvector, k_1, of the product matrix P.

The product matrix created in step six using Equation 6.2 is now factored using the method of successive squaring. See Chapter 4, Section 4.4, for the instructions of how this is done.

$$\mathbf{s}_{(1)} = \mathbf{P}1 = \begin{bmatrix} 0.5838 & -0.1652 & 0.0930 & -0.2133 \\ -0.1652 & 0.2814 & -0.1667 & 0.0507 \\ 0.0930 & -0.1667 & 0.1838 & 0.0900 \\ -0.2133 & 0.0507 & 0.0900 & 0.6674 \end{bmatrix}\begin{bmatrix} 1 \\ 1 \\ 1 \\ 1 \end{bmatrix} = \begin{bmatrix} 0.2983 \\ 0.0001 \\ 0.2011 \\ 0.5948 \end{bmatrix}$$

$$\mathbf{k}_{(1)} = \frac{\mathbf{s}_{(1)}}{\max\left(\mathbf{s}_{(1)}\right)} = \begin{bmatrix} 0.5016 \\ 0.0001 \\ 0.3364 \\ 1.0000 \end{bmatrix}$$

$$\mathbf{s}_{(2)} = \mathbf{P}^2 1 = \begin{bmatrix} 0.422290 & -0.169282 & 0.079782 & -0.266901 \\ -0.169282 & 0.136843 & -0.088376 & 0.068325 \\ 0.079782 & -0.088376 & 0.078335 & 0.048285 \\ -0.266901 & 0.068325 & 0.048285 & 0.501627 \end{bmatrix}\begin{bmatrix} 1 \\ 1 \\ 1 \\ 1 \end{bmatrix} = \begin{bmatrix} 0.065889 \\ -0.052490 \\ 0.118026 \\ 0.351337 \end{bmatrix}$$

$$\mathbf{k}_{(2)} = \frac{\mathbf{s}_{(2)}}{max\left(\mathbf{s}_{(2)}\right)} = \begin{bmatrix} 0.1875 \\ -0.1494 \\ 0.3359 \\ 1.0000 \end{bmatrix}$$

$$\mathbf{s}_{(4)} = \mathbf{P}^4 1 = \begin{bmatrix} 0.284586 & 0.119938 & 0.042014 & -0.254308 \\ -0.119938 & 0.059861 & -0.029223 & 0.084538 \\ 0.042014 & -0.029223 & 0.022643 & 0.000671 \\ -0.254308 & 0.084538 & 0.000671 & 0.329865 \end{bmatrix}\begin{bmatrix} 1 \\ 1 \\ 1 \\ 1 \end{bmatrix} = \begin{bmatrix} -0.047646 \\ -0.004762 \\ 0.036105 \\ 0.160767 \end{bmatrix}$$

$$\mathbf{k}_{(4)} = \frac{\mathbf{s}_{(4)}}{\max\left(\mathbf{s}_{(4)}\right)} = \begin{bmatrix} -0.2964 \\ -0.0296 \\ 0.2246 \\ 1.0000 \end{bmatrix}$$

$$\mathbf{s}_{(8)} = \mathbf{P}^8 1 = \begin{bmatrix} 0.161812 & -0.064039 & 0.016242 & -0.166371 \\ -0.064039 & 0.025969 & -0.007393 & 0.063428 \\ 0.016242 & -0.007393 & 0.003132 & -0.012918 \\ -0.166371 & 0.063428 & -0.012918 & 0.180631 \end{bmatrix}\begin{bmatrix} 1 \\ 1 \\ 1 \\ 1 \end{bmatrix} = \begin{bmatrix} -0.052355 \\ 0.017965 \\ -0.000937 \\ 0.064770 \end{bmatrix}$$

$$\mathbf{k}_{(8)} = \frac{\mathbf{s}_{(8)}}{\max\left(\mathbf{s}_{(8)}\right)} = \begin{bmatrix} -0.8083 \\ 0.2774 \\ -0.0145 \\ 1.0000 \end{bmatrix}$$

$$\mathbf{s}_{(16)} = \mathbf{P}^{16}\mathbf{1} = \begin{bmatrix} 0.058227 & -0.022698 & 0.005302 & -0.061244 \\ -0.022698 & 0.008853 & -0.002075 & 0.023854 \\ 0.005302 & -0.002075 & 0.000495 & -0.005545 \\ -0.061244 & 0.023854 & -0.005545 & 0.064497 \end{bmatrix} \begin{bmatrix} 1 \\ 1 \\ 1 \\ 1 \end{bmatrix} = \begin{bmatrix} -0.020413 \\ 0.007934 \\ -0.001823 \\ 0.021561 \end{bmatrix}$$

$$\mathbf{k}_{(16)} = \frac{\mathbf{s}_{(16)}}{\max(\mathbf{s}_{(16)})} = \begin{bmatrix} -0.9467 \\ 0.3680 \\ -0.0845 \\ 1.0000 \end{bmatrix}$$

$$\mathbf{s}_{(32)} = \mathbf{P}^{32}\mathbf{1} = \begin{bmatrix} 0.007685 & -0.002994 & 0.000698 & -0.008087 \\ -0.002994 & 0.001167 & -0.000272 & 0.003151 \\ 0.000698 & -0.000272 & 0.000063 & -0.000735 \\ -0.008087 & 0.003151 & -0.000735 & 0.008510 \end{bmatrix} \begin{bmatrix} 1 \\ 1 \\ 1 \\ 1 \end{bmatrix} = \begin{bmatrix} -0.002699 \\ 0.001052 \\ -0.00245 \\ 0.002840 \end{bmatrix}$$

$$\mathbf{k}_{(32)} = \frac{\mathbf{s}_{(32)}}{\max(\mathbf{s}_{(32)})} = \begin{bmatrix} 0.9502 \\ 0.3703 \\ -0.0863 \\ 1.0000 \end{bmatrix}$$

$$\mathbf{s}_{(64)} = \mathbf{P}^{64}\mathbf{1} = \begin{bmatrix} 0.000134 & -0.000052 & 0.000012 & -0.000141 \\ -0.000052 & 0.000020 & -0.000005 & 0.000055 \\ 0.000012 & -0.000005 & 0.000001 & -0.000013 \\ -0.000141 & 0.000055 & -0.000013 & 0.000148 \end{bmatrix} \begin{bmatrix} 1 \\ 1 \\ 1 \\ 1 \end{bmatrix} = \begin{bmatrix} -0.000047 \\ 0.000018 \\ -0.000004 \\ 0.000049 \end{bmatrix}$$

$$\mathbf{k}_{(64)} = \frac{\mathbf{s}_{(64)}}{\max(\mathbf{s}_{(64)})} = \begin{bmatrix} 0.9502 \\ 0.3703 \\ -0.0863 \\ 1.0000 \end{bmatrix}.$$

The eigenvalue converges on the 64th power of the product matrix, where the trial eigenvector $\mathbf{k}_{(64)}$ is seen to be equivalent to the preceding trial eigenvector $\mathbf{k}_{(64)}$ to four place accuracy. This trial eigenvector is therefore taken to be $\mathbf{k}_1$, the true eigenvector of the product matrix.

$$\mathbf{k}_1 = \begin{bmatrix} 0.9502 \\ 0.3703 \\ -0.0863 \\ 1.0000 \end{bmatrix}.$$

Step 7: Obtain the corresponding eigenvalue λ_1 for the first eigenvector $\mathbf{k}_1$, and take the square root of it to obtain the first canonical correlation coefficient r_1.

When the product matrix **P** is postmultiplied by its first eigenvector $\mathbf{k}_1$, the resultant vector contains the first eigenvalue as its largest value (corresponding to the 1.0000 in the $\mathbf{k}_1$ vector), which in this case is $\lambda_1 = 0.8811$.

$$\mathbf{Pk}_1 = \begin{bmatrix} 0.584 & -0.165 & 0.093 & -0.213 \\ -0.165 & 0.281 & -0.167 & 0.051 \\ 0.093 & -0.167 & 0.184 & 0.090 \\ -0.213 & 0.051 & 0.090 & 0.667 \end{bmatrix} \begin{bmatrix} 0.9502 \\ 0.3703 \\ -0.0863 \\ 1.0000 \end{bmatrix} = \begin{bmatrix} -0.8373 \\ 0.3263 \\ -0.0761 \\ 0.8811 \end{bmatrix}.$$

This eigenvalue is interpreted like an r^2 value. It gives the proportion of variance accounted for in the Y set of latent variables by the X set of latent variables (or vice versa). In other words, it is the square of the first canonical correlation coefficient as expressed in Equation 6.3.

$$r_j = \sqrt{\lambda_j} \tag{6.3}$$

The first canonical correlation coefficient is therefore 0.9387, the square root of 0.8811.

Step 8: Create four additional eigenvectors from $\mathbf{k}_1$: the $\mathbf{q}_1$ eigenvector (normalized to 1), the $\mathbf{f}_1$ eigenvector (normalized to the eigenvalue), the $\mathbf{c}_1$ eigenvector (standardized coefficients for the X set), and the $\mathbf{d}_1$ eigenvector (standardized coefficients for the Y set).

There are five forms of the eigenvector that are used in canonical correlation analysis. The first is the $\mathbf{k}_1$ vector that was obtained in step 6. The second is $\mathbf{q}_1$, the $\mathbf{k}_1$ eigenvector normalized to one. This is done by dividing each entry in $\mathbf{k}_1$ by the square root of its current sum of squares (as explained in Chapter 3, Section 3.11). The sum of squares of $\mathbf{k}_1$ is 2.0475, the square root of which is 1.4309.

$$\mathbf{q}_1 = \left(\frac{1}{1.4309}\right)\mathbf{k}_1 = \left(\frac{1}{1.4309}\right)\begin{bmatrix} 0.9502 \\ 0.3703 \\ -0.0863 \\ 1.0000 \end{bmatrix} = \begin{bmatrix} -0.6641 \\ 0.2588 \\ -0.0603 \\ 0.6989 \end{bmatrix}.$$

The third form is vector $\mathbf{f}_1$, the eigenvector normalized to λ_1. This is done by multiplying the normalized eigenvector, $\mathbf{q}_1$, by the square root of the eigenvalue, which has already been determined to be 0.9387.

$$\mathbf{f}_1 = (0.9387)\mathbf{q}_1 = (0.9387)\begin{bmatrix} -0.6641 \\ 0.2588 \\ -0.0603 \\ 0.6989 \end{bmatrix} = \begin{bmatrix} -0.6234 \\ 0.2429 \\ -0.0566 \\ 0.6560 \end{bmatrix}.$$

We are now ready to calculate the regression coefficients for the X set of variables, which is c_1, a fourth form of the eigenvector. This is accomplished with Equation 6.4.

$$c_j = R_{xx}^{-1/2} q_j \tag{6.4}$$

$$c_1 = R_{xx}^{-1/2} q_1 = \begin{bmatrix} 1.0000 & 0.7569 & -1.5580 & 0.1318 \\ 0 & 1.2542 & -.2660 & -0.4241 \\ 0 & 0 & 1.7557 & 0.1155 \\ 0 & 0 & 0 & 1.1583 \end{bmatrix} \begin{bmatrix} -0.6641 \\ 0.2588 \\ -0.0603 \\ 0.6989 \end{bmatrix} = \begin{bmatrix} -0.2803 \\ 0.0442 \\ -0.0252 \\ 0.8094 \end{bmatrix}$$

The d_1 eigenvector, the fifth form, is calculated according to Equation 6.5.

$$d_j = \left(\frac{1}{r_j} \right) R_{yy}^{-1} R_{yx} \, c_j \tag{6.5}$$

With "j" set to 1 (for the first canonical correlation), the Y set regression coefficients d_1 are calculated from Equation 6.5.

$$d_1 = \left(\frac{1}{r_1} \right) R_{yy}^{-1} R_{yx} \, c_1$$

$$= \left(\frac{1}{0.9387} \right) \begin{bmatrix} 280.371 & 14.460 & 33.963 & -171.693 & 87.309 \\ 14.460 & 2.028 & 1.116 & -8.815 & -4.621 \\ 33.963 & 1.116 & 5.920 & -19.796 & -10.699 \\ -171.693 & -8.815 & -19.796 & 113.775 & 45.923 \\ -87.309 & -4.621 & -10.699 & 45.923 & 35.153 \end{bmatrix}$$

$$\begin{bmatrix} 0.595 & -0.495 & 0.496 & -0.498 \\ -0.347 & 0.576 & -0.328 & 0.316 \\ -0.156 & 0.353 & -0.169 & 0.219 \\ 0.588 & -0.441 & 0.486 & -0.396 \\ 0.588 & -0.428 & 0.447 & -0.622 \end{bmatrix} \begin{bmatrix} -0.2803 \\ 0.0442 \\ -0.0252 \\ 0.8094 \end{bmatrix}$$

$$= \begin{bmatrix} -6.1160 \\ 0.1163 \\ -1.0016 \\ 4.8737 \\ 0.0540 \end{bmatrix}.$$

Step 9: Calculate the first-factor product matrix G_1 by postmultiplying f by its transpose $G_1 = ff'$. Subtract this from the product matrix of step 5 to obtain the matrix of first factor residuals, $P_1 = P - G_1$.

The purpose of the f_1 eigenvector is to create a matrix of the amount of variance and covariance accounted for by the first pair of latent variables. This is done according to Equation 6.6, by postmultiplying $\mathbf{f}_1$ by its transpose to obtain matrix $\mathbf{G}_1$

$$\mathbf{G}_j = \mathbf{f}_j \, \mathbf{f}_j' \tag{6.6}$$

$$\mathbf{G}_1 = \mathbf{f}_1 \, \mathbf{f}_1' = \begin{bmatrix} -0.6234 \\ 0.2429 \\ -0.0566 \\ 0.6560 \end{bmatrix} \begin{bmatrix} -0.6234 & 0.2429 & -0.0566 & 0.6560 \end{bmatrix}$$

$$= \begin{bmatrix} 0.389 & -0.151 & 0.035 & -0.409 \\ -0.151 & 0.059 & -0.014 & 0.159 \\ 0.035 & -0.014 & 0.003 & -0.037 \\ -0.409 & 0.159 & -0.037 & 0.430 \end{bmatrix}.$$

The $\mathbf{G}_1$ matrix is subtracted from the product matrix P to obtain matrix $\mathbf{P}_1$, the matrix of first factor residuals. This removes all of the variance and covariance accounted for by the first factor from the product matrix $\mathbf{P}$, so that an orthogonal second pair of latent variables can be obtained by factoring $\mathbf{P}_1$.

$$\mathbf{P}_1 = \mathbf{P} - \mathbf{G}_1$$

$$= \begin{bmatrix} 0.584 & -0.165 & 0.093 & -0.213 \\ -0.165 & 0.281 & -0.167 & 0.051 \\ 0.093 & -0.167 & 0.184 & 0.090 \\ -0.213 & 0.051 & 0.090 & 0.667 \end{bmatrix} - \begin{bmatrix} 0.389 & -0.151 & 0.035 & -0.409 \\ -0.151 & 0.059 & -0.014 & 0.159 \\ 0.035 & -0.014 & 0.003 & -0.037 \\ -0.409 & 0.159 & -0.037 & 0.430 \end{bmatrix}$$

$$= \begin{bmatrix} 0.195 & -0.014 & 0.058 & 0.196 \\ -0.014 & 0.222 & -0.153 & -0.109 \\ 0.058 & -0.153 & 0.181 & 0.127 \\ 0.196 & -0.109 & 0.127 & 0.237 \end{bmatrix}.$$

Step 10: Factor the first factor residual matrix $\mathbf{P}_1$ by the method of successive squaring to obtain the second eigenvector, $\mathbf{k}_2$.

This is the same process of step 6, but on the residual matrix $\mathbf{P}_1$, rather than the original product matrix. The result will be the second eigenvector $\mathbf{k}_2$. The iteration process requires six steps this time, to obtain trial eigenvector $\mathbf{k}_{(32)}$.

$$\mathbf{s}_{(1)} = \mathbf{P}_1 \mathbf{1} = \begin{bmatrix} 0.195231 & -0.013818 & 0.057748 & 0.195616 \\ -0.013818 & 0.222374 & -0.152976 & -0.108676 \\ 0.057748 & -0.152976 & 0.180603 & 0.127108 \\ 0.195616 & -0.108676 & 0.127108 & 0.237095 \end{bmatrix} \begin{bmatrix} 1 \\ 1 \\ 1 \\ 1 \end{bmatrix} = \begin{bmatrix} 0.4348 \\ -0.0531 \\ 0.2125 \\ 0.4511 \end{bmatrix}$$

$$\mathbf{k}_{(1)} = \frac{\mathbf{s}_{(1)}}{\max(\mathbf{s}_{(1)})} = \begin{bmatrix} 0.9637 \\ -0.1177 \\ 0.4710 \\ 1.0000 \end{bmatrix}$$

$$\mathbf{s}_{(2)} = \mathbf{P}_1^2 \mathbf{1} = \begin{bmatrix} 0.079907 & -0.035863 & 0.048682 & 0.093412 \\ -0.035863 & 0.084853 & -0.076257 & -0.072080 \\ 0.048682 & -0.076257 & 0.075510 & 0.081014 \\ 0.093412 & -0.072080 & 0.081014 & 0.122446 \end{bmatrix} \begin{bmatrix} 1 \\ 1 \\ 1 \\ 1 \end{bmatrix} = \begin{bmatrix} 0.1861 \\ -0.0993 \\ 0.1289 \\ 0.2248 \end{bmatrix}$$

$$\mathbf{k}_{(2)} = \frac{\mathbf{s}_{(2)}}{\max(\mathbf{s}_{(2)})} = \begin{bmatrix} 0.8280 \\ -0.4420 \\ 0.5736 \\ 1.0000 \end{bmatrix}$$

$$\mathbf{s}_{(4)} = \mathbf{P}_1^4 \mathbf{1} = \begin{bmatrix} 0.018767 & -0.016354 & 0.017868 & 0.025431 \\ -0.016354 & 0.019497 & -0.019814 & -0.024470 \\ 0.017868 & -0.019814 & 0.020450 & 0.026081 \\ 0.025431 & -0.024470 & 0.026081 & 0.035478 \end{bmatrix} \begin{bmatrix} 1 \\ 1 \\ 1 \\ 1 \end{bmatrix} = \begin{bmatrix} 0.0457 \\ -0.0411 \\ 0.0446 \\ 0.0625 \end{bmatrix}$$

$$\mathbf{k}_{(4)} = \frac{\mathbf{s}_{(4)}}{\max(\mathbf{s}_{(4)})} = \begin{bmatrix} 0.7312 \\ -0.6581 \\ 0.7131 \\ 1.0000 \end{bmatrix}$$

$$\mathbf{s}_{(8)} = \mathbf{P}_1^8 \mathbf{1} = \begin{bmatrix} 0.001586 & -0.001602 & 0.001688 & 0.002246 \\ -0.001602 & 0.001639 & -0.001722 & -0.002278 \\ 0.001688 & -0.001722 & 0.001810 & 0.002398 \\ 0.002246 & -0.002278 & 0.002398 & 0.003184 \end{bmatrix} \begin{bmatrix} 1 \\ 1 \\ 1 \\ 1 \end{bmatrix} = \begin{bmatrix} 0.0039 \\ -0.0040 \\ 0.0042 \\ 0.0056 \end{bmatrix}$$

$$\mathbf{k}_{(8)} = \frac{\mathbf{s}_{(8)}}{\max(\mathbf{s}_{(8)})} = \begin{bmatrix} 0.7058 \\ -0.7140 \\ 0.7521 \\ 1.0000 \end{bmatrix}$$

$$\mathbf{s}_{(16)} = \mathbf{P}_1^{16} \mathbf{1} = \begin{bmatrix} 0.000013 & -0.000013 & 0.000014 & 0.000018 \\ -0.000013 & 0.000013 & -0.000014 & -0.000019 \\ 0.000014 & -0.000014 & 0.000015 & 0.000020 \\ 0.000018 & -0.000019 & 0.000020 & 0.000026 \end{bmatrix} \begin{bmatrix} 1 \\ 1 \\ 1 \\ 1 \end{bmatrix} = \begin{bmatrix} 0.000032 \\ -0.000033 \\ 0.000034 \\ 0.000046 \end{bmatrix}$$

$$\mathbf{k}_{(16)} = \frac{\mathbf{s}_{(16)}}{\max(\mathbf{s}_{(16)})} = \begin{bmatrix} 0.7047 \\ -0.7164 \\ 0.7538 \\ 1.0000 \end{bmatrix}$$

$$\mathbf{s}_{(32)} = \mathbf{P}_1^{32}\mathbf{1} = \begin{bmatrix} 8.74E-10 & -8.88E-10 & 9.35E-10 & 1.24E-09 \\ -8.88E-10 & 9.03E-10 & -9.50E-10 & -1.26E-09 \\ 9.35E-10 & -9.50E-10 & 1.00E-09 & 1.33E-09 \\ 1.24E-09 & -1.26E-09 & 1.33E-09 & 1.76E-09 \end{bmatrix} \begin{bmatrix} 1 \\ 1 \\ 1 \\ 1 \end{bmatrix}$$

$$= \begin{bmatrix} 2.16E-09 \\ -2.20E-09 \\ 2.31E-09 \\ 3.06E-09 \end{bmatrix}. \qquad \mathbf{k}_{(32)} = \frac{\mathbf{s}_{(32)}}{\max(\mathbf{s}_{(32)})} = \begin{bmatrix} 0.7047 \\ -0.7164 \\ 0.7538 \\ 1.0000 \end{bmatrix}.$$

Trial eigenvector $\mathbf{k}_{(32)}$ now becomes the second eigenvector $\mathbf{k}_2$ of this canonical correlation analysis.

$$\mathbf{k}_2 = \begin{bmatrix} 0.7047 \\ -0.7164 \\ 0.7538 \\ 1.0000 \end{bmatrix}.$$

Step 11: Obtain the corresponding eigenvalue λ_2 for the second eigenvector $\mathbf{k}_2$, and also obtain the second canonical correlation coefficient r_2 by taking the square root of λ_2.

The second eigenvalue is obtained by the same process used in step 7.

$$\mathbf{P}_1\mathbf{k}_2 = \begin{bmatrix} 0.195 & -0.014 & 0.058 & 0.196 \\ -0.014 & 0.222 & -0.153 & -0.109 \\ 0.058 & -0.153 & 0.181 & 0.127 \\ 0.196 & -0.109 & 0.127 & 0.237 \end{bmatrix} \begin{bmatrix} 0.7047 \\ -0.7164 \\ 0.7538 \\ 1.0000 \end{bmatrix} = \begin{bmatrix} 0.3866 \\ -0.3930 \\ 0.4135 \\ 0.5486 \end{bmatrix}.$$

The second eigenvalue is the largest value of this product vector $\lambda_i = 0.5486$. The second canonical correlation coefficient is obtained by Equation 6.3.

$$r_2 = \sqrt{\lambda_2} = \sqrt{0.5486} = 0.7407.$$

Step 12: Create three additional eigenvectors from $\mathbf{k}_2$: the $\mathbf{q}_2$ eigenvector (normalized to 1), the $\mathbf{c}_2$ eigenvector (standardized coefficients for the X set), and the $\mathbf{d}_2$ eigenvector (standardized coefficients for the Y set).

The $\mathbf{q}_2$ eigenvector is obtained by normalizing $\mathbf{k}_2$ to 1.

$$\mathbf{q}_2 = \left(\frac{1}{1.6056}\right)\mathbf{k}_2 = \left(\frac{1}{1.6056}\right) \begin{bmatrix} 0.7047 \\ -0.7164 \\ 0.7538 \\ 1.0000 \end{bmatrix} = \begin{bmatrix} 0.4389 \\ -0.4462 \\ 0.4695 \\ 0.6228 \end{bmatrix}.$$

The $\mathbf{c}_2$ eigenvector, the regression coefficients for the X set of variables, is obtained using Equation 6.4 from step 8.

$$\mathbf{c}_2 = \mathbf{R}_{xx}^{-1/2}\mathbf{q}_2 = \begin{bmatrix} 1.0000 & 0.7569 & -1.5580 & 0.1318 \\ 0 & 1.2542 & -0.2660 & -0.4241 \\ 0 & 0 & 1.7557 & 0.1155 \\ 0 & 0 & 0 & 1.1583 \end{bmatrix} \begin{bmatrix} 0.4389 \\ -0.4462 \\ 0.4695 \\ 0.6228 \end{bmatrix} = \begin{bmatrix} -0.5622 \\ -0.9486 \\ 0.8962 \\ 0.7214 \end{bmatrix}.$$

The $\mathbf{d}_2$ eigenvector, regression coefficients for the Y set of variables, is obtained using Equation 6.5 from step 8.

$$\mathbf{d}_2 = \left(\frac{1}{r_2}\right)\mathbf{R}_{yy}^{-1}\,\mathbf{R}_{yx}\,\mathbf{c}_2$$

$$= \left(\frac{1}{0.7407}\right) \begin{bmatrix} 280.371 & 14.460 & 33.963 & -171.693 & 87.309 \\ 14.460 & 2.028 & 1.116 & -8.815 & -4.621 \\ 33.963 & 1.116 & 5.920 & -19.796 & -10.699 \\ -171.693 & -8.815 & -19.796 & 113.775 & 45.923 \\ -87.309 & -4.621 & -10.699 & 45.923 & 35.153 \end{bmatrix}$$

$$\begin{bmatrix} 0.595 & -0.495 & 0.496 & -0.498 \\ -0.347 & 0.576 & -0.328 & 0.316 \\ -0.156 & 0.353 & -0.169 & 0.219 \\ 0.588 & -0.441 & 0.486 & -0.396 \\ 0.588 & -0.428 & 0.447 & -0.622 \end{bmatrix} \begin{bmatrix} -0.5622 \\ -0.9486 \\ 0.8962 \\ 0.7214 \end{bmatrix}$$

$$= \begin{bmatrix} 6.1254 \\ -0.1916 \\ 0.8308 \\ -1.6146 \\ -3.9197 \end{bmatrix}.$$

Step 13: Adjoin the individual $\mathbf{c}$ and $\mathbf{d}$ eigenvectors to create matrix $\mathbf{C}$, the matrix of standardized coefficients for the X set, and matrix $\mathbf{D}$, the matrix of standardized coefficients for the Y set.

The matrix of regression coefficients $\mathbf{C}$ for creating canonical variate scores (latent variables) for the X set is obtained by adjoining $\mathbf{c}_1$ from step eight and $\mathbf{c}_2$ from step 12.

$$\mathbf{C} = \mathbf{c}_1 \mid \mathbf{c}_2 = \begin{bmatrix} -0.2803 & -0.5622 \\ 0.0442 & -0.9486 \\ -0.0252 & 0.8962 \\ 0.8094 & 0.7214 \end{bmatrix}.$$

The corresponding matrix of regression coefficients $\mathbf{D}$ for creating canonical variate scores for the Y set is obtained by adjoining $\mathbf{d}_1$ from step 8 and $\mathbf{d}_2$ from step 12.

$$\mathbf{D} = \mathbf{d}_1 \mid \mathbf{d}_2 = \begin{bmatrix} -6.1160 & 6.1254 \\ 0.1163 & -0.1916 \\ -1.0016 & 0.8308 \\ 4.8737 & -1.6146 \\ 0.0540 & -3.9197 \end{bmatrix}.$$

Step 14: Obtain the matrix of canonical variates χ *(chi), the latent variables for the* X *set, by postmultiplying* $\mathbf{Z}_x$ *by its matrix of standardized coefficients* **C**.

The X side matrix of latent variables, that is, the matrix of canonical variates χ is obtained by postmultiplying the $\mathbf{Z}_x$ scores by matrix **C**, according to Equation 6.7.

$$\chi = \mathbf{Z}_x \mathbf{C}. \tag{6.7}$$

$$\chi = \mathbf{Z}_x \mathbf{C} = \begin{bmatrix} 0.90 & -1.30 & 0.71 & -0.53 \\ -0.05 & 0.33 & 0.07 & -0.63 \\ -1.71 & 1.00 & -1.11 & 0.03 \\ 0.05 & 0.94 & 1.00 & -0.13 \\ 1.66 & -0.79 & 1.25 & -0.94 \\ -1.09 & -0.51 & -1.61 & 0.50 \\ -0.19 & 0.73 & -0.11 & -0.16 \\ -0.57 & 0.94 & -1.72 & 0.19 \\ -0.10 & -1.60 & -0.14 & -0.44 \\ 0.71 & 0.06 & 0.14 & -0.28 \\ -0.95 & 1.45 & -0.36 & 2.98 \\ -0.24 & -0.09 & 0.57 & -0.91 \\ 1.57 & -1.15 & 1.29 & 0.31 \end{bmatrix} \begin{bmatrix} -0.2803 & -0.5622 \\ 0.0442 & -0.9486 \\ -0.0252 & 0.8962 \\ 0.8094 & 0.7214 \end{bmatrix}$$

$$= \begin{bmatrix} -0.761 & 0.983 \\ -0.482 & -0.678 \\ 0.577 & -0.956 \\ -0.099 & -0.111 \\ -1.295 & 0.253 \\ 0.731 & 0.024 \\ -0.039 & -0.792 \\ 0.397 & -1.971 \\ -0.397 & 1.130 \\ -0.429 & -0.534 \\ 2.755 & 0.988 \\ -0.689 & 0.075 \\ -0.268 & 1.589 \end{bmatrix}.$$

Step 15: Obtain the matrix of canonical variates η (eta), the latent variables for the Y set, by postmultiplying Z_y by its matrix of standardized coefficients D.

The matrix of canonical variates η is obtained according to Equation 6.8 by postmultiplying the Z_y scores by matrix D.

$$\eta = Z_y D \tag{6.8}$$

$$\eta = Z_y D = \begin{bmatrix} 0.00 & -1.19 & -0.44 & -0.29 & 0.16 \\ -0.86 & -1.36 & 0.22 & -1.01 & -0.66 \\ -1.86 & 0.34 & 1.63 & -1.88 & -1.80 \\ 0.14 & 1.19 & 0.65 & 0.29 & 0.33 \\ 1.29 & -1.02 & 0.22 & 1.30 & 1.31 \\ -0.43 & 0.00 & -0.76 & -.58 & -0.66 \\ 0.43 & 1.36 & 0.22 & 0.43 & 0.82 \\ -0.71 & 1.69 & 1.09 & -0.58 & -0.16 \\ 1.14 & -0.85 & -0.44 & 1.30 & 0.98 \\ 1.43 & -0.68 & -1.42 & 1.45 & 1.15 \\ -1.00 & 0.17 & 0.11 & -0.72 & -1.48 \\ 0.86 & -0.17 & -1.96 & 0.58 & 0.66 \\ -0.43 & 0.51 & 0.87 & -2.9 & -0.66 \end{bmatrix} \begin{bmatrix} -6.1160 & 6.1254 \\ 0.1163 & -0.1916 \\ -1.0016 & 0.8308 \\ 4.8737 & -1.6146 \\ 0.0540 & -3.9197 \end{bmatrix}$$

$$= \begin{bmatrix} -1.102 & -0.311 \\ -.102 & -0.604 \\ 0.503 & 0.024 \\ 0.037 & -0.562 \\ -1.787 & 1.006 \\ 0.531 & 0.247 \\ -0.523 & -1.369 \\ 0.647 & -2.218 \\ -0.257 & 0.842 \\ -0.289 & 0.869 \\ 2.424 & 0.886 \\ -0.445 & 0.149 \\ 0.363 & 1.040 \end{bmatrix}.$$

Now that we have given a full numerical demonstration of the formulas and steps in canonical correlation analysis, we will verify our matrix algebra calculations using Stata. To run this canonical correlation analysis in Stata, we first enter the data. Suppose that the **Zx**|**Zy** data are recorded on a spreadsheet as shown below.

zx1	zx2	zx3	zx4	zy1	zy2	zy3	zy4	zy5
−0.90	−1.30	0.71	−0.53	0.00	−1.19	−0.44	−0.29	0.16
−0.05	0.33	0.07	−0.63	−0.86	−1.36	0.22	−1.01	−0.66
−1.71	1.00	−1.11	0.03	−1.86	0.34	1.63	−1.88	−1.80
−0.05	0.94	1.00	−0.13	0.14	1.19	0.65	0.29	0.33
1.66	−0.79	1.25	−0.94	1.29	−1.02	0.22	1.30	1.31
−1.09	−0.51	−1.61	0.50	−0.43	0.00	−0.76	−0.58	−0.66
−0.19	0.73	−0.11	−0.16	0.43	1.36	0.22	0.43	0.82
−0.57	0.94	−1.72	0.19	−0.71	1.69	1.09	−0.58	−0.16
−0.10	−1.60	−0.14	−0.44	1.14	−0.85	−0.44	1.30	0.98
0.71	0.06	0.14	−0.28	1.43	−0.68	−1.42	1.45	1.15
−0.95	1.45	−0.36	2.98	−1.00	0.17	0.11	−072	−1.48
−0.24	−0.09	0.57	−0.91	0.86	−0.17	−1.96	0.58	0.66
1.57	−1.15	1.29	0.31	−0.43	0.51	0.87	−0.29	−0.66

We copy the entire 13×9 matrix of data, including the labels at the top from the spreadsheet, and paste it directly into the Stata data editor (the leftmost of the two editor buttons on the Stata menu at the top).

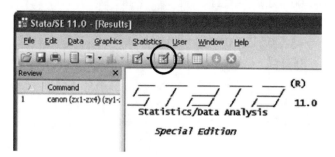

(StataCorp, 2009. Stata: Release 11. Statistical Software. College Station, TX: StataCorp LP.)

In the Stata command line, the following code is entered, and it runs the canonical correlation analysis on these data.

```
canon (zx1-zx4) (zy1-zy4), first(2)
```

The Stata instruction format for the "canon" procedure requires that one put the X set variables in parentheses, and then the Y set variables in parentheses. We enter the option "first(2)" at the end (following the comma), which indicates that Stata should extract only the first two canonical correlation coefficients. The following results will appear in the output screen of Stata.

Canonical correlation analysis   Number of obs =   13

Raw coefficients for the first variable set

	1	2
x1	0.2797	-0.5604
x2	-0.0419	-0.9477
x3	0.0250	0.8960
x4	-0.8113	0.7195

Raw coefficients for the second variable set

	1	2
y1	6.1104	6.1175
y2	-0.1152	-0.1933
y3	0.9997	0.8298
y4	-4.8748	-1.6105
y5	-0.0494	-3.9162

Canonical correlations:
   0.9385   0.7401   0.4956   0.2066

Tests of significance of all canonical correlations

	Statistic	df1	df2	F	Prob>F
Wilks' lambda	.0389544	20	14.2164	1.1803	0.3808 a
Pillai's trace	1.71682	20	28	1.0527	0.4419 a
Lawley-Hotelling trace	8.96525	20	10	1.1207	0.4437 a
Roy's largest root	7.38384	5	7	10.3374	0.0039 u

e = exact, a = approximate, u = upper bound on F

(StataCorp, 2009. Stata: Release 11. Statistical Software. College Station, TX: StataCorp LP.)

The canonical correlation coefficients reported here by Stata are very close to the two coefficients we obtained (0.9387 for canonical correlation coefficient one in step 7, and 0.7407 for canonical correlation coefficient two in step 11). Stata prints out all four canonical correlation coefficients (equivalent to the number of variables in the smaller of the two input matrices), even though we have only requested the canonical analysis corresponding to the first two.

What Stata refers to as the "raw coefficients for the first variable set," and also the ones they give for the second variable set in the output above, are the same (within rounding error) as the **C** and **D** matrices of regression coefficients we obtained in step 13 above.[6] If we had submitted the raw data matrix **X|Y** for analysis rather than the standardized **Zx|Zy** matrix, these would indeed have been the raw regression coefficients (to be used for constructing latent variables from raw data), but since we have entered a **Z** score matrix for the input data, these coefficient are equivalent to standardized regression coefficients, notwithstanding their label in Stata. The regression coefficients are used to obtain canonical variate scores as we did in step 15 by postmultiplying the X set and Y

---

[6] In a demonstration like this, our computational precision is not as high as in the statistical packages, but even they will have small differences like this among themselves.

set **Z** scores by them. Although the standard Stata output does not give canonical variate scores, it is a simple matter to create them using the matrix algebra capabilities of a spreadsheet like Excel, drawing upon the regression coefficients from Stata output and the original **Z** score data (as shown in step fourteen).

Step 16: Concatenate the matrices of observed variables, $\mathbf{Z}_x$ and $\mathbf{Z}_y$, with the matrices of latent variables, chi and eta, to create an adjoined total matrix of **Z** *scores, $\mathbf{Z}_t = \mathbf{Z}_x|\chi|\mathbf{Z}_y|\eta$. Calculate an adjoined total matrix of correlation coefficients $\mathbf{R}_t$ from this, by $\mathbf{R}_t = (1/df)\mathbf{Z}_t'\mathbf{Z}_t$.*

It is substantially easier to obtain canonical variate scores (the chi and eta latent variable score matrices) in SAS than in Stata. The following SAS code (with the "all" option and also the "out=DEMOout" instruction in the first line of the PROC CANCOR module) creates essentially all of the canonical correlation output one would want, and then some.

```
*****************************************************************************
***     Program: CCdemo.sas
***     Canonical correlation of demonstration dataset
*****************************************************************************/;
data demo;
input zx1-zx4 zy1-zy5;
 datalines;
  .903   -1.302   .715   -.534    .000  -1.186  -.436   -.289    .164
 -.048    .333    .071   -.628   -.857  -1.355   .218  -1.012   -.656
-1.710    .999  -1.108    .031  -1.857   .339   1.633  -1.880  -1.804
  .048    .939   1.001   -.126    .143   1.186   .653    .289    .328
 1.663   -.787   1.251   -.942   1.286  -1.017   .218   1.301   1.312
-1.093   -.515  -1.608    .503   -.429    .000  -.762   -.578   -.656
 -.190    .727   -.107   -.157    .429   1.355   .218    .434    .820
 -.570    .939  -1.715    .188   -.714   1.694  1.089   -.578   -.164
 -.095  -1.605   -.143   -.440   1.143   -.847  -.436   1.301    .984
  .713    .061    .143   -.283   1.429   -.678 -1.416   1.446   1.148
 -.950   1.453   -.357   2.984  -1.000    .169   .109   -.723  -1.476
 -.238   -.091    .572   -.911    .857   -.169 -1.960    .578    .656
 1.568  -1.150   1.287    .314   -.429    .508   .871   -.289   -.656
 ;
run;

proc cancorr data=demo all out=DEMOout corr
  vprefix=DV vname='Independent Variables'
  wprefix=IV wname='Dependent Variables';
 var zx1-zx4;
 with zy1-zy5;
 title 'Canonical correlation of demonstration data';
run;

proc print data=DEMOout;
run;
```

The biggest problem with the SAS output is making sense of a large and confusing array of results—about 20 separate modules. There is a way to simplify and organize the output. A simplified way of describing what we have done so far is that we have derived two pairs of latent variables, one for the X set (chi) and one for the Y set (eta). In the process, we have determined the correlation coefficient for each pair (canonical correlation coefficients one and two), and we have obtained the regression coefficients (matrix **C** and matrix **D**) that are used to create the latent variable scores. To see all of the output matrices in context, we will now do two things—we will combine the observed and the latent scores into a total **Z** score matrix, $\mathbf{Z}_t$, and from that matrix, we will obtain a total correlation matrix that contains all of the relationships among the four kinds of **Z** scores (raw X set, chi latent variables for the X set, raw Y set, and eta latent variables for the Y set).

We created the $\mathbf{Z}_t$ matrix on a spreadsheet by concatenating the four matrices in the order shown in the spreadsheet screen capture below.[7] The number of rows (observations) in this $\mathbf{Z}_t$ matrix is $n = 13$. Interestingly, it accidently happens that the number of columns is also 13, since we have four $\mathbf{Z}_x$ variables, two chi[8] latent variables, five $\mathbf{Z}_y$ variables, and two eta latent variables, which sums to 13.

$\mathbf{Z}_t =$	x1	x2	x3	x4	chi1	chi2	y1	y2	y3	y4	y5	eta1	eta2
	.903	-1.302	.715	-.534	-.761	.983	.000	-1.186	-.436	-.289	.164	-1.102	-.311
	-.048	.333	.071	-.628	-.482	-.678	-.857	-1.355	.218	-1.012	-.656	-.102	-.604
	-1.710	.999	-1.108	.031	.577	-.956	-1.857	.339	1.633	-1.880	-1.804	.503	.024
	.048	.939	1.001	-.126	-.099	-.111	.143	1.186	.653	.289	.328	.037	-.562
	1.663	-.787	1.251	-.942	-1.295	.253	1.286	-1.017	.218	1.301	1.312	-1.787	1.006
	-1.093	-.515	-1.608	.503	.731	.024	-.429	.000	-.762	-.578	-.656	.531	.247
	-.190	.727	-.107	-.157	-.039	-.792	.429	1.355	.218	.434	.820	-.523	-1.369
	-.570	.939	-1.715	.188	.397	-1.971	-.714	1.694	1.089	-.578	-.164	.647	-2.218
	-.095	-1.605	-.143	-.440	-.397	1.130	1.143	-.847	-.436	1.301	.984	-.257	.842
	.713	.061	.143	-.283	-.429	-.534	1.429	-.678	-1.416	1.446	1.148	-.289	.869
	-.950	1.453	-.357	2.984	2.755	.988	-1.000	.169	.109	-.723	-1.476	2.424	.886
	-.238	-.091	.572	-.911	-.689	.075	.857	-.169	-1.960	.578	.656	-.445	.149
	1.568	-1.150	1.287	.314	-.268	1.589	-.429	.508	.871	-.289	-.656	.363	1.040

We copy the values of this 13×13 $\mathbf{Z}_t$ matrix from the spreadsheet and paste them directly into the Stata spreadsheet for data entry. Once the 13 variables have been entered (with the same labels shown here), the correlation matrix is obtained with the Stata command:

`correlate x1-x4 chi1 chi2 y1-yt eta1 eta2`

The correlation matrix will be given by Stata in lower triangular matrix form which can be copied and pasted directly into a spreadsheet and reworked into the form most useable.

We have re-organized it into the spreadsheet format (shown in the screen capture of $\mathbf{R}_t$ below) in a way that will facilitate identifying all of the major items in the SAS canonical correlation output results. The large bold letters on each matrix partition below are used to identify the corresponding section of SAS output.

[7] We have used the symbols x1, x2, y1, y2, etc., for these for simplification of labeling even though they are in fact $\mathbf{Z}_x$ and $\mathbf{Z}_y$ scores.

[8] The variable names chi and eta are used in this spreadsheet table of data rather than the Greek symbols because Stata does not deal well with Greek letters in the variable names.

$R_t = Z_t' Z_t =$

	x1	x2	x3	x4	chi1	chi2	y1	y2	y3	y4	y5	eta1	eta2
x1	1.000												
x2	-.604	1.000											
x3	.813	-.394	1.000										
x4	-.416	.474	-.337	1.000									
chi1	-.664	.607	-.543	.955	1.000								
chi2	.439	-.621	.570	.204	.000	1.000							
y1	.595	-.495	.496	-.498	-.604	.221	1.000						
y2	-.347	.576	-.328	.316	.387	-.417	-.309	1.000					
y3	-.156	.353	-.169	.219	.241	-.240	-.657	.466	1.000				
y4	.588	-.441	.486	-.396	-.517	.237	.985	-.229	-.572	1.000			
y5	.588	-.428	.447	-.622	-.699	.028	.957	-.195	-.519	.935	1.000		
eta1	-.623	.570	-.510	.897	.939	.000	-.643	.412	.256	-.551	-.744	1.000	
eta2	.325	-.460	.422	.151	.000	.741	.298	-.563	-.324	.321	.038	.000	1.000

Block labels: **a** R_{xx}; **b** R_{yy}; **c**; **d**; **e** I; **f**; R_{xy}; I

The list of titles of the 20 SAS output modules from the SAS code above (with corresponding identifying letters for the screen capture) is as follows:

(1) Means and Standard Deviations

(2) Correlations Among the Independent Variables (matrix partition "a" on the screen capture)

(3) Correlations Among the Dependent Variables ("b")

(4) Correlations Between the Independent Variables and the Dependent Variables ("c")

(5) Canonical Correlation Analysis (This is a large section that includes the canonical correlation coefficients, "d," and the maximum likelihood significance tests for each.)

(6) Multivariate Statistics and F Approximations (the four multivariate statistics for the whole group of coefficients, as shown in Section 6.2 on Whitley's data)

(7) Raw Canonical Coefficients for the Independent Variables

(8) Raw Canonical Coefficients for the Dependent Variables

(9) Standardized Canonical Coefficients for the Independent Variables (matrix **C** of step 13)

(10) Standardized Canonical Coefficients for the Dependent Variables (matrix **D** of step 13)

(11) Correlations Between the Independent Variables and Their Canonical Variables ("e"): These are what Whitley calls canonical variate loadings in his Table 2, and are the basis of the canonical correlation summary tables that we will construct in step seventeen.

(12) Correlations Between the Dependent Variables and Their Canonical Variables ("f"): The canonical variate loadings for the Y set.

The remaining eight modules of SAS output give additional correlation matrices (X set observed variables with Y set latent variables, etc.) which we will not use here. They also give "variance accounted for" statistics that are more conveniently and understandably presented through the canonical correlation summary tables of step 17.

Step 17: From the correlation coefficients in the $R_{x\chi}$ partition of the total R_t matrix, create the canonical correlation summary table for the X set of variables. From the correlation coefficients in the $R_{y\eta}$ partition of the R_t matrix, create the canonical correlation summary table for the Y set of variables.

We need two sets of canonical variate loadings, one for the X set and one for the Y set. Each is the matrix of correlation coefficients between the observed variables and the corresponding latent variables. Each correlation matrix is a partition of the R_t matrix shown in the screen capture above. For the X set, the correlation matrix partition is $R_{x\chi}$, and, for the Y set, the correlation matrix partition is $R_{y\eta}$, as shown below.

$$\mathbf{R}_{x\chi} = \begin{bmatrix} -0.664 & 0.439 \\ 0.607 & -0.621 \\ -0.543 & 0.570 \\ 0.955 & 0.204 \end{bmatrix}$$

$$\mathbf{R}_{y\eta} = \begin{bmatrix} -0.643 & 0.298 \\ 0.412 & -0.563 \\ 0.256 & -0.324 \\ -0.551 & 0.321 \\ -0.744 & 0.038 \end{bmatrix}.$$

From these two correlation matrix partitions we create the linked canonical correlation summary table displayed as Table 6.3.

Canonical correlation analysis can be conducted on either a correlation matrix, or a covariance matrix, but the correlation matrix approach is more common.[9] We have used the correlation matrix approach here. In the foregoing, we have given a simplified demonstration of the underlying mathematics of the canonical correlation process coordinated with the results obtained from SAS and Stata. We try to accomplish three things in this chapter. In the first section, Section 6.2, canonical correlation is introduced as it has been usefully applied in published research. In this section, (Section 6.3), the process is given of how you can analyze your own data, together with an explanation of the conceptual and mathematical basis of the results that will be obtained from the statistical packages.

If the data of this section were actual empirical data, we would complete the demonstration process by interpreting the canonical correlation summary table just given and create and interpret a joint vector plot (like Fig. 6.1 of Whitley data) and create and demonstrate comparative scatter plots of individual observations. Instead, we will now close this mathematical demonstration and give those interpretative demonstrations using an empirical data set of 650 market trading days on nine market index variables in the X set, and nine TIAA-CREF mutual fund performance variables in the Y set.

6.4 ILLUSTRATIONS OF CANONICAL CORRELATION TABLES AND GRAPHICS WITH FINANCE DATA

On March 9 and 10 of 2000, the Nasdaq closing price exceeded 5000 (5046.86 and 5048.62 on those 2 days). For only those two days, the Nasdaq closing price exceeded one half the closing price of the Dow Jones Industrial Average, which was hovering around 10000. Within less than a month, however, the

---

[9] As was mentioned in Chapter 4 in conjunction with principal components of a covariance matrix, canonical correlation of a covariance matrix only makes sense when the variables are commensurate, that is, share a scale of measurement that is in some way comparable.

TABLE 6.3. Canonical Correlation Summary Table with the _Y_ Set of Cognitive Performance Test Variables on Top and the _X_ Set of Neurological Measurement Variables on the Bottom Artificial Demonstration Data

	Loadings		Squared Loadings			Uniqueness
	LV1	LV2	LV1	LV2	Total	
Canonical Variate Loadings for Five Cognitive Performance Tests (Y set)						
Y1 cognitive test	−.643	.298	.414	.089	.503	.497
Y2 cognitive test	.412	−.563	.170	.317	.487	.513
Y1 cognitive test	.256	−.324	.066	.105	.171	.829
Y2 cognitive test	−.551	.321	.304	.103	.407	.593
Y1 cognitive test	−.744	.038	.554	.001	.555	.445
Sum of squares by columns:			1.507	.615	2.123	2.877
Percents of sums of squares:			30%	12%	42%	58%
Canonical Variate Loadings for Four Neurological Measurement Variables (X set)						
X1 neurological measurement	−.664	.439	.441	.193	.634	.366
X2 neurological measurement	.607	−.621	.369	.385	.754	.246
X3 neurological measurement	−.543	.570	.295	.325	.620	.380
X4 neurological measurement	.955	.204	.913	.041	.954	.046
Sum of squares by columns:			2.017	.945	2.962	1.038
Percents of sums of squares:			50%	24%	74%	26%
			Coefficient	_p_ Value		
First Canonical Correlation			.9387	n.s.		
Second Canonical Correlation			.7407	n.s.		

Nasdaq had receded from this high to a level of 3321.29, about a 34% drop. It continued a fairly steady decline until, by the summer of 2001, it had dropped below the 2000 level as shown in Figure 6.2.

Of course, there is a close causal relationship between the performance of market indices and the current value of mutual funds. Hendrix et al. (2002) used these historical market events to illustrate the application of canonical correlation analysis to financial data of this kind—to examine the link between several market index variables and several mutual fund prices. In particular,

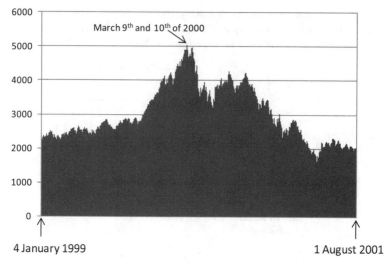

Figure 6.2 Histogram of Nasdaq closing prices for the 650 market trading days from January 4, 1999 to August 1, 2001 (data from Hendrix et al., 2002).

they examined the daily closing values of nine mutual funds of TIAA-CREF (stock fund, money market fund, bond fund, social choice fund, global equities fund, growth fund, equities fund, inflation-linked bond fund, and TIAA real estate fund) over the 650 trading days from January 4, 1999 to August 1, 2001. In concert with these mutual fund daily values, they also obtained corresponding daily market performance statistics with regard to the Dow Jones Industrial Average and the Nasdaq index. For both the Dow and also for the Nasdaq, they obtained four statistics: daily closing price, daily percent price gain or loss, percentile location of the closing price within the daily spread (thought by some to be a predictor of tomorrow's performance), and daily spread as a percentage of the closing price. The ninth statistic was the market volume for the day.

Since these are certainly not commensurate measures (with daily volume rising to over 18 million on occasion, and several market measures being measured in percents), the data were first standardized and then analyzed by Hendrix et al. using canonical correlation analysis. The results are given in Table 6.4, the linked canonical correlation summary table.

The first thing we notice from this table is that the latent variables from the X set and from the Y set are very highly correlated with one another, particularly for the first two pairs of latent variables, with the first canonical correlation being 0.9948, and the second being 0.9477. Actually, the first five canonical correlation coefficients are all highly significant. However, we quit extracting canonical factors at three in order to be able to investigate the relationships graphically. Figure 6.3 shows the nine vectors from the canonical variate loadings of market indices (bold vectors) superimposed on the nine vectors from the TIAA-CREF mutual funds (light vectors).

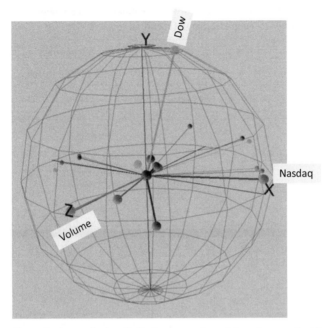

Figure 6.3   Canonical correlation linked vector plot of the nine market indices, the X set, and the nine TIAA-CREF mutual funds, the Y set. The larger vectors show the nine market indices, with Nasdaq closing price being the vector corresponding to the x-axis. The Dow closing price is the large vector pointing upward, and the daily volume is the large vector that points off to the left coinciding with the z-axis. (From Hendrix et al., 2002. Data by permission of authors. Image courtesy of (Metrika.org), © Appzilla, LLC.)

We notice that three market index vectors are "bounding vectors" in the sense that the other six market index variables lie in between them. In fact, they coincide quite closely with the axes for the three latent variables, with the Nasdaq corresponding to latent variable 1, the x-axis (pointing to the right), the Dow corresponding to latent variable 2, the vertical axis y, and the vector for market volume pointing down and to the left in correspondence with latent variable 3, the z-axis in this three-dimensional sphere.

In Table 6.4 (the bottom half showing the X set), we confirm that these three market indices corresponding very closely to the three latent variables, with all three (the circled loadings entries) having loadings on their respective latent variable at 0.96 or above. We also see from the row total sums of squares that the Dow closing price and the Nasdaq closing price are 100% accounted for within this X set latent space.

Several additional things can be observed from Table 6.4. The latent variable space accounts for substantially more variance in the mutual fund variables of the Y set (92%). The corresponding latent variable space for market index variables, the X set, only accounts for 40%. In other words, there is much more unique variance (60%) in the market indices—many other things going

TABLE 6.4. Canonical Correlation Summary Tables Showing a Y Set of Variables for Mutual Fund Performance on Top, and an X Set of Market Index Variables on the Bottom. Taken from an Analysis of Measurements on 650 Market Trading Days from January 4, 1999 to August 1, 2001 (Hendrix et al., 2002)

	Loadings			Squared Loadings				Uniqueness
	LV1	LV2	LV3	LV1	LV2	LV3	Total	
Canonical Variate Loadings for Nine Mutual Fund Performance Variables (Y set)								
CREF Stock Fund	.89	.39	.00	.79	.15	.00	.94	.06
CREF Money Market Fund	−.17	.29	.91	.03	.08	.83	.94	.06
CREF Bond Fund	−.45	.14	.86	.20	.02	.73	.96	.04
CREF Social Choice Fund	.56	.53	.41	.32	.28	.17	.76	.24
CREF Global Equities	.97	.21	.08	.94	.04	.01	.99	.01
CREF Growth Fund	.95	.15	−.14	.90	.02	.02	.94	.06
CREF Equities Index	.85	.42	.06	.73	.17	.00	.90	.10
CREF Inflation Linked Bond	−.37	.22	.87	.14	.05	.75	.94	.06
TIAA Real Estate Fund	−.17	.30	.90	.03	.09	.81	.93	.07
Sum of squares by columns:				4.07	.91	3.33	8.31	.69
Percents of sums of squares:				45%	10%	37%	92%	8%
Canonical Variate Loadings for Nine Market Index Variables (X set)								
DJIA Closing Price	.26	⟨.96⟩	−.04	.07	.93	.00	1.00	.00
DJIA Percent Gain	−.01	.11	−.08	.00	.01	.01	.02	.98
DJIA Percentile Closing	−.10	.07	−.03	.01	.00	.00	.02	.98
DJIA Spread for the Day	.34	−.22	.40	.12	.05	.16	.33	.67
DJIA Volume	.02	.02	⟨.96⟩	.00	.00	.93	.93	.07
NASD Closing Price	⟨.99⟩	.11	−.01	.99	.01	.00	1.00	.00
NASD Percent Gain	.03	.05	−.10	.00	.00	.01	.01	.99
NASD Percentile Closing	.03	.03	−.14	.00	.00	.02	.02	.98
NASD Spread for the Day	.07	−.03	.50	.01	.00	.25	.25	.75

TABLE 6.4. (*Continued*)

Sum of squares by columns:	1.19	1.01	1.38	3.58	5.42
Percents of sums of squares:	13%	11%	15%	40%	60%

Canonical Correlation Analysis Summary Statistics

	Coefficient	*p* Value
First Canonical Correlation	.9948	<.0001
Second Canonical Correlation	.9477	<.0001
Third Canonical Correlation	.6722	<.0001
Fourth Canonical Correlation	.4207	<.0001
Fifth Canonical Correlation	.2770	<.0001

on that are unrelated to mutual fund performance. On the other hand, the mutual fund variables have very little uniqueness (8%). They are for the most part dominated by market index performance.

It is interesting to note in the top half of Table 6.4 and in the vector plot that none of the mutual funds is highly related to the latent variable corresponding to the Dow, the vertical latent variable. The highest is the social choice fund. The other mutual funds seem to go in fours. Four are highly related to the Nasdaq closing price: stock fund, global equities, growth fund, and equities index. Four are highly related to volume (in other words simply tracking overall market growth): money market, bond fund, inflation-linked bond, and real estate fund. Probably none of these particular observations would be surprising to an economist or a finance professor, but a "hard numbers" data set of this kind is particularly useful in investigating the operating characteristics of the instrument, canonical correlation analysis.

Of particular interest are the two linked scatter plots of individual observations, created from the canonical variate matrices, χ and η. Before examining the time-series scatter plots of these two, we will first examine Figure 6.4, which shows the overall topology of the time course of the 650 trading days through this latent variable space.

The overall appearance of Figure 6.4 is like a three-legged stool or two croquet rings, in the shape of a V from above. The market indices start at the beginning of 1999 at the bottom of the back leg of the stool and rise vertically for the first part of 1999, indicating a rise in the Dow. The plot then turns horizontal and moves to the right, indicating a rise in the Nasdaq while the Dow stays about where it is. For the last part of 1999, the time-series plot line drops with the Nasdaq being at its high while the Dow has dropped. Then, in 2000, the plot line begins to rise again, signaling a rise in the Dow. This soon turns left, however, indicating that the Nasdaq is beginning to drop. The Dow stays up and the Nasdaq drops until the spring of 2001, when the Dow also drops, creating the third leg of the three-legged stool.

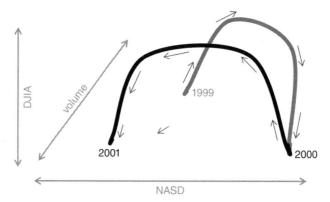

Figure 6.4 Topology of the time-series path of 650 trading days in the space defined by the Y set of variables—mutual fund performance.

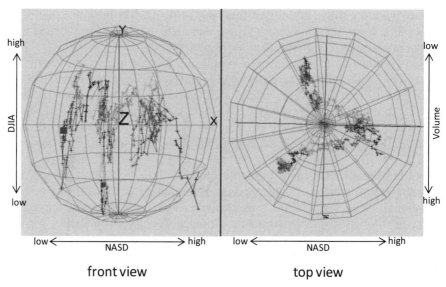

front view   top view

Figure 6.5 Front view and top view of the time line of 650 trading days within the three-dimensional space defined by the first three canonical variates of the Y set of observed variables, the financial performance of nine mutual funds. (From Hendrix et al., 2002. Data by permission of authors. Images courtesy of Metrika (metrika.org), © Appzilla, LLC.)

We now examine this same topology in the latent variable space of the nine mutual funds of TIAA-CREF, as shown in Figure 6.5, in front view and in top view. On the right of this figure, in the top view, one has the positive end of the z-axis pointing downward, and the negative end of it pointing upward. The V shape of the time line is very clear here, with 1999 starting at the top of this "top view figure" and moving over to the right (increase on the Nasdaq), then turning and going back left (decrease on the Nasdaq), all the while the market increases in volume (which, since the z-axis points down, means that the points progress toward the bottom of the chart from start to finish).

The front view shows the three-legged stool shown in Figure 6.4 and the accompanying discussion, with the middle leg (corresponding to the start of 1999) being in the background, moving forward and to the right, and then continuing to move forward toward the viewer as it goes back to the left (decreasing Nasdaq prices) and finally drops to the bottom of the third leg (Dow also having declined). It should be emphasized that what we are seeing here is the space of the prices of mutual funds, the Y set, not the actual market indices. The corresponding figure for the market indices, the X set, is Figure 6.6.

From the front view of Figure 6.6, one can see the three-legged stool, but in the top view, the V shape cannot be seen. This is because volume swings back and forth quite rapidly (in terms of days) rather than progressing slowly as the prices on the Dow and Nasdaq do. In Figure 6.5, we do not see volume directly. We see it only through its more long-term effects on mutual fund prices, and this smoothes the time line into a recognizable pattern.

Again, it should be emphasized that the point of this is not to learn finance in that most of what we are observing is probably already known to an

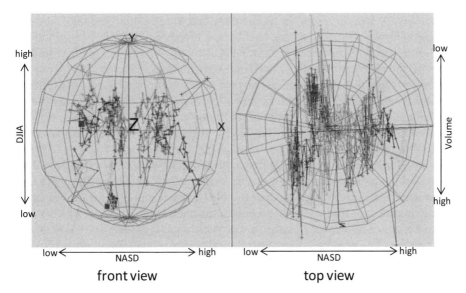

front view   top view

Figure 6.6 Front view and top view of the time line of 650 trading days within the three-dimensional space defined by the first three canonical variates of the X set of observed variables, the daily values of nine market indices. (From Hendrix et al., 2002. Data by permission of authors. Images courtesy of Metrika (metrika.org), © Appzilla, LLC.)

economist, but to see the strong performance characteristics of canonical correlation analysis in apprehending the holistic pattern in two linked sets of multivariate data.

6.6   SUMMARY AND CONCLUSIONS

It has been argued in this chapter that canonical correlation potentially has much to offer as a method of understanding one's data holistically. We have seen in Chapter 4 that factor analysis has played a major role in the behavioral science of the past century. We have also have seen in Chapter 5 that factor analysis and related methods have much to offer in the process of examining data graphically. Most of the multivariate graphical methods are based in one way or another on factoring.

On the other hand, multiple regression has the capability of demonstrating how well performance on some measure can be predicted by a combination of other measures. Canonical correlation combines the holistic organizational strengths of factoring with the regression capability of identifying predictability of one variable by a combination of others. There is an important synergy from canonical correlation analysis comparing two sides of a predictive equation. If a factor analysis summary table is useful in identifying the structure in a set of variables (such things as how the variables relate to one another, which have the most common variance with the rest of the set), canonical correlation does this for two data sets simultaneously and at the same time enables comparison between the two tables. Likewise, if a vector plot from factor loadings is useful in understanding the relationships among variables, canonical doubles this by showing the pattern of two sets, with the interactive effect of comparisons between the two on the same set of persons or observation units. If multivariate scatter plots in a reduced latent variable space are of value in understanding the internal structure of data, again, canonical correlation multiplies this by two, and, in addition, has the comparative information, as shown with the market trading data.

There is another way of dealing with complex sets of multivariate data, structural equations modeling and confirmatory factor analysis. These are very powerful later developments (over the past three decades or so) in the factor analytic tradition that constitute a computational general case of which factor analysis and canonical correlation analysis are special cases. Whereas principal component analysis, canonical correlation analysis, and some ways of using factor analysis are strong in enabling one to gain a holistic understanding of complex data and in visually getting a better apprehension of one's data, structural equations models enable one to test specific scientific models against complex empirical data. Both approaches begin with correlation matrices or covariance matrices, but the ways in which they use them and the kinds of information they derive from them are quite different. The strength of structural equations modeling is in testing specific deductive scientific hypotheses, while the major strength of the multivariate graphical approach is in exploratory data analysis and finding holistic patterns.

STUDY QUESTIONS

A. Essay Questions

1. What are some reasonable and useful applications of canonical correlation?

2. How is canonical correlation related to calculating a Pearson correlation? How is it related to multivariate multiple regression?

3. Explain the interpretation of canonical correlation as a "double factor analysis" and how "summary tables" can help with interpretation of the output.

4. Explain how canonical variates and canonical coefficients compare with factor loadings, transformation matrix coefficients, and factor scores.

5. Eigenvectors are defined "only to an arbitrary multiplier." Therefore, an eigenvector can appear in various forms, defined according to the value to which that eigenvector is normalized. Enumerate and explain the various forms in which the eigenvectors appear in canonical correlation and the uses for each.

6. Explain "left-handed" and "right-handed" eigenvectors (computational step 5) and the part they play in canonical correlation.

7. Explain each 2×2 submatrix in the canonical correlation output matrix below:

$$
R_o = \begin{bmatrix}
\begin{bmatrix} 1.000 & -.161 \\ -.161 & 1.000 \end{bmatrix} & \begin{bmatrix} 0.736 & 0.677 \\ -0.787 & 0.617 \end{bmatrix} & \begin{bmatrix} 0.758 & -.341 \\ 0.110 & 0.857 \end{bmatrix} & \begin{bmatrix} 0.673 & 0.516 \\ -0.719 & 0.471 \end{bmatrix} \\
\begin{bmatrix} 0.736 & -0.787 \\ 0.677 & 0.617 \end{bmatrix} & \begin{bmatrix} 1.000 & 0.000 \\ 0.000 & 1.000 \end{bmatrix} & \begin{bmatrix} 0.399 & -0.801 \\ 0.686 & 0.367 \end{bmatrix} & \begin{bmatrix} 0.914 & 0.000 \\ 0.000 & 0.762 \end{bmatrix} \\
\begin{bmatrix} 0.758 & 0.110 \\ -0.341 & 0.857 \end{bmatrix} & \begin{bmatrix} 0.399 & 0.686 \\ -0.801 & 0.367 \end{bmatrix} & \begin{bmatrix} 1.000 & 0.051 \\ 0.051 & 1.000 \end{bmatrix} & \begin{bmatrix} 0.436 & 0.900 \\ -0.876 & 0.482 \end{bmatrix} \\
\begin{bmatrix} 0.673 & -0.719 \\ 0.516 & 0.471 \end{bmatrix} & \begin{bmatrix} 0.914 & 0.000 \\ 0.000 & 0.762 \end{bmatrix} & \begin{bmatrix} 0.436 & -0.876 \\ 0.900 & 0.482 \end{bmatrix} & \begin{bmatrix} 1.000 & 0.000 \\ 0.000 & 1.000 \end{bmatrix}
\end{bmatrix}
$$

$$
= \begin{bmatrix}
R_{xx} & R_{x\chi} & R_{xy} & R_{x\eta} \\
R_{\chi x} & R_{\chi\chi} & R_{\chi y} & R_{\chi\eta} \\
R_{yx} & R_{y\chi} & R_{yy} & R_{y\eta} \\
R_{\eta x} & R_{\eta\chi} & R_{\eta y} & R_{\eta\eta}
\end{bmatrix}.
$$

B. Computational Questions

1. Compute a canonical correlation analysis of the following simplest case data all the way to the scores on canonical variates.

X Set		Y Set	
X1	X2	Y1	Y2
9	13	6	10
10	12	2	7
7	10	3	9
12	10	12	6
8	9	10	4
4	9	5	12
6	7	11	8

2. Compute a canonical correlation analysis of the following standardized data from all the way to the scores on canonical variates.

X Set		Y Set	
X1	X2	Y1	Y2
1.64	0	0.32	−0.61
0	1.44	1.29	1.22
0	0.96	0.65	0
0.55	0	0	0.61
0.55	−0.96	−0.194	0.61
−1.64	−1.44	−0.32	−1.83

3. From the data in questions 1 and 2, plot the vectors for the two manifest variables in each set within the space of the latent variables and compare the X set to the Y set.

C. Data Analysis

1. Use SAS (or SPSS or Stata) to compute a canonical correlation analysis on the data of computational (part B) question 1 above. Use the "all" command to get complete output tables. Create an output file with the "out" command, and print it using PROC PRINT. Compare your results to those from your computations from part B.

2. Plot the results of the computations in questions 1 of part B with three "generalized draftsman's displays" (among canonical variates, within the X set and within the Y set) using Excel's graphing function or some other graphing method.

3. Use SAS (or SPSS or Stata) to compute a canonical correlation analysis on the data of computational (part B) question 2 above. Use the "all" command to get complete output tables. Create an output file with the "out" command, and print it using PROC PRINT. Compare your results with those from your computations from Part B.

4. Use SAS (or SPSS or Stata) to compute a canonical correlation analysis on the following simplest case data that involves more variables (four in the X set and five in the Y set), and more observations (13 of them). Use the "all" command to get complete output tables. Create an output file with the "out" command, and print it using PROC PRINT. Assume that the four X set variables are market indices and the five Y set variables are monthy performance values for a bank. From the SAS output create a canonical correlation summary table like Table 6.4, a linked vector plot like Figure 6.3, and linked time series scatter plots like Figures 6.5 and 6.6, and interpret the results.

x1	x2	x3	x4	y1	y2	y3	y4	y5
9	13	6	10	9	9	4	1	3
10	12	2	7	6	2	7	2	9
7	10	3	9	7	12	6	9	8
12	10	12	6	3	8	3	3	5
8	9	10	4	7	11	1	7	6
4	9	5	12	1	3	3	6	6
6	7	11	8	2	4	4	7	5
10	13	4	10	8	10	6	2	8
12	12	9	7	4	9	4	7	5
9	10	7	9	4	7	7	5	3
7	9	10	6	2	6	3	8	9
6	9	6	4	7	6	1	4	5
4	7	6	12	5	4	3	4	6

5. Use the SAS program at the end of Section 6.3 to run the full canonical correlation analysis on the demonstration data from that section. Go through the output and identify each of the many correlation matrices and identify which each is of those in the $\mathbf{R}_t$ matrix from the end of the demonstration in Section 6.3.

REFERENCES

Hendrix, K. A., Brown, B. L., and Hendrix, S. B. 2002. Canonical correlation: The underused method. *American Statistical Association 2002 Proceedings of the Section on Statistical Graphics* (1430–1454). Alexandria, Virginia: American Statistical Association.

Hotelling, H. 1935. The most predictable criterion. *Journal of Educational Psychology, 26,* 139–142.

Hotelling, H. 1936. Relations between two sets of variables. *Biometrika, 28,* 321–377.

Johnson, R. A., and Wichern, D. W. 2007. *Applied Multivariate Statistical Analysis, Sixth Edition.* Upper Saddle River, NJ: Pearson Prentice-Hall.

Lattin, J., Carroll, J. D., and Green, P. E. 2003. *Analyzing Multivariate Data.* Belmont, CA: Brooks/Cole.

Rencher, A. C. 2002. *Methods of Multivariate Analysis, Second Edition.* New York: Wiley.

Tabachnick, B. G., and Fidell, L. S. 2007. *Using Multivariate Statistics.* Boston: Pearson.

Van de Geer, J. P. 1971. *Introduction to Multivariate Analysis for the Social Sciences.* San Francisco, CA: W. H. Freeman and Company.

Whitley, B. E. 1999. Right-wing authoritarianism, social dominance orientation, and prejudice. *Journal of Personality and Social Psychology, 77*(1), 126–134.

CHAPTER SEVEN

HOTELLING'S T^2 AS THE SIMPLEST CASE OF MULTIVARIATE INFERENCE

7.1 INTRODUCTION

Harold Hotelling was one of the major contributors to modern statistical methods (Olkin and Sampson, 2001). Besides his T-squared test (T^2), Hotelling created canonical correlation analysis. In addition to being a theoretical statistician, he was also an applied statistician and is well known for his work in economic theory.

In creating the T^2 test, Hotelling explicitly set out to create a multivariate generalization of the Student's t-distribution (Gossett, 1908, 1943), as the titles of his two major papers on the subject (Hotelling, 1931, 1951) attest. It has become a useful and indispensable component in contemporary statistical method and theory. Although many still conduct multiple t-tests on multivariate data without using a multivariate omnibus test, such as Hotelling's T^2 test, they do so at their own peril. Such practices invite problems with alpha inflation. Rencher (2002) gives four arguments as to why a multivariate approach to hypothesis testing is desirable.

Multivariate Analysis for the Biobehavioral and Social Sciences: A Graphical Approach,
First Edition. Bruce L. Brown, Suzanne B. Hendrix, Dawson W. Hedges, Timothy B. Smith.
© 2012 John Wiley & Sons, Inc. Published 2012 by John Wiley & Sons, Inc.

RENCHER'S (2002) FOUR ARGUMENTS FOR A MULTIVARIATE APPROACH TO HYPOTHESIS TESTING

1. The use of p univariate tests inflates the Type I error rate.
2. Univariate tests do not model the correlations among the dependent variables, whereas the multivariate test does.
3. The multivariate test is in many cases more powerful.
4. Many multivariate significance tests obtain an optimal descriptive linear combination, such as a discriminant function, as a by-product.

It is generally expected that when someone conducts a multivariate hypothesis test, it will often be accompanied by or followed up with a univariate hypothesis test, if for no other reason than to discover which particular variables are responsible for the effect. Rencher's second reason we should consider adopting a multivariate approach to hypothesis testing is because univariate tests do not take into account the covariance structure of the data. In addition, the multivariate test in many cases is more powerful than the univariate test.

Taking into account the covariance structure was an important issue to Hotelling, and also to his contemporary, Mahalanobis. At the very time Hotelling was developing the multivariate generalization of the Student's t-test, nearly 80 years ago, Mahalanobis (1930, 1936) was contributing to the development of statistically based distance functions (now referred to as Mahalonobis distance). There is much in common between Hotelling's T^2 development and Mahalanobis distance-function work. In fact, when Mahalanobis distance is applied to means and multiplied by a constant reflecting sample size, it becomes the formula for Hotelling's T^2 test.

Consider the progression of equations below. The first is the Pythagorean theorem, which forms the basis of Euclidean geometry. The second is a matrix expression of the generalized Pythagorean theorem, which expands the capacity to deal with a higher number of dimensions than three.

$$c^2 = a^2 + b^2$$

$$d^2 = (\mathbf{y}_1 - \mathbf{y}_2)'(\mathbf{y}_1 - \mathbf{y}_2).$$

These distance functions are geometrical, but they are not statistical. There is no adjustment for the variances or the covariances. For a statistical distance, we standardize by inserting the inverse of the covariance matrix, as shown in Equation 7.1, the Mahalanobis distance function. Notice that the formulas for Hotelling's three types of T^2 analysis, the one for a single mean (Eq. 7.2), the one for two independent samples (Eq. 7.3), and the one for two correlated samples (Eq. 7.4), only differ from Mahalanobis' formula by the n-related multiplicative constants and the incorporation of means rather than raw data.

$$d^2 = (\mathbf{y}_1 - \mathbf{y}_2)' \mathbf{S}^{-1} (\mathbf{y}_1 - \mathbf{y}_2) \tag{7.1}$$

$$T^2 = n' (\bar{\mathbf{y}} - \boldsymbol{\mu}_0) \, \mathbf{S}^{-1} (\bar{\mathbf{y}} - \boldsymbol{\mu}_0) \tag{7.2}$$

$$T^2 = \left(\frac{n_1 n_2}{n_1 + n_2} \right) (\bar{\mathbf{y}}_1 - \bar{\mathbf{y}}_2)' \mathbf{S}_{pl}^{-1} (\bar{\mathbf{y}}_1 - \bar{\mathbf{y}}_2) \tag{7.3}$$

$$T^2 = n \bar{\mathbf{d}}' \mathbf{S}_d^{-1} \bar{\mathbf{d}}. \tag{7.4}$$

Notice that all four of these are expressions of the "quadratic form" introduced in Chapter 3, Section 3.10.

7.2   AN APPLIED EXAMPLE OF HOTELLING'S T^2 TEST: FAMILY FINANCES AND RELATIONAL AGGRESSION

Many studies have investigated the causes of conflict in marriage and the family dynamics that lead to it (Schwartz, 2000). It has been shown (Papp et al., 2009) that the most problematic and pervasive form of marital conflict is over finances. However, an alternative explanation is that rather than finances being the source of marital conflict, finances may be just one more battleground for a relationship that is already conflicted. This is the position taken in a recent paper by McBride et al. (2011). The authors hold that finances are just one more aspect of a pattern of conflict and aggressive strategies, including relational aggression (Carroll et al., 2010), rather than being the source of conflict.

From a group of 325 couples participating in the Flourishing Families Project introduced in Chapter 5, Section 5.5, two groups are selected—the high financial conflict group being those with a combined score of nine[1] or higher on a financial conflict measure, and the low financial conflict group being those with a combined score of two or lower. A Hotelling's T^2 test (for independent groups) is used to determine how well high financial conflict group versus low financial conflict group accounts for the pattern in couple ratings on three scales: (1) the Gottman Couple Communication Scale, an indicator of destructive communication strategies (gotcomw and gotcomh for wife and husband, respectively); (2) the Relational Aggression Scale, an indicator of the extent to which couples engage in strategies of subtle mutual offenses, including

---

[1] A nine is a very high score. The question is "indicate the amount of conflict you have in each of these areas," with one of the areas being finances. A five is "we always have conflict about this." To obtain their couple score, both her score and his score were combined. That means that a 10 would be both of them giving the maximum score of five, but a nine would be that one puts four and one puts five.

social sabotage, love withdrawal, rumors and gossip, mutual defamation, and so forth (relagw and relagh); and (3) a Financial Wellbeing Scale constructed from a collection of nine variables that measure satisfaction and agreement/disagreement on financial matters (allfinw and allfinh). It was hypothesized that financial conflict would be *at least* as related to the Gottman Scale and the Relational Aggression Scale as to the Financial Wellbeing Scale.

The 87×7 matrix of data is entered into the Stata data editor. The 87 rows are the 42 high financial conflict couples plus the 45 low financial conflict couples. The seven columns are the scores of both husband and wife on the three scales, plus a column entitled "bgroup" that contains the binary grouping variable. The Stata instruction to run the Hotelling procedure on the data is:

hotelling relagw gotcomw allfinw relagh gotcomh allfinh, by(bgroup)

Figure 7.1 gives the results of the Hotelling independent groups T^2 analyses for these data. Stata gives quite detailed output from the analysis, with a table

```
. hotelling relagw gotcomw allfinw relagh gotcomh allfinh, by( bgroup)

-> bgroup = 1
    variable |      Obs        Mean    Std. Dev.       Min       Max

      relagw |       42    32.16667    14.09881        12        66
     gotcomw |       42    35.90476     8.447281       22        54
     allfinw |       42    23.57143     6.443537       10        35
      relagh |       42    36.7619     11.98131        16        74
     gotcomh |       42    37.5         6.608383       25        59

      allfinh |      42    24.88095     5.747544       12        35

-> bgroup = 2
    variable |      Obs        Mean    Std. Dev.       Min       Max

      relagw |       45    19.46667     8.817854       12        53
     gotcomw |       45    25.4         5.193527       16        37
     allfinw |       45    35.97778     5.006763       23        44
      relagh |       45    20.28889     6.265425       12        32
     gotcomh |       45    28.06667     7.212363       13        49

      allfinh |      45    34.84444     4.6902         23        44
```

```
2-group Hotelling's T-squared = 167.2275
F test statistic: ((87–6–1)/(87–2)(6)) x 167.2275 = 26.231764

H0: Vectors of means are equal for the two groups
          F(6,80) =    26.2318
   Prob > F(6,80) =     0.0000
```

scale	Hi Conflic	Lo Conflic
relagw	2.681	1.622
gotcomw	2.762	1.954
relagh	3.063	1.691
gotcomh	2.885	2.159
allfinw	2.946	4.497
allfinh	3.110	4.356

Figure 7.1 Hotelling T^2 tabular output from Stata for the Flourishing Families anaysis of Relational Agression Scale, Gottman Communication Scale, and the combined financial variables, tested for difference between the high financial conflict group and the low financial conflict group. The insert at the bottom right has the scale scores from the analysis transformed into average scale rating to facilitate interpretation. (StataCorp, 2009. Stata: Release 11. Statistical Software. College Station, TX: StataCorp LP.)

for each of the two groups showing the mean ratings, the standard deviation, and the minimum and the maximum values. At the bottom of the output, it gives the obtained T^2 value, which is 167.2275, and the probability value associated with it, $P = 0.0000$. The result is highly significant and in the direction the investigators hypothesized. That is, there is a substantially higher level of conflict on the Gottman scale and the Relational Aggression Scale for the high-financial conflict couples than for the low-financial conflict couples. The T^2 value is quite large.

The little adjusted means insert in the lower right of Figure 7.1 is intended to make the results more interpretable in absolute terms. There were 13 rating scale responses combining together to obtain the Gottman scale score and 12 for the Relational-Aggression scale score. These mean scale scores were divided by 13 and 12, respectively, to get the figures in this small table, which indicate the average 7-point scale score for each group on the Relational Aggression measure and the average five-point scale score for the Gottman. The collection of financial questions were a mix of 5- and 4-point scales and are not therefore as interpretable, but from the Gottman and the Relation Aggression Means, it can be seen that even the high financial conflict group have scores below the scale midpoint on both the Gottman (midpoint of three) and also Relational Aggression (midpoint of four).

7.3 MULTIVARIATE VERSUS UNIVARIATE SIGNIFICANCE TESTS

Two kinds of situations are of interest with regard to the trade-off between univariate and multivariate hypothesis tests. The first is when the multivariate test indicates that there is statistical significance showing that the null hypothesis can be rejected, but there is no corresponding significant univariate statistic. The second is the reverse: when there is a statistically significant univariate test, but the corresponding multivariate test is not significant. In this chapter, we present examples of artificial datasets that illustrate each of these situations.

In Section 7.4 in connection with a demonstration of the two-sample independent groups Hotelling's T^2, we have the first case—highly significant multivariate test but with neither univariate test being significant. For this particular configuration of data, it just so happens that the basis of group separation is in between the two variables. That is, it requires the combined information of both variables to adequately separate the two groups, and discriminant analysis fulfills that function—to reveal the optimal combination of the variables. The second case is where the multivariate test is not significant but a univariate t-test indicates significance.

Figure 7.2 (reprinted from Rencher, 2002) provides a way for thinking clearly about situations in which the significance differs between the multivariate and the univariate tests. We have included a 3D inset to remind the reader that what looks like an ellipse arranged at a diagonal and intersecting

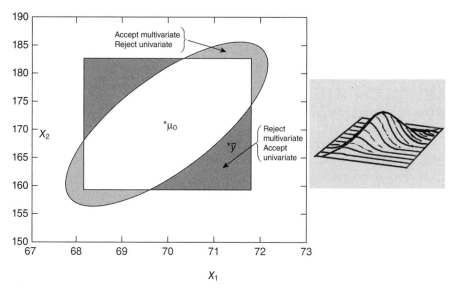

Figure 7.2 Acceptance and rejection regions for univariate and multivariate tests, (Reproduced from page 116 of Rencher, A. C. 2002. *Methods of Multivariate Analysis,* Second Edition. New York: Wiley. Copyright © 2002 by John Wiley & Sons, Inc. All rights reserved.)

a rectangle is meant to represent an actually existing "hill" of data—a multivariate normal recording of the recording of the density of numerical occurrence at each point on the X_1 and X_2 plane.

What we are seeing from Rencher's top view is a topological ring around the "hill" of data that has exactly 5% of the integrated density of this bivariate normal distribution outside of it. It is not a matter of debating about whether the univariate rectangle of decision or the elliptical topographical contour around the actually existing distribution is the correct nature of things— obviously it is the actual distribution. In other words, the multivariate decision is always right. The dark-shaded area inside the rectangle but outside that ellipse that Rencher has labeled "reject multivariate, accept univariate" shows us the region where the univariate decision procedure is inadequate because it lacks power. It fails to identify differences that truly are significant.

On the other hand, the lighter-shaded areas in Rencher's figure, inside the tips of the ellipse but outside the rectangle, are where the univariate tests are guilty of alpha inflation. In actuality, those points are not outside the 95% majority of the multivariate normal distribution, but our myopic univariate decision strategy leads us to wrongly think they are and to reject the null hypothesis while declaring a statistically significant result where there is none. This phenomenon has been referred to as Rao's paradox (for further discussion, see Rao, 1966; Healy, 1969; and Morrison, 1990, 174). The lesson in all of this is that we should only believe a univariate test when it has the decision of a corresponding multivariate test to back it up. The Bonferroni adjustment method (Bonferroni, 1936) and other methods of controlling for alpha inflation are really "protectionist" measures. They simply reduce the size of the rectangle by making the tests more conservative, but without taking into account the covariance structure, thus weakening the power of the test. In contrast, Hotelling's T^2 prevents alpha inflation without loss of power.

7.4 THE TWO SAMPLE INDEPENDENT GROUPS HOTELLING'S T^2 TEST

We now return to the subject of the z-tests discussed in Chapter 2 (Section 2.7), and the relationship of the z-tests to Z-scores. Consider a random variable X. It might be the girth size of each peach in the harvest of 326,429 peaches in a small Midwestern town in the fall of 2012.[2] The measurement of girth is in principle continuous, but to simplify, suppose that our measurements are taken in whole numbers, whole centimeters.

Ordinarily, the girth size and weight of peaches in this small town can be thought of as somewhat random, but this year, there is a new factory in town, and the residents who love their peaches crop are afraid that their crop will be harmed by industrial pollution. Subjectively, it seems that the peaches close to the factory are smaller and also more dense (slightly higher weight).

As an experiment, they make comparative measurements of a sample of peaches close to the factory and a sample safely distant from the factory. They take measurements on peach girth and also on peach weight in grams. Since there are two measures to be considered, we need a multivariate test. In particular, we will use the Hotelling's T^2 procedure for two independent samples (Eq. 7.3) to test the difference. Table 7.1 shows the input data for this analysis.

The Hotelling's T^2 procedures are among the simplest and most straightforward of all of the multivariate methods. If there were no statistical packages, we could still do these analyses well with only a handheld calculator. We first demonstrate the mathematics and concepts with matrix notation and then show that the Stata results agree with ours. We will use Equation 7.3, the one for use with independent groups data.

---

[2] For hypothetical data, the peach example is memorable, concrete, and pedagogically attractive even though it's not a behavioral or social science example, although it does fit the "bio" part of the "biobehavioral sciences." We've tried to give the storyline a political science spin.

TABLE 7.1. Input Data for Independent Groups Hotelling's T^2 Demonstration

Sample	Girth	Grams
1	15	130
1	17	139
1	17	164
1	19	155
1	20	172
2	17	105
2	17	122
2	19	122
2	19	138
2	21	156
2	23	173
2	21	122

Sample 1 is from near the factors. Sample 2 is far from the factory.

$$T^2 = \left(\frac{n_1 n_2}{n_1 + n_2}\right)(\bar{\mathbf{y}}_1 - \bar{\mathbf{y}}_2)' \mathbf{S}_{pl}^{-1}(\bar{\mathbf{y}}_1 - \bar{\mathbf{y}}_2). \qquad (7.3)$$

In the first group, the ones that come from the affected area, we have $n_1 = 5$ bushels of peaches sampled (for which we have these averaged measurements).[3] In the second group, we have $n_2 = 7$ bushels of peaches sampled. We will simplify the formula somewhat by defining a quantity w.

$$w = \frac{n_1 n_2}{n_1 + n_2} = \frac{(5)(7)}{5 + 7} = 2.917.$$

We obtain means vectors for sample 1 and sample 2 with the mean of the first variable, girth, in the first column of each vector, and the mean of the second variable, grams, in the second column of each vector.

---

[3] We obviously could do the analysis on individual peaches instead, in which case there would be thousands of units, but that would ruin the simplicity of the demonstration and almost certainly assure that we would have statistically significant results with such a large N, even if they are not practically meaningful.

$$\bar{y}_1 = [17.600 \quad 152.00]$$

$$\bar{y}_2 = [19.571 \quad 134.00].$$

We subtract the second means vector from the first to obtain a vector of differences between the means.

$$(\bar{y}_1 - \bar{y}_2) = [1.971 \quad -18].$$

The pooled SSCP matrix for within group error, $\mathbf{S}_{pl}$, is obtained using Equation 7.5, where $\mathbf{T}_j$ is the matrix of group totals, and $\mathbf{M}_j$ is the matrix of group means, and where the df_1 for the first sample is 4, and df_2 for the second sample is 6.

$$
\mathbf{S}_{pl} = \left(\frac{1}{df_1 + df_2}\right)[\mathbf{X'X} - \mathbf{T}_j\mathbf{M}_j] = \left(\frac{1}{10}\right)\left[\begin{bmatrix} 4275 & 32102 \\ 32102 & 245712 \end{bmatrix} - \begin{bmatrix} 4230.09 & 31734 \\ 31734 & 241212 \end{bmatrix}\right]
$$

$$
= \left(\frac{1}{10}\right)\begin{bmatrix} 44.91 & 368 \\ 368 & 4500 \end{bmatrix} = \begin{bmatrix} 4.491 & 36.8 \\ 36.8 & 450.0 \end{bmatrix}.
$$

(7.5)

We now need the inverse of this pooled covariance matrix for the formula. It could be done by one of the methods explained in Chapter 3, but we will instead use a spreadsheet to save some time. The Excel command to obtain the inverse of a matrix in, for example, cells C2 to D3 is "=minverse(C2:D3)". We find the inverse of $\mathbf{S}_{pl}$.

$$
\mathbf{S}_{pl}^{-1} = \begin{bmatrix} 4.491 & 36.8 \\ 36.8 & 450.0 \end{bmatrix}^{-1} = \begin{bmatrix} 0.675 & -0.055 \\ -0.055 & 0.007 \end{bmatrix}.
$$

These quantities can now be entered into Equation 7.3 to obtain the Hotelling's T^2 statistic.

$$
T^2 = \left(\frac{n_1 n_2}{n_1 + n_2}\right)(\bar{y}_1 - \bar{y}_2)'\mathbf{S}_{pl}^{-1}(\bar{y}_1 - \bar{y}_2) = w(\bar{y}_1 - \bar{y}_2)'\mathbf{S}_{pl}^{-1}(\bar{y}_1 - \bar{y}_2)
$$

$$
= (2.917)[1.971 \quad -18]'\begin{bmatrix} 0.675 & -0.055 \\ -0.055 & 0.007 \end{bmatrix}[1.971 \quad -18] = 25.436
$$

We go now to Table E of the Appendix and look up the tabled critical ratio of Hotelling's T^2 distribution, with $p = 2$ (two dependent variables), and with 10 degrees of freedom. The critical ratio for the 0.01 level is 17.826, which we exceed, indicating that we can reject the null hypothesis of no difference between peach samples at the 0.01 level. We will now confirm our work by running the same analysis on a statistical package.

Hotelling's T^2 is not available in the SAS system. Since Hotelling's T^2 is a special case of multiple analysis of variance (MANOVA), perhaps the rationale is that any user who would like a Hotelling's T^2 could run it as a MANOVA. However, the univariate t-test is more powerful than the corresponding ANOVA on two-group data, for the simple reason that it can be used in its "one-tailed" form, so MANOVA might not fulfill the need entirely. Rencher (2002), on page 130, gives formulas for converting each of the four multivariate significance tests into a T^2 value, which would enable one to use SAS's PROC GLM to obtain the MANOVA statistics and then convert them to a T^2 value.

If using Stata, we paste the data matrix from Table 7.1 (including titles at the top) into the Stata data editor and enter the following instruction:
hotelling girth grams, by(sample)
We obtain the following output for this data set.

```
. hotelling girth grams, by( sample)

-> sample = 1
    Variable |      Obs        Mean    Std. Dev.         Min         Max
       girth |        5        17.6     1.949359          15          20
       grams |        5         152     17.36376         130         172

-> sample = 2
    Variable |      Obs        Mean    Std. Dev.         Min         Max
       girth |        7    19.57143     2.225395          17          23
       grams |        7         134     23.43075         105         173

2-group Hotelling's T-squared = 25.435596
F test statistic: ((12-2-1)/(12-2)(2)) x 25.435596 = 11.446018

H0: Vectors of means are equal for the two groups
          F(2,9) =    11.4460
     Prob > F(2,9) =     0.0034
```

At the bottom of the page of the output, we find the Hotelling's T^2 value to be 25.435596, with a p-value of 0.0034, consistent with our results. We also see the mean girth and mean grams for each of the two samples of peaches, and also the standard deviations and the minimum and maximum values.

With the peaches data still in Stata, we run the two t-tests to determine whether we have significant effects on either of the two dependent variables. The instruction is
ttest girth, by(sample) and **ttest gram, by(sample)**
We obtain the following output.

```
. ttest girth, by(sample)

Two-sample t test with equal variances
```

Group	Obs	Mean	Std. Err.	Std. Dev.	[95% Conf. Interval]	
1	5	17.6	.8717798	1.949359	15.17955	20.02045
2	7	19.57143	.8411201	2.225395	17.51328	21.62958
combined	12	18.75	.6527912	2.261335	17.31322	20.18678
diff		-1.971429	1.240934		-4.736403	.7935457

```
     diff = mean(1) - mean(2)                              t =  -1.5887
Ho: diff = 0                              degrees of freedom =      10

   Ha: diff < 0              Ha: diff != 0                Ha: diff > 0
Pr(T < t) = 0.0716     Pr(|T| > |t|) = 0.1432       Pr(T > t) = 0.9284

. ttest gram, by(sample)

Two-sample t test with equal variances
```

Group	Obs	Mean	Std. Err.	Std. Dev.	[95% Conf. Interval]	
1	5	152	7.765307	17.36376	130.4401	173.5599
2	7	134	8.855991	23.43075	112.3302	155.6698
combined	12	141.5	6.422616	22.2486	127.3639	155.6361
diff		18	12.42118		-9.676114	45.67611

```
     diff = mean(1) - mean(2)                              t =   1.4491
Ho: diff = 0                              degrees of freedom =      10

   Ha: diff < 0              Ha: diff != 0                Ha: diff > 0
Pr(T < t) = 0.9110     Pr(|T| > |t|) = 0.1779       Pr(T > t) = 0.0890
```

(StataCorp, 2009, Stata: Release 11. Statistical Software. College Station, TX: StataCorp LP.)

Surprisingly, neither t-test is significant. One has a value of -1.5887 with a p-value of 0.1432, and the other has a value of 1.4491 with a p-value of 0.1779. This is like the situation described in Section 7.3 at the beginning of the chapter, where we have a highly significant multivariate relationship, but it is not reflected in the analyses of any of the dependent variables. The answer to this, we said, is a discriminant analysis of the data to find the best linear combination of the variables for predicting group membership. We will in fact do that discriminant analysis in the next and closing section of the chapter. First, let's examine the bivariate scatter plot shown in Figure 7.3 to see if we can tell what the pattern of results is and why it should be significant multivariately but not univariately.

This pattern of the two groups within the bivariate scatter plot is highly informative—a classic case of where the multivariate test is substantially more powerful than any of the univariate tests. Notice that if we were to project these two swarms of data points down to the x-axis or over to the y-axis, there would be substantial overlap across the two groups, not pulling apart enough to be statistically significant. However, if we were to draw a diagonal axis up between the encircling ellipses for the two groups and then place a line perpendicular to that axis and project the data points onto it, we would see that the two groups are perfectly separable in that projected view (as can be seen by comparing Figs. 7.3 and 7.4). This very separable projection is the discriminant function.

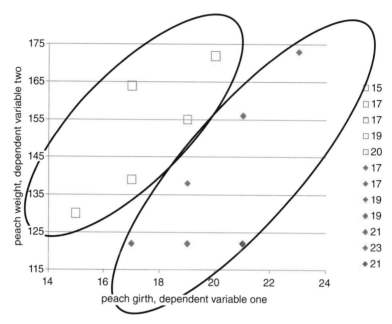

Figure 7.3   An example from our artificial peaches data about how the difference between two groups could be nonsignificant with respect to both the univariate t-test for dependent variable 1 and also the univariate t-test for dependent variable 2, and yet be highly significant in the multivariate test, with the holistic pattern having no overlap between the groups.

7.5   DISCRIMINANT ANALYSIS FROM A HOTELLING'S T^2 TEST

One of the four arguments for a multivariate approach to hypothesis testing at the beginning of the chapter is that many multivariate significance tests obtain an optimal descriptive linear combination, such as a discriminant function, as a by-product. In this final section, we will see that Hotelling's T^2 does indeed produce discriminant analysis as a by-product. In fact, in our calculations in the previous section, we were almost there. The formula for the discriminant function coefficients is given in Equation 7.6. We see that this is just the last two terms of the Hotelling's T^2 equation itself (Equation 7.3).

$$a = S_{pl}^{-1}(\bar{y}_1 - \bar{y}_2) \tag{7.6}$$

Solving this equation, we obtain the discriminant function coefficients vector, **a**.

$$a = S_{pl}^{-1}(\bar{y}_1 - \bar{y}_2) = \begin{bmatrix} 0.675 & -0.055 \\ -0.055 & 0.007 \end{bmatrix} [1.971 \quad -18] = \begin{bmatrix} 2.323 \\ -0.230 \end{bmatrix}$$

These are like regression coefficients. When we postmultiply our observed data matrix by them, according to Equation 7.7, the data are transformed to give us the projected values on the discriminant function.

$$\mathbf{z} = \mathbf{Ya} \qquad (7.7)$$

We now do the calculations.

$$\mathbf{z} = \mathbf{Ya} = \begin{bmatrix} 15 & 130 \\ 17 & 139 \\ 17 & 164 \\ 19 & 155 \\ 20 & 172 \\ 17 & 105 \\ 17 & 122 \\ 19 & 122 \\ 19 & 138 \\ 21 & 156 \\ 23 & 173 \\ 21 & 122 \end{bmatrix} \begin{bmatrix} 2.323 \\ -0.230 \end{bmatrix} = \begin{bmatrix} 4.95 \\ 7.53 \\ 1.78 \\ 8.49 \\ 6.91 \\ 15.35 \\ 11.44 \\ 16.09 \\ 12.40 \\ 12.91 \\ 13.65 \\ 20.73 \end{bmatrix}$$

In just visually examining these values, we can see that the first group of five (the first peaches sample) are quite separable from the remaining seven. When we create a multivariate plot of the transformed data (Fig. 7.4), we can see what the discriminant function accomplishes. The groups are now completely separable on the x-axis.

We will conclude with a comparative graph from Eero Simoncelli's web page[4] that shows clearly what is accomplished by a discriminant analysis, by comparing it with principal component analysis (Fig. 7.5). It is amazing how much is conveyed with just the graph itself. Principal component analysis (PCA) on the left identifies a latent variable that is the major axis of a swarm of data points (the highest possible variance), as we saw in Chapter 4. However, Fisher's discriminant function identifies for us a latent variable that will optimally separate between groups of data, as we saw that it did for our demonstration.

---

[4] This was found in an online paper entitled "least squares optimization" on Eero Simoncelli's website, referred to in Figure 7.5.

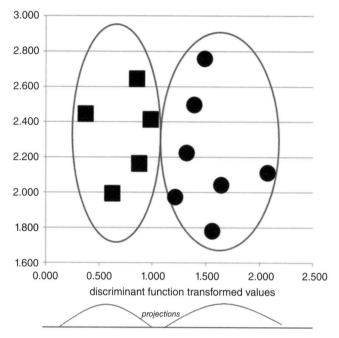

Figure 7.4   The same data shown in Figure 7.3, but rotated using discriminant analysis, such that the first dimension, the horizontal one, now has a highly significant *t*-value, and the groups separate perfectly on this linear combination variable.

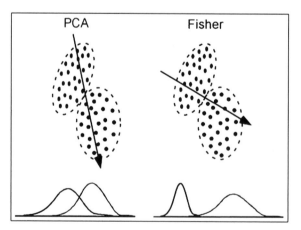

Figure 7.5   Vector projection of principal component analysis (PCA) and Fisher's discriminant analysis. (From *Least Squares Opimization*, Course Notes by Eeero P. Simoncelli, Center for Neural Science and Courant Insitute for Mathematical Sciences, New York University. July 2003. http://www.cns.nyu.edu/~eero/teaching.html.)

7.6 SUMMARY AND CONCLUSIONS

In this chapter, we have discussed several advantages of multivariate methods compared with univariate methods when analyzing multivariate data. Specifically, multivariate methods do not cause alpha inflation, a problem that plagues the use of multiple univariate tests on multivariate data. Second, multivariate tests model correlations among the dependent variables, whereas univariate tests do not. In addition, multivariate tests are in many cases more powerful than univariate tests, and finally, a discriminant function is a by-product of many multivariate tests and provides a descriptive linear combination of the outcome variables that is relevant to distinguishing the groups of interest (Fig 7.4).

The T^2 test is the multivariate analogue of the t-test and is used to compare two groups that each contains two or more dependent variables. To derive the formulas for the T^2 tests, we first convert from geometrical distance to statistical distance, standardizing the generalized Pythagorean theorem by inserting the inverse of the covariance matrix, which results in the formulas for Hotelling's three types of T^2 analyses:

$$T^2 = n(\bar{y} - \mu_0)' S^{-1} (y_1 - y_2) \tag{7.2}$$

$$T^2 = \left(\frac{n_1 n_2}{n_1 + n_2} \right) (\bar{y}_1 - \bar{y}_2)' S_{pl}^{-1} (\bar{y}_1 - \bar{y}_2). \tag{7.3}$$

$$T^2 = n \bar{d}' S_d^{-1} \bar{d}, \tag{7.4}$$

which correspond to single mean, two independent means, and two correlated samples, respectively.

In some analyses, we may encounter the case in which the multivariate test is significant, but the univariate ones are negative. In this situation, the group separation is between the variables; that is, information is required from two or more of the dependent variables to separate the groups. The univariate tasks have insufficient power to detect a difference when it exists.

The reverse can occur as well: the multivariate test is not significant, but one or more of the univariate tests is significant. In this case, the univariate test shows significance because of alpha inflation due to multiple testing. In short, we should only accept a univariate test as truly significant if the multivariate test is positive.

Another advantage to the T^2 test is that it obtains a discriminant function as a by-product. In fact, from the last two terms of the T^2 equation, we can obtain the discriminant function coefficients vector. Postmultiplying the observed data matrix by the discriminant function coefficients vector provides the projected values on the discriminant function. A plot using the transformed data shows that the discriminant function aids in interpreting the findings, particularly when the univariate tests were not significant but the multivariate ones were.

STUDY QUESTIONS

A. Essay Questions

1. Do the derivation to show that the F-ratio (for independent groups ANOVA) is indeed the square of the t-ratio (for independent groups analysis): $F = t^2$.

2. Do the derivation to show that Hotelling's T^2 statistic for independent groups (Eq. 7.3) is the square of the corresponding univariate t ratio formula, made multivariate by the expansion of the formula to deal with two or more dependent variables.

3. Do the derivation to show that Hotelling's T^2 statistic for correlated groups (Eq. 7.4) is the square of the corresponding univariate t-ratio formula, made multivariate by the expansion of the formula to deal with two or more dependent variables.

4. Explain from Figure 7.2 why the T^2 statistic is more powerful than multiple t-tests on multivariate data.

5. Why is the single sample T^2 test explained (Eq. 7.2) even though we almost never encounter single sample data?

6. Why must the degrees of freedom in a Hotelling's T^2 test be greater than P, the number of dependent variables?

B. Computational Questions

1. Calculate a Hotelling's T^2 on the data set below, assuming that it is independent groups data:

Treatment 1	Treatment 2
(8,9)	(12,5)
(6,7)	(16,9)
(5,1)	(10,7)
(5,7)	(4,8)
(4,5)	(8,11)
(2,13)	(10,8)

2. Calculate two univariate t-tests on the data of question 1 above, and compare your results to those you received from the T^2 test in question 1.

3. Now calculate the discriminant function ($z = \mathbf{Ya}$) for the data of question 1 above, according to Equation 7.7, and do the univariate t-test on this new z linear combination variable, and compare the results to the tests in questions 1 and 2.

4. Now suppose that the data of question 1 above is repeated measures data rather than independent groups, and calculate a paired observations Hotelling's T^2 test according to Equation 7.4 on those data.

5. Calculate two univariate paired observations *t*-tests on the data of question 4 and compare the results to those of the multivariate test in question 4.

C. Data Analysis Questions

1. Use Stata to run a Hotelling's T^2 (independent groups) on the data of question 1 of the computational questions above. Obtain the T^2 value and compare with what you obtained from your analysis in question 1.

2. Use Stata to run two *t*-tests (independent groups) on the data of question 1 of the computational questions above. Obtain the *t*-values and compare with what you obtained from your analysis in question 2.

3. Use Stata to run a Hotelling's T^2 (repeated measures) on the data of question 4 of the computational questions above. Obtain the T^2 value and compare with what you obtained from your analysis in question 4.

4. Use Stata to run two *t*-tests (repeated measures) on the data of question 8 of the computational questions above. Obtain the *t*-values and compare with what you obtained from your analysis in question 5.

REFERENCES

Bonferroni, C. E. 1935. Il calcolo delle assicurazioni su gruppi di teste. In *Studi in Onore del Professore Salvatore Ortu Carboni*. Rome, Italy.

Carroll, J. S., Nelson, D. A., Yorgason, J. B., Harper, J. M., Ashton, R. H., and Jensen, A. C. 2010. Relational aggression in marriage. *Aggressive Behavior*, *36*(5), 315–329.

Gossett W. S. [Student, pseud.] 1908. The probable error of a mean. *Biometrika*, *6*, 1–25.

Gossett, W. S. 1943. *Student's Collected Papers*. London: Biomtrika Office, University College.

Healy, M. J. R. 1969. Rao's paradox concerning multivariate tests of significance. *Biometrics*, *25*, 411–413.

Hotelling, H. 1931. The generalization of Student's ratio. *Annals of Mathematical Statistics*, *2*, 360–378.

Hotelling, H. 1951. A generalized t test and measure of multivariate dispersion. Proceedings of the Second Berkeley Symposium on Mathematical Statistics and Probability, *1*, 23–41.

Mahalanobis, P. C. 1930. On tests and measures of group divergence. *Journal of the Asiatic Society of Bengal*, *26*, 541–588.

Mahalanobis, P. C. 1936. On the generalized distance in statistics. *Proceedings of the National Institute of Sciences of India*, *12*, 49–55.

McBride, J., Brown, B. L., and Ostenson, J. 2011. Finances: Source of marital conflict or battleground for aggressive strategies? Paper presented at the Rocky Mountain Psychological Association Meetings, April 14–16, Salt Lake City, UT.

Morrison, D. F. 1990. *Multivariate Statistical Methods, Third Edition*. New York: McGraw-Hill.

Olkin, I., and Sampson, A. R. 2001. Harold Hotelling. In Heyde, C. C., and Seneta, E. (Eds.), *Statisticians of the Centuries*. New York: Springer.

Papp, L. M., Cummings, E. M., and Godke-Morey, M. 2009. For richer, for poorer: Money as a topic of marital conflict in the home. *Family Relations, 58*(1), 91–103.

Rao, C. R. 1966. Covariance adjustment and related problems in multivariate analysis. In Krishnaiah, P. (Ed.), *Multivariate Analysis*. New York: Academic Press.

Rencher, A. C. 2002. *Methods of Multivariate Analysis, Second Edition*. New York: Wiley.

Schwartz, T. W. 2000. The land mines of marriage: Intergenerational causes of marital conflict. *Gestalt Review, 4*(1), 47–62.

CHAPTER EIGHT

MULTIVARIATE ANALYSIS OF VARIANCE

8.1 INTRODUCTION

Chapter two ended with a demonstration of how to use matrix algebra to calculate a one-way univariate analysis of variance (ANOVA). In this chapter, we continue from that point and show how the matrix algebra approach lends itself well to extending the univariate principles to a one-way multivariate analysis of variance (MANOVA). As we shall see, the multivariate extension of ANOVA has much to offer as an analytical tool. We will begin by discussing nine particularly useful aspects of MANOVA.

PRINCIPLES AND POWERS OF MULTIVARIATE ANALYSIS OF VARIANCE

1. To avoid alpha inflation, one should not consider significant univariate results unless the multivariate test is also significant.
2. Univariate tests do not model the correlations among the dependent variables, whereas the multivariate test does.
3. The multivariate test is in many cases more powerful.

Multivariate Analysis for the Biobehavioral and Social Sciences: A Graphical Approach,
First Edition. Bruce L. Brown, Suzanne B. Hendrix, Dawson W. Hedges, Timothy B. Smith.
© 2012 John Wiley & Sons, Inc. Published 2012 by John Wiley & Sons, Inc.

4. The multivariate test in MANOVA obtains discriminant analysis coefficients as a by-product.

5. It is possible to have a significant multivariate test but no corresponding significant univariate test. In that case, a discriminant analysis will usually find a linear combination of the dependent variables that will be signficant.

6. Multivariate graphics and MANOVA summary tables help with the interpretation of the results from a complex MANOVA.

7. It is important to learn how the significance test results, the R-squared values, and the graphical patterns all relate to one another in a MANOVA. There is much power in the convergent information.

8. A multivariate graphical display of three-way means (ABC) does *not* correspond to the R-squared values and the significance levels for the three-way interaction of ABC. Rather it displays the combined information from all of the component parts, the three one-way terms of A, B, and C; the three two-way terms of AB, AC, and BC; and the three-way term ABC. The sum of all seven of these R-squared values will correspond to the means pattern.

9. In a one-way MANOVA, the Wilks' lambda statistic is equal to one minus R-squared for between the groups and the dependent variables. There are no other convincing multivariate indices of association.

The first three of these nine essential principles about MANOVA come from previously published sources. Specifically, Hummel and Sligo's (1971) Monte Carlo simulation study of MANOVA provided essential information that greatly improved its use in the social sciences. Twenty years later, Rencher and Scott (1990) conducted an extensive simulation study that validated the points made by Hummel and Sligo, but also went beyond the original conclusions in many ways. Subsequently, Rencher's two multivariate texts contained these key clarifications regarding the uses of MANOVA (Rencher, 1998; Rencher, 2002). These sources confirm that (1) one avoids alpha inflation by using MANOVA as a "filter" for univariate tests, (2) univariate tests fail to properly model the correlations among dependent variables, and (3) the multivariate tests are in many cases more powerful. Not only does MANOVA often detect significant results that a series of univariate ANOVAs would miss, but it also corrects spurious univariate results attributable to alpha inflation when conducting multiple univariate ANOVA tests on multivariate data.

As we saw in Chapter 7, Hotelling's T^2 statistic has discriminant analysis as a by-product, and the same is true of MANOVA (point 4 above). In fact, we shall see in Section 8.4 of this chapter that the discriminant function coefficients are an integral part of the MANOVA calculations. This information can help correct for the problem identified in point 5 on our list: it is not unusual to have a significant result on a multivariate test and yet no significant results on any of the corresponding univariate tests. In such a situation, a discriminant analysis will correct the problem by enabling the researcher to find a linear combination of the dependent variables that will be statistically significant, as was demonstrated in the previous chapter for Hotelling's T_2 test.

Starting in the next section, we will demonstrate point 6 with published data, showing that a MANOVA summary table and multivariate graphics can greatly clarify the results and animate data interpretation. Point 7 recommends that you diligently look for ways that all of the products of a MANOVA (R^2 values, significance test results, graphical patterns, and simple comparative tables) are complementary to one another and can thus be considered in combination with one another to give a sharper picture of the results of a MANOVA. Points 8 and 9 are higher-level concepts that are useful in integrating the results of significance tests from factorial MANOVA with the detailed topological patterns of multivariate graphs. They are best learned with demonstrations from data.

8.2   AN APPLIED EXAMPLE OF MULTIVARIATE ANALYSIS OF VARIANCE (MAV1)

An example of MANOVA in a published study comes from a paper by Myers and Siegel (1985) in the *Journal of Personality and Social Psychology* (JPSP).[1] The paper is a test of the applicability of Solomon and Corbit's (1974) opponent-process theory of acquired motivation to the human experience of infant breastfeeding. The theory had been used (primarily with animal models) to account for cycles of hedonic contrast, hedonic habituation, and withdrawal distress in addictive behaviors. However, Myers and Siegel saw the theory as being more generally applicable to a number of common behaviors that also exhibit cyclic hedonic patterns. In particular, they were interested in evaluating the hedonic content for women in their memories of six specific events that have occurred in their breastfeeding experience: (1) the very first experience with the youngest child, (2) the time just prior to a typical morning feeding,

---

[1] We were looking for an example study for this chapter either from the social psychological literature or from the cognitive neuroscience literature. There were a number of studies in the cognitive neuroscience journals that employed MANOVA, but interestingly, most of them used it as a standalone significance test rather than in concert with the separate univariate ANOVAs as is recommended. Interestingly, there were a number of social psychological studies in *JPSP* in the mid-1980s that used MANOVA, but there were no recent examples in that journal. However, the social psychological studies from the 1980s that we examined (e.g., Lemyre and Smith, 1985; Warner and Sugarman, 1986) all seemed to be aware of and followed the recommendation of Hummel and Sligo (1971) that MANOVA be followed by separate univariate ANOVA and the two be reported together.

(3) during the typical morning feeding, (4) during an enforced delay of a duration at least twice the usual interval between feedings, (5) during the feeding period following the delay, and (6) the time during weaning, which is recalled as being the most emotional.

The subjects for the study were 21 women attending a regional conference of a "prepared childbirth group" who were either currently breastfeeding or had done so recently. Nine of subjects were primiparas (first birth mothers) and 12 were multiparas (mothers with previous births). To evaluate the hedonic content of the six recall situations, the researchers used a modified short form of the Nowlis Mood Adjective Checklist and also a 7-point measure of physical discomfort. The MANOVA model used in the study is a MAV1 ($m = 6$), which is the notation we have adopted in this book. The "MAV" stands for multivariate analysis of variance, and the "1" indicates that it is a one-way MANOVA, with six treatments being compared ($m = 6$). Table 5.1 summarizes the results of the MANOVA (including the three individual analyses of variance) reported in the main study. There was also a second part to the study to control for demand characteristics.

The first thing to notice from the multivariate test is that the results are very strong. The multivariate significance test used was the Wilks' lambda, which in a one-way MANOVA can be interpreted as $1 - R^2$. This indicates that the six treatments, the recall situations, account for 84.5% of the variance in the multivariate pattern of results. We also see that all three individual dependent variables, euphoria, dysphoria, and discomfort are also highly significant statistically with unusually large F ratios. The paper did not report the univariate R^2 values, or they would have been included in this summary table. When everything is significant like this, it helps to have a graph or even a table of means to discover the meaning of the significant results. We used the Gabriel biplot from Stata to display the results, as shown in Figure 8.1.

Once the 6×3 matrix of means from the paper have been pasted (or key entered) into Stata, as shown in the lower left window of Figure 8.1, the single line of Stata instruction code shown below will initiate the biplot.

```
biplot euphoria dysphoria discomfort, std
```

This biplot is a helpful addition to the tabular results in Table 8.1. The horizontal dimension on this graph represents a bipolar hedonic dimension of euphoria (on the right) to dysphoria (on the left), with discomfort ratings going upward and to the left. We used the "std" option at the end of the Stata command to indicate that it should be a standardized biplot, since the variables are not commensurate with one another (discomfort has a 7-point scale, but the other two dependent variables have much larger scale score values).

The table informs us that the within-group error in these results is small— the effect sizes are strong and highly significant statistically, both in the

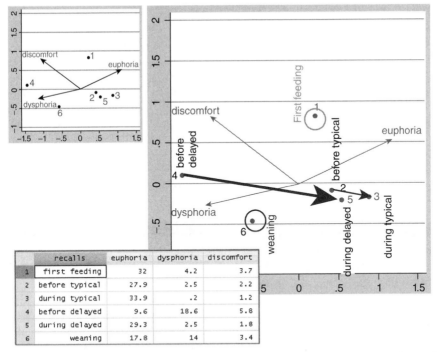

	recalls	euphoria	dysphoria	discomfort
1	first feeding	32	4.2	3.7
2	before typical	27.9	2.5	2.2
3	during typical	33.9	.2	1.2
4	before delayed	9.6	18.6	5.8
5	during delayed	29.3	2.5	1.8
6	weaning	17.8	14	3.4

Figure 8.1 Gabriel biplot showing the average ratings given to each of the six recall situations, by the 21 breastfeeder subjects, on the three dimensions of euphoria, dysphoria, and discomfort. The largest contrast difference is from the "before delayed" condition to the "during delayed" condition, as shown with the large arrow. First feeding seems to have involved discomfort, but is also rated high on euphoria. Weaning and "before delayed" are the two conditions rated high in dysphoria. The input data to Stata are shown in the lower left. (Graph created from the data of Myers and Siegel, 1985.) (StataCorp, 2009. Stata: Release 11. Statistical Software. College Station, TX: StataCorp LP.)

TABLE 8.1. Multivariate Analysis of Variance Summary Table for the Evaluations of the Six Recall Situations by Myers and Siegel's (1985) 21 Breastfeeder Subjects (Nine Primaparas and 12 Multiparas) on Measures of Euphoria, Dysphoria, and Discomfort

Dependent Variables	Test Statistic	P-Value
Euphoria	$F = 38.23$	<0.0001
Dysphoria	$F = 47.50$	<0.0001
Discomfort	$F = 28.45$	<0.0001
Multivariate test	$\Lambda = 0.155$	<0.0001
Multivariate strength of effect	$R^2 = 0.845$	

multivariate pattern and also the three individual univariate variables, but the interpretation of these results is not clear from the table alone. On the other hand, the biplot shows us with precision the topological nature of these strong effects. By mentally "drawing" perpendicular lines from the data points down to the vectors for the variables, we can put together quite a bit of information from this graph. From the "mental perpendiculars," we see that "first feeding" (1) is typified by some discomfort but is rated high on euphoria. "During typical" (3) is also high euphoria but without the discomfort of the first time. "During delayed" (5) and also "before typical" are both fairly positive on the euphoria versus dysphoria dimension and relatively low on the discomfort vector. The two negative ratings are "weaning" (6), and "before delayed" (4). Weaning is not really very high on discomfort, but it is very high on the dysphoria dimension. On the other hand, "before delayed" (4) is by far the highest of the six conditions on both of these variables, discomfort and dysphoria.

The authors mention in the paper that "the methodology used in the present study is especially vulnerable to the charge that the data are influenced by demand characteristics (Orne, 1962)." In other words, perhaps one could argue that the ratings are just a reflection of what is commonly known or supposed about the topic. To control for this, they replicated the study with 21 nulliparous females (who had neither given birth themselves nor breastfed) to whom they refer as their "simulator" respondents. They did additional multivariate statistical analyses that showed that the simulators were clearly different from the women who breastfed. Figure 8.2, which is the parallel biplot of the simulator results, also shows that the demand-characteristics argument is not tenable.

Even from the standpoint of the implicit nature in the pattern of the three dependent variables, the simulators view breastfeeding very differently from those who have actual experience. They have discomfort and euphoria as opposites rather than euphoria and dysphoria. Along that same line, they have no idea of how euphoric first feeding is, rating it primarily as a major discomfort. They do seem to have some idea from the demand characteristics of the situation that there is a large difference between "before delayed" and "during delayed," but to them it is a discomfort versus euphoria dimension, whereas for those with experience it is primarily a dysphoria versus euphoria dimension. In fact, they have lower evaluations of breastfeeding overall on the euphoria dimension (a mean rating of 21.1 as compared with 25.1) than do those who have had experience.

This study exemplifies the basic utility of MANOVA and associated graphs. Respondents' scores can greatly differ across variables, and those variables are best evaluated in multivariate space where interactions can be ascertained. Whereas it may be interesting to know that differences between groups exist (univariately), when those same differences and their interactions are projected into multivariate space, the results can strengthen study conclusions (in this case by eliminating the explanation of demand characteristics) and better answer research questions so as to inform theory.

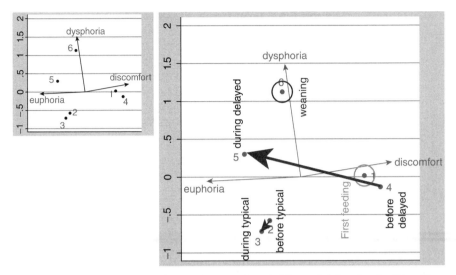

Figure 8.2   Gabriel biplot similar to the one given in Figure 8.1, but for the 21 nulliparous "simulator" subjects. On the left is shown the biplot as it appears in Stata, and on the right with markings to show comparisons of the six treatment conditions. Notice that the euphoria pattern is substantially different from and not nearly as coherent as the actual breastfeeder results in Figure 8.1. (Created from the data of Myers and Siegel, 1985.) (StataCorp, 2009. Stata: Release 11. Statistical Software. College Station, TX: StataCorp LP.)

8.3   ONE-WAY MULTIVARIATE ANALYSIS OF VARIANCE (MAV1)

The goal of this section is to walk you through the calculations of a MAV1, one-way MANOVA, so that you will know what is actually happening when you run it on one of the statistical packages. The first step will be to do the calculations and create the summary tables and graphs of results. We will then run the same analysis, first on Stata and then on SAS, and show that we obtain the same results.

We will use the contrived data from Chapter 7, Section 7.5, involving the measurement of two samples of peaches, one exposed to industrial wastes and the other not, only this time we add a third group, perhaps a group of peaches grown in a spot with even stronger pollutants. The data set, then, will be precisely the same as the one used for Hotelling's T^2 for independent groups, except that one more group is added, requiring that a MANOVA model be run rather than the Hotelling T^2 statistic, which will only handle two groups. The data are shown in Table 8.2.

We refer back now to Chapter 2. In Section 2.8 of that chapter, matrix-based formulas were given for obtaining the three sums of squares needed for a one-way independent groups ANOVA. We wish to show the continuity between those formulas and the ones we will use for MANOVA.

TABLE 8.2. Input Data for MAV1 Demonstration, Peaches

IV	Girth	Grams
1	15	130
1	17	139
1	17	164
1	19	155
1	20	172
2	17	105
2	17	122
2	19	122
2	19	138
2	21	156
2	23	173
2	21	122
3	21	122
3	24	138
3	24	147
3	25	130

$$\text{SSB} = Q_{\text{group}} - Q_{\text{total}} \text{ (Eq. 2.36 from Chapter 2)}$$

$$\text{SSW} = Q_{\text{data}} - Q_{\text{group}} \text{ (Eq. 2.37 from Chapter 2)}$$

$$\text{SSB} = Q_{\text{group}} - Q_{\text{total}} \text{ (Eq. 2.38 from Chapter 2)}$$

The formulas are written in terms of "Q" values, where Q stands for "quantity." There is one quantity Q_{data} based on the raw data (Eq. 2.35), and it is just the raw data matrix premultiplied by its own transpose. There is also a quantity Q_{groups} (Eq. 2.34) based on group means and group totals. The final quantity is Q_{total}, and it is based on the grand total multiplied by the grand mean (Eq. 2.33).

$$Q_{\text{data}} = \mathbf{x}'\mathbf{x} \text{ (Eq. 2.34 from Chapter 2)}$$

$$Q_{\text{group}} = \mathbf{t}'_j\mathbf{m}_j \text{ (Eq. 2.35 from Chapter 2)}$$

$$Q_{\text{total}} = \mathbf{t}'\mathbf{m} \text{ (Eq. 2.33 from Chapter 2)}$$

Notice that none of the Q symbols in these formulas are in bold type. That is because each represents a scalar, a single sum of squares, not an SSCP matrix. These are formulas for ANOVA not MANOVA. The corresponding MANOVA formulas change very little. The essential difference is that they

now become matrix formulas to handle multiple dependent variables simultaneously, and they therefore yield SSCP matrices (matrices of sums of squares
in the diagonal entries and cross products off the diagonals).

$$\mathbf{B} = \text{SSCP}_{\text{between}} = \mathbf{Q}_{\text{group}} - \mathbf{Q}_{\text{total}} \tag{8.1}$$

$$\mathbf{W} = \text{SSCP}_{\text{within}} = \mathbf{Q}_{\text{data}} - \mathbf{Q}_{\text{group}} \tag{8.2}$$

$$\mathbf{T} = \text{SSCP}_{\text{total}} = \mathbf{Q}_{\text{data}} - \mathbf{Q}_{\text{total}} \tag{8.3}$$

The formulas for the $\mathbf{Q}$ values are also matrix formulas. Matrix $\mathbf{X}$ is the
$n \times p$ matrix of raw data, with n being the number of observations and p being
the number of dependent variables. The symbol $\mathbf{T}_j$ represents the sums or
totals for each of the individual treatment or sample groups, and the symbol
$\mathbf{M}_j$ represents the means of those groups. The symbol $\mathbf{T}$ is used for the vector
of grand sums of all the data for each dependent variable, and $\mathbf{M}$ is the vector
of grand means for the dependent variables.

$$\mathbf{Q}_{\text{data}} = \mathbf{X}'\mathbf{X} \tag{8.4}$$

$$\mathbf{Q}_{\text{group}} = \mathbf{T}_j'\mathbf{M}_j \tag{8.5}$$

$$\mathbf{Q}_{\text{total}} = \mathbf{TM}' \tag{8.6}$$

COMPUTATIONAL STEPS IN ONE-WAY MANOVA

1. Calculate all $\mathbf{Q}$ matrices for the entire MANOVA.
2. Calculate all SSCP matrices from the $\mathbf{Q}$ matrices.
3. Calculate all the multivariate statistic(s) from the $\mathbf{H}$ and the $\mathbf{E}$ matrix,
   and calculate R^2.
4. Gather all of the univariate sum of squares statistics from the diagonals
   of the SSCP matrices.
5. Collect all of the results, univariate and multivariate, in a MANOVA
   summary table.
6. Create graphics to reveal the descriptive relationships in the data.

Step 1: Calculate all $\mathbf{Q}$ matrices for the entire MANOVA
First, we calculate $\mathbf{Q}_{\text{data}}$ by premultiplying the matrix of observed data by
its own transpose.

$$\mathbf{Q}_{data} = \mathbf{X'X} = \begin{bmatrix} 15 & 17 & 17 & 19 & 20 & 17 & 17 & 19 & 19 & 21 & 23 \\ 130 & 139 & 164 & 155 & 172 & 105 & 122 & 122 & 138 & 156 & 173 \end{bmatrix}$$

$$\begin{bmatrix} 21 & 21 & 24 & 24 & 25 \\ 122 & 122 & 138 & 147 & 130 \end{bmatrix} \begin{bmatrix} 15 & 130 \\ 17 & 139 \\ 17 & 164 \\ 19 & 155 \\ 20 & 172 \\ 17 & 105 \\ 17 & 122 \\ 19 & 122 \\ 19 & 138 \\ 21 & 156 \\ 23 & 173 \\ 21 & 122 \\ 21 & 122 \\ 24 & 138 \\ 24 & 147 \\ 25 & 130 \end{bmatrix}$$

$$= \begin{bmatrix} 6493 & 44754 \\ 44754 & 318449 \end{bmatrix}$$

We next calculate $\mathbf{Q}_{group}$ from the matrix of group totals $\mathbf{T}_j$ and the matrix of group means $\mathbf{M}_j$.

$$\mathbf{Q}_{group} = \mathbf{T'_j M_j} = \begin{bmatrix} 88 & 137 & 94 \\ 760 & 938 & 537 \end{bmatrix} \begin{bmatrix} 17.600 & 152.000 \\ 19.571 & 134.000 \\ 23.500 & 134.250 \end{bmatrix} = \begin{bmatrix} 6439.1 & 44354 \\ 44354 & 313304 \end{bmatrix}.$$

Finally, we calculate $\mathbf{Q}_{total}$ from the matrix of group totals $\mathbf{T}$ and the matrix of group means $\mathbf{M}$.

$$\mathbf{Q}_{total} = \mathbf{TM'} = \begin{bmatrix} 319 \\ 2235 \end{bmatrix} [19.938 \quad 139.688] = \begin{bmatrix} 6360.1 & 44560 \\ 45560 & 312202 \end{bmatrix}.$$

Step 2: Calculate all SSCP matrices from the $\mathbf{Q}$ matrices.

$$\mathbf{B} = SSCP_{between} = \mathbf{Q}_{group} - \mathbf{Q}_{total} = \begin{bmatrix} 6439.1 & 44354 \\ 44354 & 313304 \end{bmatrix} - \begin{bmatrix} 6360.1 & 44560 \\ 45560 & 312202 \end{bmatrix}$$

$$= \begin{bmatrix} 79.023 & -206.81 \\ -206.81 & 1102.7 \end{bmatrix}$$

$$\mathbf{W} = SSCP_{within} = \mathbf{Q}_{data} - \mathbf{Q}_{group} = \begin{bmatrix} 6493 & 44754 \\ 44754 & 318449 \end{bmatrix} - \begin{bmatrix} 6439.1 & 44354 \\ 44354 & 313304 \end{bmatrix}$$

$$= \begin{bmatrix} 53.914 & 400.5 \\ 400.5 & 4844.8 \end{bmatrix}$$

$$\mathbf{T} = SSCP_{total} = \mathbf{Q}_{data} - \mathbf{Q}_{total} = \begin{bmatrix} 6493 & 44754 \\ 44754 & 318449 \end{bmatrix} - \begin{bmatrix} 6360.1 & 44560 \\ 45560 & 312202 \end{bmatrix}$$

$$= \begin{bmatrix} 132.94 & 193.69 \\ 193.69 & 5947.4 \end{bmatrix}.$$

Step 3: Calculate all the multivariate statistics from the SSCP matrices, including the R² value.

In the next section, we will discuss the calculation of all four of the multivariate statistics, but here we will only calculate Wilks' lambda because of its simplicity. It is equal to the determinant of the within groups SSCP matrix, $\mathbf{W}$, divided by the determinant of the total SSCP matrix as shown in Equation 8.7.

$$\Lambda = \frac{|\mathbf{W}|}{|\mathbf{T}|}. \tag{8.7}$$

$$\Lambda = \frac{|\mathbf{E}|}{|\mathbf{E} + \mathbf{H}|}. \tag{8.8}$$

Equation 8.7 works fine for a one-way MANOVA, but Equation 8.8 is a more general form of the formula that will work with higher-order models. The symbol in the denominator, $|\mathbf{E} + \mathbf{H}|$, indicates the determinant of the sum of the hypothesis matrix and the error matrix. Wilks' lambda is found as the ratio of these two determinants. For a 2×2 matrix, the determinant is very simple. It is calculated for the obtained W matrix and the obtained $\mathbf{T}$ matrix; they are then entered into Equation 8.8 to obtain a Wilks' lambda value of 0.13384.

$$|\mathbf{W}| = \begin{bmatrix} 53.914 & 400.5 \\ 400.5 & 4844.8 \end{bmatrix} = (53.914)(4844.8) - (400.5)^2 = 100801.$$

$$|\mathbf{T}| = \begin{bmatrix} 132.94 & 193.69 \\ 193.69 & 5947.4 \end{bmatrix} = (132.94)(5947.4) - (193.69)^2 = 753123.$$

$$\Lambda = \frac{|\mathbf{E}|}{|\mathbf{H} + \mathbf{E}|} = \frac{|\mathbf{W}|}{|\mathbf{T}|} = \frac{100801}{753123} = 0.13384.$$

With the Wilks' lambda test statistic, smaller is better, so to be significant, our obtained lambda must be smaller than the tabled value. There are three groups, so the degrees of freedom between groups is $dfb = k - 1 = 2$. There are 16 observations, so the degrees of freedom within groups is

$n - k = 16 - 3 = 13$. The number of dependent variables is $p = 2$. We next go to Table F in the back of the book and look up the critical ratio for Wilks' lambda that has 2 and 13 degrees of freedom and $p = 2$ to find the value 0.476. Our obtained lambda is well below that, so we can reject the null hypothesis with respect to the entire multivariate pattern of the data. Whether or not we also obtain a significant univariate statistic, we do have a significant effect somewhere in the multivariate pattern. Sometimes, the univariate test will indicate the basis for the significant effect, but sometimes we will have to do it by using discriminant analysis or from a multivariate graph. We will in fact examine the pattern of the data here graphically and with discriminant analysis to demonstrate how this is done.

In a one-way MANOVA, the R^2 value is equal to one minus the lambda value, so for these artificial data, the R^2 value is $1 - 0.13384 = 0.86616$. We account for 86.6% of the variance with our three groups. Now, before the next step of our process, let's check our results against Stata. We first paste the data shown in Table 8.2 into the Stata data editor (that looks like a little spreadsheet) and then enter the instruction:

MANOVA girth grams = iv

We receive the following MANOVA summary table output:

```
. manova girth grams = iv

                    Number of obs =        16

                    W = Wilks' lambda        L = Lawley-Hotelling trace
                    P = Pillai's trace       R = Roy's largest root

        Source | Statistic     df   F(df1,     df2) =     F    Prob>F
        -------+----------------------------------------------------------
           iv  | W   0.1338     2    4.0       24.0    10.40 0.0000 e
               | P   0.9251          4.0       26.0     5.59 0.0022 a
               | L   6.0312          4.0       22.0    16.59 0.0000 a
               | R   5.9574          2.0       13.0    38.72 0.0000 u
        -------+----------------------------------------------------------
        Residual |               13
        -------+----------------------------------------------------------
           Total |               15
        -------+----------------------------------------------------------
                  e = exact, a = approximate, u = upper bound on F
```

(StataCorp, 2009. Stata: Release 11. Statistical Software. College Station, TX: StataCorp LP.)

The first of the four listed statistics here, "W" is the Wilks' lambda, and we see that it matches the one we have calculated, 0.1338. There are also three other multivariate statistics reported, Pillai's trace, the Lawley–Hotelling trace, and Roy's largest root. These will be explained in Section 8.4.

Step 4: Gather all of the univariate sum of squares statistics from the diagonals of the SSCP matrices.

The sums of squares for Between, Within, and Total are all found in the diagonal elements of the SSCP matrices.

Step 5: Collect all of the results, univariate and multivariate, in a MANOVA summary table.

TABLE 8.3. Multivariate Analysis of Variance Summary Table for the Analysis of the Three Samples of Peaches (Artificial Data), Corresponding to Figures 8.3 and 8.4

Source	dv1: ANOVA Summary Table for Girth of Peaches					dv2: ANOVA Summary Table for Weight of Peaches				
	R^2	SS	df	MS	F	R^2	SS	df	MS	F
Between	0.594	79.02	2	39.51	9.53*	0.185	1102.69	2	551.34	1.48
Within	0.405	53.91	13	4.15		0.815	4844.75	13	372.67	
Total	1.000	132.94	15			1.000	5947.44	15		
Multivariate test(s)										
Wilks' lambda = 0.13384					$R^2 = 0.866$	$P < 0.0001$				

* $P = 0.0028$.

We construct a MANOVA summary table (Table 8.3) from the statistics just calculated and gathered, with the univariate summary tables at the top and the multivariate summary table at the bottom. Notice that for the univariate tables, we only need the sum of square values (SS) and everything else in the table follows (df, MS, F, R^2). To get the R^2 values, we simply divide each of the SS values by the SS for total.

We have evidence of a statistically significant multivariate relationship in the data, with the differences among three samples of peaches accounting for 86.6% of the variance in the multivariate pattern. One of the two dependent variables, peach girth, is also found to be significant, with the three sample grouping of peaches accounting for 59.4% of the variance. When these same data for groups 1 and 2 were analyzed in Chapter 7 using Hotelling's T^2 analysis, the multivariate test was highly significant, but neither of the dependent variables had a statistically significant t test. Apparently, the addition of the third group has created more separation among the groups on the first dependent variable, peach girth.

Step 6: Create graphics to reveal the descriptive relationships in the data.

Figure 8.3 is a bivariate scatter plot showing the location of each of the sixteen bivariate observations in the demonstration data. Examining the elliptical "shells" around each of the three peach sample groups, one can see why dependent variable 2, peach weight, is not significant. If one projects the ellipses for the three groups over to the left, onto the y-axis (peach weight), it is apparent that the three groups pretty much overlap with one another on the y-axis projection without much separation. However, a similar projection down onto the x-axis (peach girth), has some separation among the groups, especially the two outer ones.

If we now do a discriminant analysis (Chapter 7, Section 7.5) on these same data and plot them in the discriminant space (Fig. 8.4), we see that the effect is essentially to rotate the data until we have optimal separation, and now there is almost no overlap on the x-axis.[2] The x-axis in this figure is the

---

[2] Notice that there is a data point on the right side of the middle group that is located directly on top of another point, one from the rightmost group. This is signified as a filled diamond inside an unfilled one. Except for this one common data point, there is no overlap among the three groups on the x-axis.

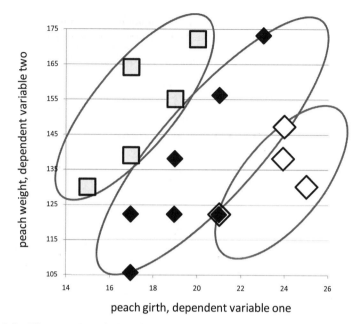

Figure 8.3   The same peach data from Figure 7.3 of Chapter 7, but now expanded to three samples of peaches, to demonstrate a MAV1, one-way multivariate analysis of variance. These data have a highly significant Wilks' lambda value of 0.1338, which is indicative of a multivariate R^2-value of 0.8662. In Figure 8.4, these data are rotated using multiple discriminant analysis, to create highly separable groups on the discriminant function.

discriminant function, which has the property of maximally separating groups.[3] This shows there is an optimal combination of girth and weight that will separate three groups of peaches nearly perfectly. In other words, even though the three "types" of peaches differ more by girth than by weight, as we saw in the univariate ANOVA tests, they differ even more according to some characteristic that is reflected in both girth and weight.

To run this MANOVA analysis in Stata requires multiple passes. Each of the univariate ANOVAs is run one at a time, as is the multivariate one. In SAS, they are all considered one analysis, and many tables, both univariate ones and multivariate ones, are produced. Before closing this section, we will run one of the univariate tests in Stata to verify our results in Table 8.3. The Stata command for executing the univariate ANOVA on the girth dependent variable is:

```
ANOVA girth iv
```

We receive the following output, and we see that the results do in fact agree with those we obtained by following the formulas and the MANOVA process.

---

[3] Obviously, this rotation is no great feat in a two-dimensional variable space. You can do it by simply tilting your head. But discriminant analysis can find the best separation function in three-dimensional space, or four, or 20, and that is useful. We are, of course, limiting ourselves to two dependent variables in the data set to make the demonstration simpler.

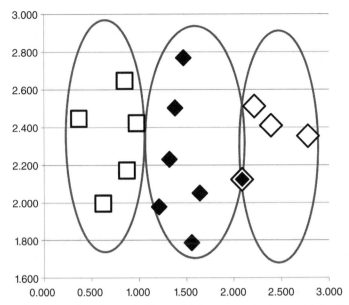

Figure 8.4   The same data shown in Figure 8.3, but rotated using multiple discriminant analysis, such that the first dimension, the horizontal one, now has a highly significant F-ratio of 38.72, and the groups separate almost perfectly on this linear combination discriminant function variable.

```
. anova girth iv

                            Number of obs =       16     R-squared      =  0.5944
                            Root MSE      = 2.03648     Adj R-squared  =  0.5320

              Source |   Partial SS    df       MS            F      Prob > F

               Model |   79.0232143     2   39.5116071        9.53     0.0028

                  iv |   79.0232143     2   39.5116071        9.53     0.0028

            Residual |   53.9142857    13   4.14725275

               Total |    132.9375     15      8.8625
```

(StataCorp, 2009. Stata: Release 11. Statistical Software. College Station, TX: StataCorp LP.)

8.4   THE FOUR MULTIVARIATE SIGNIFICANCE TESTS

The four multivariate significance tests are all calculated from the eigenvalues of the $\mathbf{E}^{-1}\mathbf{H}$ matrix. The $\mathbf{E}^{-1}\mathbf{H}$ matrix is the multivariate analog of the F-ratio. Just as the F-ratio is the mean square for the hypothesis matrix (MSB in a one-way) divided by the mean square for the error matrix (MSW), $\mathbf{E}^{-1}\mathbf{H}$ is the hypothesis SSCP matrix (**H**) divided by—or actually multiplied by the inverse of—the error SSCP matrix (**E**). The $\mathbf{E}^{-1}\mathbf{H}$ matrix is then factored, and the four multivariate statistics can be calculated from the eigenvalues (λ_i) of the $\mathbf{E}^{-1}\mathbf{H}$ matrix according to Equations 8.9–8.12.

$$\text{Wilks' lambda: } \Lambda = \Pi \frac{1}{1+\lambda_i}. \tag{8.9}$$

$$\text{Pillai: } V = \Sigma \frac{\lambda_i}{1+\lambda_i}. \tag{8.10}$$

$$\text{Lawley-Hotelling: } U = \Sigma \lambda_i. \tag{8.11}$$

$$\text{Roy's largest root: } \theta = \frac{\lambda_1}{1+\lambda_1}. \tag{8.12}$$

Rather than the laborious process of creating the $\mathbf{E}^{-1}\mathbf{H}$ matrix and then factoring it, we will run the "peaches" data in SAS and find the eigenvalues of the $\mathbf{E}^{-1}\mathbf{H}$ matrix in the printout, and then verify one or two of these four formulas on the basis of the eigenvalues. The data and SAS code for running the data are shown in Table 8.4.

This SAS code produces four pages of output, with the first pages summarizing the univariate results. The last page is the summary of the four multivariate statistics, as show in Figure 8.5.

TABLE 8.4. Input Data and SAS Code for MAV1 Analysis of the Demonstration Data

```
data peaches; input sample girth grams; datalines;
1                        15                        130
1                        17                        139
1                        17                        164
1                        19                        155
1                        20                        172
2                        17                        105
2                        17                        122
2                        19                        122
2                        19                        138
2                        21                        156
2                        23                        173
2                        21                        122
3                        21                        122
3                        24                        138
3                        24                        147
3                        25                        130
; run;
proc print; run;
proc glm data=peaches; class sample;
model girth grams=sample;
manova h=sample; lsmeans sample; run;
```

```
                              The SAS System           18:27 Wednesday, April 13, 2011   4

                                The GLM Procedure
                            Multivariate Analysis of Variance

                  Characteristic Roots and Vectors of: E Inverse * H, where
                         H = Type III SSCP Matrix for sample
                              E = Error SSCP Matrix

                Characteristic                  Characteristic Vector   V'EV=1
                      Root        Percent            girth              grams

                 5.95736036        98.78          0.21634868         -0.02020715
                 0.07388164         1.22          0.03543689          0.01124856


    MANOVA Test Criteria and F Approximations for the Hypothesis of No Overall sample Effect
                         H = Type III SSCP Matrix for sample
                              E = Error SSCP Matrix

                         S=2      M=-0.5      N=5

       Statistic                     Value     F Value   Num DF   Den DF   Pr > F

       Wilks' Lambda               0.13384405    10.40      4        24    <.0001
       Pillai's Trace              0.92506601     5.59      4        26     0.0022
       Hotelling-Lawley Trace      6.03124200    17.72      4    13.429    <.0001
       Roy's Greatest Root         5.95736036    38.72      2        13    <.0001

            NOTE: F Statistic for Roy's Greatest Root is an upper bound.
                  NOTE: F Statistic for Wilks' Lambda is exact.
```

Figure 8.5   The four multivariate statistics from SAS Output of the MAV1 demonstration data. (Created with SAS software. Copyright 2011, SAS Institute Inc. Cary, NC. All rights reserved. Reproduced with permission of SAS Institute Inc. Cary, NC.)

In the fourth line down in this output, we read "Characteristic Roots and Vectors of: E Inverse*H." Characteristic root is another name for eigenvalue, and the first two numbers on the left of this output fragment (Fig. 8.5) are indeed the eigenvectors of the $\mathbf{E}^{-1}\mathbf{H}$ matrix. Right next to them is their percent, that is, the percent of variance accounted for by each latent root of the $\mathbf{E}^{-1}\mathbf{H}$ matrix. Since the first root accounts for 98.78% of the variance in the matrix, this indicates that this is essentially a one-dimensional $\mathbf{E}^{-1}\mathbf{H}$ matrix.

We will first use Equation 8.9 to calculate the Wilks' lambda value. The Greek capital letter pi at the beginning of the expression on the right is like the capital sigma sign that is used for summation; however, the pi operator indicates that one should multiply what follows for all values of i. So we multiply the ratio of $1/(1 + \lambda_i)$ for the first eigenvalue times that same ratio for the second eigenvalue and so on for all eigenvalues of the $\mathbf{E}^{-1}\mathbf{H}$ matrix. This particular one only has two.

$$\Lambda = \Pi \frac{1}{1+\lambda_i} = \left(\frac{1}{1+5.95736}\right)\left(\frac{1}{1+0.07388}\right) = (0.14373)(0.93120) = 0.13384.$$

This is indeed the value of Wilks' lambda that we obtained in our direct calculations using Equation 8.8 in the demonstration above, and it also agrees with the output of results from Stata and SAS. Interestingly, Wilks' lambda can be calculated in either of these two ways: Equation 8.8, which uses the ratio of determinants, or Equation 8.9 that we just used, which is based upon the eigenvalues of the $\mathbf{E}^{-1}\mathbf{H}$ matrix.

Let's do a calculator check of one more of the four multivariate statistics, this time using Equation 8.10, which is the formula for Pillai's trace.

$$V = \Sigma \frac{\lambda_i}{1+\lambda_i} = \left(\frac{5.95736}{1+5.95736}\right) + \left(\frac{0.07388}{1+0.07388}\right) = 0.856267 + 0.068797 = 0.925065.$$

This value also agrees with the output of results from Stata and SAS.

The first question that comes up is why there should be four multivariate significance tests. It would seem that one would suffice. A paragraph from Rencher (2002) on this issue is instructive:

> Why do we use four different tests? All four are exact tests; that is, when H_0 is true, each test has probability α of rejecting H_0. However, the tests are not equivalent, and in a given sample they may lead to different conclusions even when H_0 is true; some may reject H_0 while others accept H_0. This is *due to the multidimensional nature of the space in which the mean vectors* $\mu_1, \mu_2, \ldots \mu_k$ *lie*. . . . (italics ours).

On pages 176–178, Rencher (2002) gives a comparison and discussion of the relative power of each of the tests under different circumstances. Roy's greatest root only depends on the first eigenvalue of the $\mathbf{E}^{-1}\mathbf{H}$ matrix and therefore is the most powerful of the four tests when the multivariate means line up in one dimension, which happens when the first eigenvalue accounts for a high percentage of the variance in the matrix. This is an area of possible research in that the particular spatial configurations and the specific conditions under which the other three tests will be more powerful are not yet clear. The common statistical packages (SAS, SPSS, Stata, and others) routinely report all four, and it is good to consider all four in making a determination. It is common practice to consider any one of them to be a proper hedge against alpha inflation in deciding whether the corresponding univariate effects can be considered. Usually, the four tests reach a similar verdict. In those cases where they do not, Rencher recommends examining the covariance matrices and the eigenvalues of $\mathbf{E}^{-1}\mathbf{H}$ to evaluate possible reasons.

8.5   SUMMARY AND CONCLUSIONS

In this chapter, we have generalized the discussion of ANOVA at the end of Chapter 2 to include MANOVA, the multivariate form of ANOVA. MANOVA has several important advantages over univariate analysis of multivariate data. MANOVA avoids alpha inflation when it is used as a filter for univariate tests—if the MANOVA does not show significance, significant univariate tests are assumed to be due to chance from multiple testing. Further, univariate tests fail to properly model the correlations among dependent variables. Multivariate tests are also in many cases more powerful than are univariate ones. Or, in other words, MANOVA can detect significant results that a series of univariate ANOVAs may have missed when the overall statistical structure shows group differences, but the univariate analyses mask those differences. Discriminant analysis will also help in these situations.

In interpreting a one-way MANOVA, it is helpful to realize that the Wilks' lambda is equal to $1 - R^2$, where R^2 is the proportion of variance accounted for by the model. Biplot graphs can be useful for interpreting statistical significance with this type of complex multivariate data with correlated outcome variables, particularly when there is considerable significance.

There are six primary computational steps in a one-way MANOVA. The first is to calculate all the **Q** matrices for the entire MANOVA. The second step is to calculate all SSCP matrices from the **Q** matrices. Next, multivariate statistics, including the R^2 value, are calculated from the SCPP matrices. The fourth step is to gather all the univariate sums of squares from the diagonals of the SSCP matrices. For this, the sums of squares for between, within, and total are all found in the diagonal elements of the SSCP matrices. The fifth step consists of collecting both the univariate and multivariate results into a MANOVA summary table with the univariate summary at the top of the table and the multivariate summary at the bottom. Finally, we create graphics to elucidate the relationships in the data.

MANOVA produces four multivariate significance tests, which are all calculated from the eigenvalues of the $\mathbf{E}^{-1}\mathbf{H}$ matrix, the multivariate analog of the F-ratio (see Eqs. 8.9–8.12). Each multivariate statistic is an exact test, and each has the same probability of mistakenly rejecting the null hypothesis when it is true. However, the four tests are not equivalent. Although the four tests are often in agreement, in any one case one may show significance, whereas another may not. All four statistics should be taken into consideration when interpreting the results of a MANOVA, and differences among them can often inform the interpretation of the results. MANOVA can be a powerful tool, particularly when used in concert with multivariate graphical devices that clarify the relationships and the results.

STUDY QUESTIONS

A. Essay Questions

1. Explain the advantages of MANOVA over a series of univariate analyses of variance on data with multiple dependent variables.

2. Explain how you can have significance on the multivariate test in MANOVA but not on any of the univariate tests.

3. Describe the six steps to compute a one-way MANOVA.

4. Explain the basic formula for each of the four multivariate significance tests of MANOVA and how they relate to the $E^{-1}H$ matrix.

5. Explain how the four multivariate significance tests are related to Hotelling's T squared. (Chapter 7)

6. Outline the rationale of following up significant MANOVA results with corresponding univariate tests.

B. Computational Questions

1. Compute a MAV1 analysis of the data given below (using matrix algebra and the "Q" quantities computational approach). Use only Wilks' lambda for the multivariate test (since it is easiest).

Treatment 1	Treatment 2
(8,9)	(9,9)
(6,7)	(6,2)
(5,1)	(7,12)
(5,7)	(3,8)
(4,5)	(7,11)
(2,13)	(1,3)

2. Compute a one-way MANOVA (MAV1) on the following data (using matrix algebra and the "Q" quantities computational approach). Use only Wilks' lambda for the multivariate test (since it is easiest).

Treatment 1	Treatment 2
(8,9)	(12,5)
(6,7)	(16,9)
(5,1)	(10,7)
(5,7)	(4,8)
(4,5)	(8,11)
(2,13)	(10,8)

3. To better understand the mathematics involved in a one-way MANOVA, re-compute the MAV1 analysis of question 1 above using simple definitional formulas for SS between, within and total on both dependent variables, and then create similar definitional SP formulas and construct the H, E, and T matrices. Which is easier computationally, the question 1 method of computations or this definitional method of computations? Which is easier to understand conceptually?

C. Data Analysis Questions

1. Run the data in questions 1 and 2 of section B above in SAS, SPSS, or Stata. Compare results.

2. Use an Excel spreadsheet to compute a one-way MANOVA (MAV1) on the following data. Use matrix algebra and the "Q" quantities computational approach. Use only Wilks' lambda for the multivariate test (since it is easiest). Then re-run the data in SAS, SPSS, or Stata. Compare results.

Treatment 1	Treatment 2
(5,6)	(6,11)
(9,6)	(8,11)
(5,7)	(9,12)
(11,9)	(13,14)
(8,2)	(10,5)
(10,5)	(18,6)
(9,7)	(15,8)
(7,6)	(17,5)

3. Now redo the same analysis of question 2 above, but standardize the data before accomplishing your computations. Notice that Wilks' lambda comes out the same whether or not you standardize the data before beginning. (The other three multivariate statistics, Pilai's, Hotelling–Lawley, and Roy's would also come out the same by both computations.) What do you learn from this about MANOVA?

REFERENCES

Hummel, T. J., and Sligo, J. R. 1971. Empirical comparison of univariate and multivariate analysis of variance procedures. *Psychological Bulletin*, 76(1), 49–57.

Lemyre, L., and Smith, P. M. 1985. Intergroup discrimination and self-esteem in the minimal group paradigm. *Journal of Personality and Social Psychology*, 49(3), 660–670.

Myers, H. H., and Siegel, P. S. 1985. The motivation to breastfeed: A fit to the opponent-process theory? *Journal of Personality and Social Psychology*, 49(1), 188–193.

Orne, M. T. 1962. On the social psychology of the psychological experiment: With particular reference to demand characteristics and their implications. *The American Psychologist*, 17, 776–783.

Rencher, A. C. 1998. *Multivariate Statistical Inference and Applications*. New York: Wiley.

Rencher, A. C. 2002. *Methods of Multivariate Analysis, Second Edition*. New York: Wiley.

Rencher, A. C., and Scott, D. T. 1990. Assessing the contribution of individual variables following rejection of a multivariate hypothesis. *Communication in Statistics: Simulation and Computation*, 19(2), 535–553.

Solomon, R., and Corbit, J. 1974. An opponent-process theory of motivation: I. Temporal dynamics of affect. *Psychological Review, 81*, 119–145.

Warner, R. M., and Sugarman, D. B. 1986. Attributions of personality based on physical appearance, speech, and handwriting. *Journal of Personality and Social Psychology, 50*(4), 792–799.

CHAPTER NINE

MULTIPLE REGRESSION AND THE GENERAL LINEAR MODEL

9.1 INTRODUCTION

In this final chapter, we present one of the most elementary of methods, and yet one of the most elegant and profound—multiple regression. It is also quite old. Carl Friedrich Gauss first described the least squares method in about 1794 or 1795.[1]

There is something quite central about regression. Certainly it is central to a broad array of disciplines as one of the most common of quantitative tools. But there is also something mathematically fundamental about it and its relationship to our broader numerical system, particularly linear algebra. It is, for example, closely allied with the inverse of a matrix. As demonstrated by Equation 4.21 of Chapter 4, the inverse of a matrix contains within each of its

---

[1] Priority in the discovery of the method of least squares is a matter of some dispute. Indeed, Stigler (1981) refers to it as "the most famous priority dispute in the history of statistics" (p. 465). Legendre (1805) published a lucid account of the method in 1805 complete with a worked example, but there is some evidence that Gauss had actually devised the method as early as 1794, and perhaps had successfully made astronomical predictions using it. Seal (1967) argues that "the linear regression model owes so much to Gauss that we believe it should bear his name." Stigler (1981) maintains that "if there was any single scientist who first put the method within the reach of the common man, it was Legendre," but he also states that "when Gauss did publish on least squares, he went far beyond Legendre in both conceptual and technical development" (p. 472). See Plackett (1972) for a detailed summary of the evidence on each side of the dispute.

Multivariate Analysis for the Biobehavioral and Social Sciences: A Graphical Approach,
First Edition. Bruce L. Brown, Suzanne B. Hendrix, Dawson W. Hedges, Timothy B. Smith.
© 2012 John Wiley & Sons, Inc. Published 2012 by John Wiley & Sons, Inc.

diagonal elements the complete numerical information of a multiple regression for predicting the variable of that element from the other variables in the matrix. This property is truly remarkable. Multiple regression, rather than being a statistical invention, is an integral part of the mathematics of linear algebra. Thus, multiple regression provides the foundation from which a variety of more complex analyses can be performed. As we shall see in this chapter, it is one of the most universal, useful, and powerful of quantitative methods (see also Rao et al., 1999).

The featured "case study" data set for this chapter will not be introduced until Section 9.5, "Linear Contrasts and Complex Designs." The data are not from a published paper but rather from an unpublished master's thesis entitled "Foot-in-the-door technique: Mediating effects of material payment, social reinforcement, and familiarity" (Waranusuntikule, 1985). This particular study was chosen because of the highly unusual statistical methods design—a three-way analysis of variance plus a control group—that will be used to illustrate the power of the linear contrasts strategy for creating tailor-made analyses for unusual situations.

9.2 THE FUNDAMENTAL METHOD OF MULTIPLE REGRESSION

In this section, we describe the matrix algebra process for calculating a multiple regression analysis. This will provide a simple review of the matrix methods of Chapter 3 applied to the construction of covariance matrices, and it will also provide a conceptual foundation for introducing in the remaining sections of this chapter several higher-level applications of multiple regression: analysis of covariance (ANCOVA), factorial analysis of variance (ANOVA) with unbalanced data, and logistic regression.

For the initial demonstration, we use simplest case data with seven observations and three variables. Two of the variables are predictors, X_1 and X_2, and one of the variables is a criterion variable, Y, as shown at the left.

Y	X_1	X_2
9	13	6
10	12	2
7	10	3
12	10	12
8	9	10
4	9	5
6	7	11

Step 1: Calculate a correlation matrix $\mathbf{R}$ from the data matrix.
The simplest way to do this is to temporarily forget about the X and Y structure of this matrix and treat is as a single $\mathbf{X}$ matrix with seven rows and

three columns. When we have obtained the correlation matrix, the X and Y identities can then be restored.

- The means of the three columns of the matrix (three variables) are obtained and found to be $\bar{Y} = 8$; $\bar{X}_1 = 10$; $\bar{X}_2 = 7$. The corresponding sums are $T_y = 56$; $T_{x1} = 70$; $T_{x2} = 49$. The means are collected into a means vector $\mathbf{m}' = \begin{bmatrix} 8 & 10 & 7 \end{bmatrix}$, and the totals into a totals vector $\mathbf{t}' = \begin{bmatrix} 56 & 70 & 49 \end{bmatrix}$.
- A deviation score matrix (also called a centered matrix) $\mathbf{X}_c$ is obtained from the raw score matrix $\mathbf{X}$ by subtracting the matrix of means $\mathbf{M}$ from each score.[2] This is the process of centering the data.

$$\mathbf{X}_c = \mathbf{X} - \mathbf{M} = \begin{bmatrix} 9 & 13 & 6 \\ 10 & 12 & 2 \\ 7 & 10 & 3 \\ 12 & 10 & 12 \\ 8 & 9 & 10 \\ 4 & 9 & 5 \\ 6 & 7 & 11 \end{bmatrix} - \begin{bmatrix} 8 & 10 & 7 \\ 8 & 10 & 7 \\ 8 & 10 & 7 \\ 8 & 10 & 7 \\ 8 & 10 & 7 \\ 8 & 10 & 7 \\ 8 & 10 & 7 \end{bmatrix} = \begin{bmatrix} 1 & 3 & -1 \\ 2 & 2 & -5 \\ -1 & 0 & -4 \\ 4 & 0 & 5 \\ 0 & -1 & 3 \\ -4 & -1 & -2 \\ -2 & -3 & 4 \end{bmatrix}.$$

- A **CSSCP** matrix is created by premultiplying the centered scores $\mathbf{X}_c$ by the transpose of the raw scores. Please note that the three diagonal elements of this matrix are the sums of squares of variable Y, variable X_1, and variable X_2, respectively. The sums of squares of variables X_1 and X_2 ($SS_{x1} = 24$, $SS_{x2} = 96$) will be used in Equations 9.1 and 9.2 at the end of this section.

$$\mathbf{CSSCP} = \mathbf{X}'\mathbf{X}_c = \begin{bmatrix} 9 & 10 & 7 & 12 & 8 & 4 & 6 \\ 13 & 12 & 10 & 10 & 9 & 9 & 7 \\ 6 & 2 & 3 & 12 & 10 & 5 & 11 \end{bmatrix} \begin{bmatrix} 1 & 3 & -1 \\ 2 & 2 & -5 \\ -1 & 0 & -4 \\ 4 & 0 & 5 \\ 0 & -1 & 3 \\ -4 & -1 & -2 \\ -2 & -3 & 4 \end{bmatrix}$$

$$= \begin{bmatrix} 42 & 17 & 13 \\ 17 & 24 & -26 \\ 13 & -26 & 96 \end{bmatrix}$$

[2] The matrix $\mathbf{M}$ is a conformable 7×3 matrix of means of the three columns of data. The matrix equation for creating it, as explained in Chapter 3, Section 3.3.7, is to premultiply the $\mathbf{X}$ matrix by a conformable 7×7 $\mathbf{J}$ matrix of ones, which gives the $\mathbf{T}$ matrix, and then divide each entry of $\mathbf{T}$ by N to get the $\mathbf{M}$ matrix:

$$\mathbf{T} = \mathbf{JX} \text{ and } \mathbf{M} = \left(\frac{1}{N}\right)\mathbf{T}.$$

- Alternately, a **CSSCP** matrix can be created by the formula **CSSCP = X′X − tm′**.

CSSCP = X′X − tm′

$$= \begin{bmatrix} 9 & 10 & 7 & 12 & 8 & 4 & 6 \\ 13 & 12 & 10 & 10 & 9 & 9 & 7 \\ 6 & 2 & 3 & 12 & 10 & 5 & 11 \end{bmatrix} \begin{bmatrix} 9 & 13 & 6 \\ 10 & 12 & 2 \\ 7 & 10 & 3 \\ 12 & 10 & 12 \\ 8 & 9 & 10 \\ 4 & 9 & 5 \\ 6 & 7 & 11 \end{bmatrix} - \begin{bmatrix} 56 \\ 70 \\ 49 \end{bmatrix} \begin{bmatrix} 8 & 10 & 7 \end{bmatrix}$$

$$= \begin{bmatrix} 42 & 17 & 13 \\ 17 & 24 & -26 \\ 13 & -26 & 96 \end{bmatrix}.$$

- From the **CSSCP** matrix, a covariance matrix is obtained by dividing each entry by the degrees of freedom (*df*).

$$\mathbf{C} = \left(\frac{1}{N-1}\right)\mathbf{CSSCP} = \left(\frac{1}{6}\right)\begin{bmatrix} 42 & 17 & 13 \\ 17 & 24 & -26 \\ 13 & -26 & 96 \end{bmatrix} = \begin{bmatrix} 7.00 & 2.83 & 2.17 \\ 2.83 & 4.00 & -4.33 \\ 2.17 & -4.33 & 16.00 \end{bmatrix}.$$

- Create matrix **D** by diagonalizing the covariance matrix **C**.

$$\mathbf{D} = \text{diag}\begin{bmatrix} 7.00 & 2.83 & 2.17 \\ 2.83 & 4.00 & -4.33 \\ 2.17 & -4.33 & 16.00 \end{bmatrix} = \begin{bmatrix} 7.00 & 0 & 0 \\ 0 & 4.00 & 0 \\ 0 & 0 & 16.00 \end{bmatrix}$$

- Create the standardizing matrix **E** by taking the inverse of the square root of matrix **D**.

$$\mathbf{E} = \text{inv}\left(\text{square root}\begin{bmatrix} 7.00 & 0 & 0 \\ 0 & 4.00 & 0 \\ 0 & 0 & 16.00 \end{bmatrix}\right) = \text{inv}\begin{bmatrix} 2.65 & 0 & 0 \\ 0 & 2.00 & 0 \\ 0 & 0 & 4.00 \end{bmatrix}.$$

$$= \begin{bmatrix} 0.378 & 0 & 0 \\ 0 & 0.500 & 0 \\ 0 & 0 & 0.250 \end{bmatrix}$$

- Create the correlation matrix **R** by pre- and postmultiplying the covariance matrix **C** by the standardizing matrix **E**.

$$\mathbf{R = ECE} = \begin{bmatrix} 0.378 & 0 & 0 \\ 0 & 0.500 & 0 \\ 0 & 0 & 0.250 \end{bmatrix} \begin{bmatrix} 7.00 & 2.83 & 2.17 \\ 2.83 & 4.00 & -4.33 \\ 2.17 & -4.33 & 16.00 \end{bmatrix} \begin{bmatrix} 0.378 & 0 & 0 \\ 0 & 0.500 & 0 \\ 0 & 0 & 0.250 \end{bmatrix}$$

$$= \begin{bmatrix} 1.000 & 0.535 & 0.205 \\ 0.535 & 1.000 & -0.542 \\ 0.205 & -0.542 & 1.000 \end{bmatrix}.$$

*Step 2: Organize the correlation matrix **R** into four partitions (a matrix, two vectors, and a scalar), according to the dependent variable Y and the two independent variables, X_1 and X_2.*

R =	r_{yy}	r_{yi}	=	1.000	0.535	0.205
	r_{iy}	R_{ii}		0.535	1.000	-0.542
				0.205	-0.542	1.000

*Step 3: Obtain the inverse of the R_{ii} correlation matrix of the independent variables using the methods of Chapter 3 or the **minverse** function of an Excel spreadsheet.*

$$\mathbf{R_{ii}^{-1}} = inverse \begin{bmatrix} 1.000 & -0.542 \\ -0.542 & 1.000 \end{bmatrix} = \begin{bmatrix} 1.415 & 0.767 \\ 0.767 & 1.415 \end{bmatrix}$$

Step 4. Obtain the standardized regression coefficients β by postmultiplying the inverse of the R_{ii} matrix by the r_{iy} vector, Equation 9.1.

$$\beta = \mathbf{R_{ii}^{-1} r_{iy}} = \begin{bmatrix} 1.415 & 0.767 \\ 0.767 & 1.415 \end{bmatrix} \begin{bmatrix} 0.535 \\ 0.205 \end{bmatrix} = \begin{bmatrix} 0.915 \\ 0.700 \end{bmatrix}. \tag{9.1}$$

*Step 5: Obtain unstandardized regression coefficients **b** by multiplying β, the standardized coefficients, by the ratio of standard deviations.*[3]

$$\mathbf{B} = \beta \left(\frac{s_y}{s_x} \right) = \begin{bmatrix} (0.915)\left(\dfrac{2.65}{2} \right) \\ (0.700)\left(\dfrac{2.65}{4} \right) \end{bmatrix} = \begin{bmatrix} 1.210 \\ 0.463 \end{bmatrix}$$

Step 6: Obtain the squared multiple regression coefficient R^2 by premultiplying the standardized regression coefficients β by the r_{yi} column vector, Equation 9.2.

---

[3] We first obtain the standard deviations by taking the square root of the variance of each of the three variables in the covariance matrix and find them to be: s_y = sqrt(7) = 2.65, s_{x1} = sqrt(4) = 2, s_{x2} = sqrt(16) = 4.

$$\mathbf{R}^2 = \mathbf{r}_{yi}\boldsymbol{\beta} = [0.535 \quad 0.205]\begin{bmatrix} 0.915 \\ 0.700 \end{bmatrix} = 0.6331. \tag{9.2}$$

Step 7: Solve for a, *the* Y *intercept.*

$$a = \bar{Y} - b_1\bar{X}_1 - b_2\bar{X}_2 = 8 - (1.210)(10) - (0.463)(7) = -7.343.$$

Step 8. Create the regression equation $\hat{\mathbf{Y}} = \mathbf{XB} + \mathbf{A}$, *using unstandardized coefficients and the intercept a.*

$$\hat{\mathbf{Y}} = 1.210\mathbf{X}_1 + .463\mathbf{X}_2 - 7.343.$$

Step 9. Use the regression equation in matrix form to obtain the predicted scores $\hat{\mathbf{Y}}$.

$$\hat{\mathbf{Y}} = \mathbf{XB} + \mathbf{A} = \begin{bmatrix} 13 & 6 \\ 12 & 2 \\ 10 & 3 \\ 10 & 12 \\ 9 & 10 \\ 9 & 5 \\ 7 & 11 \end{bmatrix}\begin{bmatrix} 1.210 \\ 0.463 \end{bmatrix} + \begin{bmatrix} -7.343 \\ -7.343 \\ -7.343 \\ -7.343 \\ -7.343 \\ -7.343 \\ -7.343 \end{bmatrix} = \begin{bmatrix} 11.167 \\ 8.104 \\ 6.147 \\ 10.316 \\ 8.179 \\ 5.864 \\ 6.222 \end{bmatrix}.$$

Step 10: Calculate the total $(T_{y'})$ *and the mean* $(\bar{\hat{Y}}')$ *of the predicted scores* $(\hat{\mathbf{Y}})$. *They should be equal to the mean and the total for the observed* **Y** *scores calculated in step 1, which they are.*

$$T_{\hat{y}} = \sum \hat{Y}_i = 56 \quad \bar{\hat{Y}} = \frac{T_{\hat{y}}}{N} = \frac{56}{7} = 8$$

Step 11: Calculate SSR, the sum of squares regression, which is the sum of squared deviations of the predicted scores from their mean.

$$\mathrm{SSR} = \hat{\mathbf{Y}}'\hat{\mathbf{Y}} - T_{\hat{y}}\bar{\hat{Y}}$$

$$= [11.167 \quad 8.104 \quad 6.147 \quad 10.316 \quad 8.179 \quad 5.864 \quad 6.222]\begin{bmatrix} 11.167 \\ 8.104 \\ 6.147 \\ 10.316 \\ 8.179 \\ 5.864 \\ 6.222 \end{bmatrix}.$$

$$- (56)(8) = 474.592 - 448 = 26.592$$

Step 12: Calculate the errors of prediction by subtracting the predicted scores $\hat{\mathbf{Y}}$ *from the actual scores* $\mathbf{Y}$.

$$\mathbf{e} = \mathbf{Y} - \hat{\mathbf{Y}} = \begin{bmatrix} 9 \\ 10 \\ 7 \\ 12 \\ 8 \\ 4 \\ 6 \end{bmatrix} - \begin{bmatrix} 11.167 \\ 8.104 \\ 6.147 \\ 10.316 \\ 8.179 \\ 5.864 \\ 6.222 \end{bmatrix} = \begin{bmatrix} -2.167 \\ 1.896 \\ 0.853 \\ 1.684 \\ -0.179 \\ -1.864 \\ -0.222 \end{bmatrix}$$

Step 13. Calculate SSE, the sum of squares of the error scores, and confirm that SST = SSR + SSE, and that R^2 = SSR/SST. SST is equal to 42, the deviation sum of squares, obtained for the dependent variable Y in step 1.

The error scores have a mean of zero, so they are already in deviation score form, thus simplifying the calculation of SSE.

$$\text{SSE} = \mathbf{e}'\mathbf{e} = \begin{bmatrix} -2.167 & 1.896 & 0.853 & 1.684 & -0.179 & -1.864 & -0.222 \end{bmatrix} \begin{bmatrix} -2.167 \\ 1.896 \\ 0.853 \\ 1.684 \\ -0.179 \\ -1.864 \\ -0.222 \end{bmatrix}.$$

$$= 15.408$$

We see that SST is indeed the sum of SSR and SSE.

$$\text{SST} = \text{SSR} + \text{SSE}.$$

$$42 = 26.592 + 15.408.$$

We also see that the multiple R-square value, R^2, is equal to SSR divided by SST.

$$R^2 = \text{SSR/SST} = 26.592/42 = 0.6331.$$

Step 14. Calculate MST, MSR, and MSE by dividing SST and SSE by their respective df. The df for MST is the number of observations n minus one (n − 1), the df for regression is the number of regression parameters k (two predictors and one intercept) minus one, and the df for MSE is n − P − 1, where P is the number of regressors (independent variables) in the analysis. Use these quantities to calculate the adjusted R-Square as R^2_{adj} = 1 − MSE / MST, and to calculate the "root MSE" as the square root of MSE.

$$MST = \frac{SST}{n-1} = \frac{42}{7-1} = \frac{42}{6} = 7$$

$$MSR = \frac{SSR}{k-1} = \frac{26.592}{3-1} = \frac{26.592}{2} = 13.296$$

$$MSE = \frac{SSE}{n-p-1} = \frac{15.408}{7-2-1} = \frac{15.408}{4} = 3.852$$

$$\text{root } MSE = \sqrt{MSE} = \sqrt{3.852} = 1.963$$

$$R_{adj}^2 = 1 - \frac{MSE}{MST} = 1 - \frac{3.852}{7} = 0.4497$$

The F-ratio for the regression is calculated as MSR divided by MSE.

$$F = \frac{MSR}{MSE} = \frac{13.296}{3.852} = 3.452.$$

In the F table (Table 4 of the Appendix), we find that the critical F-ratio for 2 and 13 df is 3.80. Our obtained F-ratio falls short of this, so the results of the regression analysis are not significant.

We will now collect all of these calculations in a multiple regression summary table, Table 9.1. This table has three sections. In the top section, labeled "Analysis of Variance," we have the three SS values (SSR, meaning SS regression; SSE, meaning SS error; and SST, meaning SS total, respectively) listed in the first column. These SS values are also known as "variation" values. In the second column, we have the three values of df (df regression, df error, and df total). In the third column, we have the three variances, MSR, MSE, and MST. In the fourth column, we have the F-ratio that was just calculated as MSR/

TABLE 9.1. Multiple Regression Summary Table for Demonstration Data

Analysis of Variance				
Source	SS	df	MS	F	P
Model	26.592	2	13.296	3.452	0.1346
Error	15.408	4	3.852		
Corrected total	42.000	6	7.000		

$R^2 = 0.6331$
Adjusted $R^2 = 0.4497$
Root MSE = 1.963

Parameter Estimates				
Variable	Parameter	df	Std Error	t-Value	P
Intercept	−7.3428	1	5.8872	−1.25	0.2804
X1	1.2101	1	0.4766	2.54	0.0640
X2	0.4631	1	0.2383	1.94	0.1239

MSE. In the fifth column, we have the probability value for that F-ratio. You can find exact probability values for F-ratios on a spreadsheet. In Excel, for example, if you enter into a cell the command, "FDIST(3.452, 2, 4)" (where the first parameter of the function is the F-ratio, the second is the numerator df, and the third is the denominator df), the answer will come back as 0.1346.

In the middle section is given the two R^2 values, the unadjusted one and also the adjusted one, and the square root of MSE. The unadjusted R^2 (0.6331) is the percent of "variation" (that is, sum of squares units) that can be explained by the regression equation. The adjusted R^2 (0.4497) is the percent of variance (variation divided by its df) that can be explained, which means that it takes into account sample size and the number of predictor variables, and is therefore more comparable across models and analyses. The "root MSE" (the square root of the mean squares error) indicates the standard deviation of actual data points about the regression plane, measured in the same units as the Y, the dependent variable.

The bottom section entitled "Parameter Estimates" gives information about the effects of each of the individual predictor variables in the prediction process. In the first column, "Parameter," are three quantities we have calculated above: the Y intercept (-7.3428), the unstandardized predictor coefficient for predictor X_1 (1.2101), and the unstandardized predictor coefficient for predictor X_2 (0.4631). Since these coefficients are unstandardized, they are in the units of the observed variables and have a direct interpretation. In bivariate regression (with a single predictor variable), the coefficient indicates how much the dependent variable is expected to increase when the predictor variable increases by one unit. In multiple regression (with multiple predictor variables), the coefficient indicates how much the dependent variable is expected to increase when the particular predictor variable increases by one unit, while the other predictor variables are held constant. Each of these parameters has one degree of freedom. The next three columns in this part of the table have not yet been calculated. The standard error for the parameter estimate of X_1 is calculated by Equation 9.3, and the standard error for the parameter estimate of X_2 is calculated by Equation 9.4,[4] where SS_{x1} is the deviation sum of squares for the first predictor variable, and SS_{x2} is the deviation sum of squares for the second predictor variable, both calculated in step 1 of the demonstration above. The value r_{12}^2 is the square of the correlation coefficient between the two predictor variables that appears in the correlation matrix among the three variables at the end of step 1.

$$S_{b1} = \sqrt{\frac{\text{MSE}}{SS_{x1}(1-r_{12}^2)}}$$

$$S_{b1} = \sqrt{\frac{3.852}{24(1-(-0.542)^2)}} = \sqrt{\frac{3.852}{16.9497}} = \sqrt{0.2273} = 0.4766.$$

(9.3)

[4] These two equations will only work for the case in which there are exactly two predictors. The formula for only one predictor is simpler than these (see Pedhazur, 1982, 28), and the formula for three or more predictors is slightly more complex (Pedhazur, 1982, 59).

$$s_{b2} = \sqrt{\frac{MSE}{SS_{x2}(1-r_{12}^2)}}.$$

$$= \sqrt{\frac{3.852}{96(1-(-0.542)^2)}} = \sqrt{\frac{3.852}{67.7987}} = \sqrt{0.0568} = 0.2383.$$

(9.4)

The t-values to test each of these parameter estimates for significance are obtained from Equation 9.5—by simply dividing each unstandardized regression coefficient by its standard error. The df for this t-test are the df for MSE, 4 in this case, and it is a two-tailed test. To obtain the exact probability for the t-value for X_1 using Excel, for example, you enter "TDIST(2.54, 4, 2)," and the answer will come back as 0.0640.

$$t_i = \frac{b_i}{s_{bi}}$$

(9.5)

$$t_1 = \frac{b_1}{s_{b1}} = \frac{1.2101}{0.4766} = 2.54$$

$$t_2 = \frac{b_2}{s_{b2}} = \frac{0.4631}{0.2383} = 1.94.$$

We can now verify our calculations of these regression results by running the analysis of this simple data set, first in SAS and then in Stata. The SAS program is given below, including both the data statement and the code for running PROC REG (the most basic of SAS's wide variety of regression procedures) on this data set.

```
data basicMR;
  input y x1 x2;
  datalines;
    9      13     6
   10      12     2
    7      10     3
   12      10    12
    8       9    10
    4       9     5
    6       7    11
;
run;

proc print data=basicMR;
run;

proc reg data=basicMR;
  model y = x1 x2;
  title 'Basic Multiple Regression Demonstration';
run;
```

The SAS output from this program fits into one page as shown below.

Number of Observations Read				7		
Number of Observations Used				7		
Model	2	26.59214	13.29607	3.45	0.1346	
Error	4	15.40786	3.85197			
Corrected Total	6	42.00000				
Root MSE	1.96264	R-Square	0.6331			
Dependent Mean	8.00000	Adj R-Sq	0.4497			
Coeff Var	24.53303					
Intercept	1	-7.34275	5.88724	-1.25	0.2804	
x1	1	1.21007	0.47660	2.54	0.0640	
x2	1	0.46314	0.23830	1.94	0.1239	

This output is virtually identical to the multiple regression summary table that we created in Table 9.1, thus verifying that our calculations are correct, but more importantly, showing that we now understand the origin of each of the values that appear in the SAS output.

To run the regression analysis in Stata, you first paste the matrix of Y, X_1, and X_2 (shown in the introduction before step 1 of this demonstration) into the Stata editor, and then enter **reg y x1 x2**. The following output is obtained from Stata.

```
. reg y x1 x2

      Source |       SS       df       MS              Number of obs =       7
-------------+------------------------------           F(  2,     4) =    3.45
       Model | 26.5921376      2  13.2960688           Prob > F      =  0.1346
    Residual | 15.4078624      4   3.8519656           R-squared     =  0.6331
-------------+------------------------------           Adj R-squared =  0.4497
       Total |         42      6           7           Root MSE      =  1.9626

------------------------------------------------------------------------------
           y |      Coef.   Std. Err.      t    P>|t|     [95% Conf. Interval]
-------------+----------------------------------------------------------------
          x1 |   1.210074   .4765951     2.54   0.064    -.1131666    2.533314
          x2 |    .463145   .2382976     1.94   0.124    -.1984752    1.124765
       _cons |  -7.342752   5.887239    -1.25   0.280    -23.68835    9.002843
------------------------------------------------------------------------------
```

(StataCorp, 2009. Stata: Release 11. Statistical Software. College Station, TX: StataCorp LP.)

The reported quantities are essentially the same ones we obtained from SAS, and are therefore consistent with our demonstration calculations. One notable difference between Stata and SAS is that Stata gives us the confidence intervals for each of the three parameters that were estimated: the intercept (or "constant" as it is called in Stata), and the two unstandardized regression coefficients.

Before leaving this multiple regression demonstration, we will display the results in linear models form. The observed scores in matrix **Y** can be thought of as a linear sum of **XB** (the predictors multiplied by the unstandardized coefficients), **A** (the intercept), and **e** (error).

$$\mathbf{Y} = \mathbf{XB} + \mathbf{A} + \mathbf{e}. \tag{9.6}$$

Equation 9.6 is illustrated with the demonstration data.

Y		X		B		A		e
9	=	13	6	1.210	+	−7.343	+	−2.167
10		12	2	0.463		−7.343		1.896
7		10	3			−7.343		0.853
12		10	12			−7.343		1.684
8		9	10			−7.343		−0.179
4		9	5			−7.343		−1.864
6		7	11			−7.343		−0.222

After multiplying **XB** to create a vector, it can be verified that any entry in the **Y** vector is the sum of the entries in the three vectors to the right of the equal sign.

Y		XB		A		e
9	=	18.510	+	−7.343	+	−2.167
10		15.447		−7.343		1.896
7		13.490		−7.343		0.853
12		17.658		−7.343		1.684
8		15.522		−7.343		−0.179
4		13.206		−7.343		−1.864
6		13.565		−7.343		−0.222

We can now sum **XB** and **A** to simplify further.

Y		XB+A		e
9	=	11.167	+	−2.167
10		8.104		1.896
7		6.147		0.853
12		10.316		1.684
8		8.179		−0.179
4		5.864		−1.864
6		6.222		−0.222
mean= 8.000		mean= 8.000		mean= 0.000

The mean of vector **Y** is 8.00, and the mean of the **XB + A** vector is, of course, also 8.00, since it is predicting **Y**. We will use these means to center vector **Y** and vector **XB + A** (vector **e** is already centered) to calculate sums of squares of deviation scores.

The mean is subtracted from the **Y** vector and the **XB + A** vector to create centered scores

Y		XB+A		e
1	=	3.167	+	−2.167
2		0.104		1.896
−1		−1.853		0.853
4		2.316		1.684
0		0.179		−0.179
−4		−2.136		−1.864
−2		−1.778		−0.222

SST= 42.000		SSR= 26.592		SSE= 15.408

The sum of squares of the **Y** vector is 42, which is SST. The sum of squares of the **XB + A** vector is 26.592, which is SSR; and the sum of squares of the error vector is 15.408, which is SSE. This summarizes in basic terms the nature of the linear model for multiple regression.

9.3 TWO-WAY ANALYSIS OF VARIANCE (AV2) USING MULTIPLE REGRESSION

Cohen (1968) was the first to make a strong argument within the behavioral sciences for multiple regression as the general case method of which analysis of variance (ANOVA) could be considered a special case, in a *Psychological Bulletin* paper entitled "Multiple regression as a general data-analytic system." His initial insights struck a responsive chord with many behavioral scientists, and was followed up several years later with a nontechnical, highly accessible textbook on the subject (Cohen and Cohen, 1975, 1983).

In the 1970s, there was also much interest among statisticians in using the general linear model as an over-arching data analysis strategy, with a parallel development of computer software that employed multiple regression as a way of dealing with nonorthogonal analysis of variance designs (Speed, 1969; Bryce, 1970; Lawson, 1971; Searle, 1971; Francis, 1973; Bryce and Carter, 1974; Neter and Wasserman, 1974; Speed et al., 1978; Searle et al., 1981). All of this was experimental and avant-garde in the 1970s, but has now been adopted as the standard way of dealing with ANOVA in the major statistical packages. The rationale and calculations of the general linear models approach to regression and ANOVA is now well articulated in a number of books (see e.g., Pedhazur, 1997; West et al., 2003; Kutner et al., 2005; Rencher and Schaalje, 2008).

We now illustrate how ANOVA can be considered to be a special case of multiple regression/correlation analysis (MRC) by first demonstrating a simple two-way ANOVA by the traditional method of calculating sums of squares. We then move on to an intermediate step of showing how a two-way ANOVA can be seen as one of several possible sets of linear contrasts on a one-way ANOVA. Finally, we will use these orthogonal contrasts to cast ANOVA within the framework and computational method of multiple regression/ correlation analysis demonstrated in Section 9.2, which is the modern way of dealing with ANOVA in the standard statistical packages.

9.3.1   AV2 by the Sums of Squares Method

We now demonstrate two-way ANOVA with a set of hypothetical data.[5] Suppose that we are doing a social psychological experiment on methods of persuasion. In particular, we wish to test the effects of two factors—social reinforcement (the row treatment) and familiarity of the person (the column treatment)—on the effectiveness of the persuasive appeal. We test five persons in each of the four conditions (socially reinforced and familiar, socially reinforced and unfamiliar, not socially reinforced and familiar, and not socially reinforced and unfamiliar) and obtain 20 effectiveness scores shown in the 2×2 data matrix.

	Column 1 (Familiar)	Column 2 (Unfamiliar)
Row 1 (Socially reinforced)	9	7
	8	8
	4	4
	6	1
	3	5
Row 2 (Not socially reinforced)	2	2
	4.5	5
	1	1
	3	1
	2	1

As shown in the table below, the two-way ANOVA has five terms for which sums of squares must be obtained: rows, columns, row by column interaction, within groups, and total. The *df* are given for each of the five terms, both algebraically and also numerically. The symbol R is used to represent the number of row treatments, which in this case is two, so $R - 1$, the *df* for rows is one. Similarly, C is the number of columns, which is also two, so the *df* value for

---

[5] The data presented here are simplified, but not altogether fictitious. They are a subset of the data presented in Section 9.6 that are constructed to be consistent with the findings of the Waranusuntikule (1985) social persuasion dissertation study.

columns is also one. The df formulas are used to indicate the Q-values that are used in calculating the sums of squares. As shown in Equation 9.7, the SS formula for rows is Q_{rows} (which corresponds to "R" in the df formula) minus Q_{total} (which corresponds to the "1" in the df, since there is only one grand mean). The SS formula for columns, in like manner, is $Q_{columns}$ minus Q_{total}, as given in Equation 9.8.

Source	degrees of freedom	SS formulas	
Rows	$R-1=1$	$Q_{rows} - Q_{total}$	(9.7)
Columns	$C-1=1$	$Q_{columns} - Q_{total}$	(9.8)
R×C Interaction	$(R-1)(C-1) = RC-R-C+1=1$	$Q_{rc} - Q_{rows} - Q_{columns} + Q_{total}$	(9.9)
Within	$RC(n-1) = RCn - RC = N - RC = 16$	$Q_{data} - Q_{rc}$	(9.10)
Total	$N-1=19$	$Q_{data} - Q_{total}$	(9.11)

The df value for an interaction is always the product of the df for each source entering into that interaction, so for this two-way interaction, the RC df is $(R-1)$ times $(C-1)$, which numerically also comes out to be equal to one. We have expanded the $(R-1)(C-1)$ binomial to its product of $RC - R - C + 1$ to indicate each component (signified by Q) of which the two-way interaction is composed. The sum of squares for the RC interaction is found by Q_{rc} minus Q_{rows}, minus $Q_{columns}$, plus Q_{total}, as shown in Equation 9.9.

The df for within cells is RC multiplied by $n - 1$, where n is the number of observations within each cell (five in this case). In other words, it is the number of cells (four) times the number of df per cell (also four), for the df within cells of 16. The SS formula for within cells is equal to Q_{data} minus Q_{rc}, as shown in Equation 9.10. Likewise, the SS formula for total variation is equal to Q_{data} minus Q_{total}, as shown in Equation 9.11. The df from each of these sources are additive (as are the SS values). The first four df values do indeed sum to 19, which is $N - 1$, the df for corrected total. We will now calculate the Q-values for each data partition.

In preparation for calculating the Q-values for each data partition, we first obtain the totals within each cell (T_{rc}), the totals for each row (T_r) and each column (T_c), and the grand total (T), and also the corresponding means.

Totals	c1	c2	
r1	30	25	55
r2	12.5	10	22.5
	42.5	35	77.5

Means	c1	c2	
r1	6	5	5.5
r2	2.5	2	2.25
	4.25	3.5	3.875

From these totals and means, we construct the appropriate vectors of totals and vectors of means for the Q formulas.

$$\mathbf{t}_{\text{rows}} = \begin{bmatrix} 55 \\ 22.5 \end{bmatrix} \quad \mathbf{t}_{\text{columns}} = \begin{bmatrix} 42.5 \\ 35 \end{bmatrix} \quad \mathbf{t}_{\text{rc}} = \begin{bmatrix} 30 \\ 25 \\ 12.5 \\ 10 \end{bmatrix}$$

$$\mathbf{m}_{\text{rows}} = \begin{bmatrix} 5.5 \\ 2.25 \end{bmatrix} \quad \mathbf{m}_{\text{columns}} = \begin{bmatrix} 4.25 \\ 3.5 \end{bmatrix} \quad \mathbf{m}_{\text{rc}} = \begin{bmatrix} 6 \\ 5 \\ 2.5 \\ 2 \end{bmatrix}.$$

The formula for obtaining Q_{data}, the quantity reflecting the variance of observed data, is given by Equation 9.12 (which is also Eq. 2.34 of Chapter 2).

$$Q_{\text{data}} = \mathbf{X}'\mathbf{X}$$

$$Q_{\text{data}} = \begin{bmatrix} 9 & 8 & 4 & 6 & 3 & 7 & 8 & 4 & 1 & 5 & 2 & 4.5 & 1 & 3 & 2 & 2 & 5 & 1 & 1 & 1 \end{bmatrix} \begin{bmatrix} 9 \\ 8 \\ 4 \\ 6 \\ 3 \\ 7 \\ 8 \\ 4 \\ 1 \\ 5 \\ 2 \\ 4.5 \\ 1 \\ 3 \\ 2 \\ 2 \\ 5 \\ 1 \\ 1 \\ 1 \end{bmatrix}.$$

$$= 431.25$$

(9.12)

The formula for obtaining Q_{total}, Equation 9.13 (Eq. 2.33 of Chapter 2), is simply the product of the grand total and the grand mean.

$$Q_{total} = \mathbf{t'm}$$
$$Q_{total} = \mathbf{t'm} = (77.5)(3.875) = 300.3125.$$

(9.13)

The remaining three formulas have not been given before, but are new to the two-way ANOVA.

$$Q_{rows} = \mathbf{t'_{rows} m_{rows}}.$$

(9.14)

$$Q_{columns} = \mathbf{t'_{columns} m_{columns}}.$$

(9.15)

$$Q_{rc} = \mathbf{t'_{rc} m_{rc}}.$$

(9.16)

We calculate each of these three quantities using the appropriate vectors of totals and of means.

$$Q_{rows} = \mathbf{t'_{rows} m_{rows}} = [55 \quad 22.5]\begin{bmatrix} 5.5 \\ 2.25 \end{bmatrix} = 353.125.$$

$$Q_{columns} = \mathbf{t'_{columns} m_{columns}} = [42.5 \quad 35]\begin{bmatrix} 4.25 \\ 3.5 \end{bmatrix} = 303.125.$$

$$Q_{rc} = \mathbf{t'_{rc} m_{rc}} = [30 \quad 25 \quad 12.5 \quad 10]\begin{bmatrix} 6 \\ 5 \\ 2.5 \\ 2 \end{bmatrix} = 356.25.$$

From these five Q-values, we calculate the five sums of squares needed for the ANOVA summary table. These SS values are expressed in Equations 9.7–9.11 (derived in the table on p. 387, from the degrees of freedom formulas).

$$SSR = Q_{rows} - Q_{total}$$
$$= 353.125 - 300.3125 = 52.8125.$$

(9.7)

$$SSC = Q_{columns} - Q_{total}$$
$$= 303.125 - 300.3125 = 2.8125.$$

(9.8)

$$SSRC = Q_{rc} - Q_{rows} - Q_{columns} + Q_{total}$$
$$= 356.25 - 353.125 - 303.125 + 300.3125$$
$$= 0.3125.$$

(9.9)

$$SSW = Q_{data} - Q_{rc}$$
$$= 431.25 - 356.25 = 75.0000.$$

(9.10)

$$SST = Q_{data} - Q_{total}$$
$$= 431.25 - 300.3125 = 130.9375.$$

(9.11)

The five sum of squares values are collected, together with their respective *df*, into the ANOVA summary table, Table 9.2. The mean squares and

TABLE 9.2. Analysis of Variance Summary Table for AV2 Analysis of Persuasion Data

Source	R^2	SS	df	MS	F	P
Rows	0.4033	52.8125	1	52.8125	11.267	0.0040
Columns	0.0215	2.8125	1	2.8125	0.600	0.4499
R × C interaction	0.0024	0.3125	1	0.3125	0.067	0.7995
Within	0.5728	75.0000	16	4.6875		
Corrected total	1.0000	130.9375	19			

F-ratios are calculated and also entered into the table. One additional column is added to this table, the R^2 column to the left of the sums of squares. Each R^2 value is obtained by dividing the corresponding SS value by the total SS value. The R^2 value for rows, for example is obtained by the equation $R^2_{rows} = SSR/SST = 52.8125/130.9375 = 0.4033$. The R^2 values, the SS values, and the df values are all additive, with the values from the other four terms (rows, columns, interaction and within) summing to the corrected total value.

9.3.2 AV2 by Multiple Regression: The General Linear Model

In Table 9.1 and the corresponding SAS printout, we saw that a multiple regression analysis can be considered to consist of three parts: (1) the "Analysis of Variance" information summarizing the entire model, shown in the top third of the table; (2) the R^2 and root MSE information in the middle of the table, and (3) the detailed analysis of the separate predictors in the bottom third of the table. ANOVA can also be organized in this way. To obtain the statistics for the entire model, suppose that we do the analysis on the four treatment groups as though we were doing a one-way ANOVA. That is, we calculate a vector of totals and a vector of means for the four groups as a whole,[6] and enter these into Equation 9.17 to obtain a Q_{groups} quantity.

$$\mathbf{t}_{groups} = \begin{bmatrix} 30 \\ 25 \\ 12.5 \\ 10 \end{bmatrix} \quad \mathbf{m}_{groups} = \begin{bmatrix} 6 \\ 5 \\ 2.5 \\ 2 \end{bmatrix}.$$

$$Q_{groups} = \mathbf{t}'_{groups}\mathbf{m}_{groups} \tag{9.17}$$

$$Q_{groups} = \mathbf{t}'_{groups}\mathbf{m}_{groups} = \begin{bmatrix} 30 & 25 & 12.5 & 10 \end{bmatrix} \begin{bmatrix} 6 \\ 5 \\ 2.5 \\ 2 \end{bmatrix} = 356.25.$$

---

[6] Notice that these values for the $\mathbf{t}_{groups}$ and $\mathbf{m}_{groups}$ vectors are equivalent to the $\mathbf{t}_{rc}$ and $\mathbf{m}_{rc}$ vectors, but labeled differently for a two-way or a one-way analysis.

TABLE 9.3. Analysis of Variance AV1 Summary Table for the Entire Model of the Persuasion Data

Source	R^2	SS	df	MS	F	P
Model	0.4272	55.9375	3	18.645833	3.978	0.0271
Error	0.5728	75.0000	16	4.6875		
Corrected total	1.0000	130.9375	19			
$R^2 = 0.4272$						
Root MSE = 2.165						

TABLE 9.4. Linear Contrasts Corresponding to Row, Column, and Interaction for Persuasion Data

	r1		r2			
	c1	c2	c1	c2		
	6	5	2.5	2	Contrast	SS(Contrast)
Rows	1	1	−1	−1	6.5	52.8125
Columns	1	−1	1	−1	1.5	2.8125
RC interaction	1	−1	−1	1	0.5	0.3125

Using Equation 9.18, we now obtain SSmodel, a sum of squares for the entire model, and it has three df since there are four means. The SSW value obtained above becomes the SSerror, and it still has 16 df. The SST and dft values also remain the same.

$$SSmodel = Q_{groups} - Q_{total}$$
$$= Q_{groups} - Q_{total} = 356.25 - 300.3125 = 55.9375 \qquad (9.18)$$
$$SSerror = SSW = Q_{data} - Q_{rc} = 431.25 - 356.25 = 75.0000$$

These three sum of squares values and three df values are entered into Table 9.3, the summary table for the entire model, with the R^2 value for the entire model and the "root MSE" value entered right below the table. Root MSE, of course, is just the square root of the MSE value from the upper table: Root MSE = $\sqrt{MSE}$ = $\sqrt{4.6875}$ = 2.165. These are all collected into the ANOVA summary table for the entire model.

To obtain the detailed analysis of the separate "predictors," that is, the row, column, and row by column interaction terms, we use linear contrasts, which are also sometimes called planned comparisons. Table 9.4 shows the four cell means with three sets of weights under them, the first row of weights corresponding to the rows sum of squares, the second row corresponding to the columns sum of squares, and the last row corresponding to the interaction sum of squares.

Notice that the effect of the contrast vector for rows is to sum the first two means and subtract the sum of the last two. Each contrast is a difference between two weighted sums, and therefore has one degree of freedom per contrast vector. The rows contrast value is found to be 6.5.

$$(1)(6)+(1)(5)+(-1)(2.5)+(-1)(2)=6.5.$$

The process of calculating a contrast is perhaps clearer in matrix form, as shown in Equation 9.19.

$$C_{rows} = \mathbf{w}_{rows}\mathbf{m}$$

$$C_{rows} = \mathbf{w}_{rows}\mathbf{m} = \begin{bmatrix} 1 & 1 & -1 & -1 \end{bmatrix} \begin{bmatrix} 6 \\ 5 \\ 2.5 \\ 2 \end{bmatrix} = 6.5. \tag{9.19}$$

The contrast for columns is calculated in a similar manner and comes to 1.5.

$$C_{columns} = \mathbf{w}_{columns}\mathbf{m} = \begin{bmatrix} 1 & -1 & 1 & -1 \end{bmatrix} \begin{bmatrix} 6 \\ 5 \\ 2.5 \\ 2 \end{bmatrix} = 1.5.$$

To create the vector of four weights for the RC interaction, we multiply the four weights in the rows contrast vector by the four corresponding weights in the columns contrast vector. This is analogous to the principle that the df for an interaction are found to be the product of the individual df values of each term entering into that interaction. The contrast value for RC is calculated by adapting Equation 9.19.

$$C_{rc} = \mathbf{w}_{rc}\mathbf{m} = \begin{bmatrix} 1 & -1 & -1 & 1 \end{bmatrix} \begin{bmatrix} 6 \\ 5 \\ 2.5 \\ 2 \end{bmatrix} = 0.5.$$

Once the contrast values have been obtained, Equation 9.20 is used to obtain the sum of squares for each contrast. The symbol n in this formula is the frequency within each cell, which is 5 for this demonstration data set. The denominator for Equation 9.20 is the sum of the squared values in the vector of weights.

$$SS_{C_{rows}} = \frac{nC_{rows}^2}{\sum w_{rows}^2} \tag{9.20}$$

$$SS_{C_{rows}} = \frac{nC_{rows}^2}{\sum w_{rows}^2} = \frac{(5)(6.5^2)}{4} = \frac{211.25}{4} = 52.8125$$

Similar formulas are used for the sums of squares of the column contrast and the RC interaction contrast.

$$SS_{C_{columns}} = \frac{nC_{columns}^2}{\sum w_{columns}^2} = \frac{(5)(1.5^2)}{4} = \frac{11.25}{4} = 2.8125$$

$$SS_{C_{rc}} = \frac{nC_{rc}^2}{\sum w_{rc}^2} = \frac{(5)(0.5^2)}{4} = \frac{1.25}{4} = .3125$$

These sum of square values and df values for each of the three contrasts are collected in Table 9.5. The R^2 values are calculated by dividing each contrast sum of squares by the SST value (130.9375) from Table 9.3. The mean square values are calculated by dividing each sum of squares by its df value (which is 1 for all contrasts), and the F-ratio s are obtained by dividing each MS value by the MSE value (75.00) also found in Table 9.3. The p-values for each F ratio can be obtained using the FDIST function in Excel.

If we combine the information in Table 9.3, the model summary, and Table 9.5, the detailed analysis of the three separate "predictors" (row, column, and RC interaction), we have created essentially the same table that is reported by SAS for a two-way ANOVA, AV2. To verify this, we now set up the SAS code to conduct a two-way ANOVA of the persuasion study data, entering the data with a data statement and then analyzing the data with PROC GLM.

```
data MAV2data;
  input R C verbal;
  datalines;
    1    1    9
    1    1    8
    1    1    4
    1    1    6
    1    1    3
    1    2    7
    1    2    8
    1    2    4
    1    2    1
    1    2    5
    2    1    2
    2    1    4.5
    2    1    1
    2    1    3
    2    1    2
    2    2    2
    2    2    5
    2    2    1
    2    2    1
    2    2    1
  ;
run;

proc glm data=MAV2data;
  class R C;
  model verbal = R C R*C;
  lsmeans R C R*C;
  title 'Two-way ANOVA';
run;
```

TABLE 9.5. Detailed AV2 Analysis of Separate Predictors for Persuasion Data

Source	R^2	SS	df	MS	F	P
Rows	0.4033	52.8125	1	52.8125	11.267	0.0040
Columns	0.0215	2.8125	1	2.8125	0.600	0.4499
RxC interaction	0.0024	0.3125	1	0.3125	0.067	0.7995

The SAS output resulting from this code is contained within one page. We see that it corresponds to what we have created through the foregoing two-way ANOVA calculation process.

```
                              The GLM Procedure
Dependent Variable: verbal

                                        Sum of
        Source              DF          Squares      Mean Square   F Value    Pr > F

        Model                3       55.9375000      18.6458333      3.98     0.0271

        Error               16       75.0000000       4.6875000

        Corrected Total     19      130.9375000


                     R-Square       Coeff Var      Root MSE      verbal Mean

                     0.427208       55.87261       2.165064       3.875000


        Source              DF        Type I SS      Mean Square   F Value    Pr > F

        R                    1      52.81250000     52.81250000     11.27     0.0040
        C                    1       2.81250000      2.81250000      0.60     0.4499
        R*C                  1       0.31250000      0.31250000      0.07     0.7995


        Source              DF      Type III SS      Mean Square   F Value    Pr > F

        R                    1      52.81250000     52.81250000     11.27     0.0040
        C                    1       2.81250000      2.81250000      0.60     0.4499
        R*C                  1       0.31250000      0.31250000      0.07     0.7995
```

(Created with SAS software. Copyright 2011, SAS Institute Inc. Cary, NC. All rights reserved. Reproduced with permission of SAS Institute Inc., Cary, NC.)

It must be added, however, that SAS does not use a linear contrasts method to obtain the results we see in this output. It uses a general linear models approach similar to multiple regression. We will now demonstrate how this works by putting the MAV2 persuasion study data into the multiple regression computational form shown in Section 9.2.

Table 9.6 shows the data setup for doing this two-way ANOVA in multiple regression form. The 20 rows of data are organized with the dependent variable, "verbal compliance," on the left and then with three columns of *orthogonal coding* for the three independent variables, which we will refer to as rows, columns, and RC interaction. We demonstrated above how orthogonal linear contrasts could be used to calculate the sums of squares for rows, columns, and interaction. The same weights are used here as *orthogonal coding*[7] for a mul-

---

[7] Pedhazur (1997) describes and explains three types of coding used in analysis of variance by the general linear models approach: orthogonal coding, effect coding, and dummy coding. We have chosen to use orthogonal coding because of its similarity with linear contrasts which we also are demonstrating here.

TABLE 9.6. Persuasion Data Ready to be Analyzed with a Multiple Regression Package to Accomplish a Two-Way Analysis of Variance, AV2

Verbal	Rows	Columns	RC
9	1	1	1
8	1	1	1
4	1	1	1
6	1	1	1
3	1	1	1
7	1	−1	−1
8	1	−1	−1
4	1	−1	−1
1	1	−1	−1
5	1	−1	−1
2	−1	1	−1
4.5	−1	1	−1
1	−1	1	−1
3	−1	1	−1
2	−1	1	−1
2	−1	−1	1
5	−1	−1	1
1	−1	−1	1
1	−1	−1	1
1	−1	−1	1

tiple regression analysis of the AV2 design, but with vectors of weights 20 elements long rather than four, since the regression deals with every data point, each of the 20 observations.

If the matrix of data in Table 9.6 is pasted into the Stata editor, it can be analyzed directly with the multiple regression package using the Stata command **reg verbal rows columns rc**. The results are shown in Table 9.7. The format is somewhat different than if we were to run it as an ANOVA in SAS's GLM or Stata's **anova** command, but with a careful look, we can see that we have essentially all of the same information. The holistic ANOVA information summarizing the entire model (found in the upper third of SAS's GLM output) is all present. We have the SS values, the df values, and the mean squares for the model, for residual, and for total, and also the F-ratio and its associated P-value. From the middle section of the SAS GLM output, we also have the R^2 value for the entire model (plus an adjusted R^2 value not found in the GLM output), and the root MSE.

We also have essentially all of the information from the bottom third of the GLM output, the detailed analysis of the separate predictors, but it is in a somewhat different form. The P-values for rows, columns and RC interaction are all present and in agreement with the GLM output, but they are accomplished with a t-ratio (the one shown in Eq. 9.5 of the Section 9.2 multiple regression demonstration), and we do not have the sums of squares, df, or

TABLE 9.7. Output from Stata Command "reg verbal rows columns RC" Applied to Table 9.6 Input Data to Accomplish a Two-Way Analysis of Variance with a Regression Package

```
. reg verbal  rows columns rc

     Source |      SS       df       MS              Number of obs =      20
------------+------------------------------         F( 3,    16) =    3.98
      Model |  55.9375        3  18.6458333         Prob > F      =  0.0271
   Residual |       75       16     4.6875         R-squared     =  0.4272
------------+------------------------------         Adj R-squared =  0.3198
      Total | 130.9375       19  6.89144737         Root MSE      =  2.1651


     verbal |     Coef.   Std. Err.      t    P>|t|     [95% Conf. Interval]
------------+----------------------------------------------------------------
       rows |     1.625   .4841229     3.36   0.004     .5987053    2.651295
    columns |      .375   .4841229     0.77   0.450    -.6512947    1.401295
         rc |      .125   .4841229     0.26   0.800    -.9012947    1.151295
      _cons |     3.875   .4841229     8.00   0.000     2.848705    4.901295
```

(StataCorp, 2009. Stata: Release 11. Statistical Software. College Station, TX: StataCorp LP.)

TABLE 9.8. Partitioned Correlation Matrix for the Persuasion Data of Table 9.6, Showing Matrix R_{xx}, Vector r_{xy} and Vector r_{yx}

	Y	X1	X2	X3
Y: verbal	1	0.6351	0.1466	0.0489
X1: rows	0.6351	1	0	0
X2: columns	0.1466	0	1	0
X3: RC	0.0489	0	0	1

mean squares for these individual effects. They can, however, be readily calculated from the regression coefficients and the total sum of squares given in the upper third. In other words, all of the necessary information is present in the regression output, but in some cases expressed in a different form. A comparison of the two forms of output (and also across SAS and Stata) helps to gain insight into the characteristics of both kinds of analysis. The regression approach is more general and would allow, for example, other sets of orthogonal weights besides those that correspond to the row, column, and interaction structure of an AV2 analysis. This will become clearer in Sections 9.3.4 and 9.5 where we introduce other possible models.

Table 9.8 shows the partitioned correlation matrix for these data. Note in particular that matrix R_{xx} is an identity matrix, because the three *orthogonal coding* vectors are independent of one another. Since ANOVA designs are usually orthogonal like this, the multiple regression analysis for an ANOVA is usually a simplified model, as is this one.

Equations 9.1 and 9.2 from the beginning of the chapter can be combined to create Equation 9.21, which obtains the multiple R^2 value from the R_{xx} matrix premultiplied by r_{yx} vector and postmultiplied by r_{xy} vector.

$$\beta = \mathbf{R}_{ii}^{-1}\mathbf{r}_{iy} \tag{9.1}$$

$$\mathbf{R}^2 = \mathbf{r}_{yi}\beta \tag{9.2}$$

$$\mathbf{R}^2 = \mathbf{r}_{yi}\beta = \mathbf{r}_{yi}\mathbf{R}_{ii}^{-1}\mathbf{r}_{iy}$$

$$\mathbf{R}^2 = \mathbf{r}_{yi}\mathbf{R}_{ii}^{-1}\mathbf{r}_{iy} \tag{9.21}$$

With $\mathbf{R}_{xx}$ as an identity matrix in the orthogonal case, the multiple R-square becomes simply the sum of the squared values of the $\mathbf{r}_{yx}$ correlations.

$$\mathbf{R}^2 = \mathbf{r}_{yx}\mathbf{R}_{xx}^{-1}\mathbf{r}_{xy} = [0.6351 \quad 0.1466 \quad 0.0489] \begin{bmatrix} 1 & 0 & 0 \\ 0 & 1 & 0 \\ 0 & 0 & 1 \end{bmatrix}^{-1} \begin{bmatrix} 0.6351 \\ 0.1466 \\ 0.0489 \end{bmatrix}$$

$$= [0.6351 \quad 0.1466 \quad 0.0489] \begin{bmatrix} 0.6351 \\ 0.1466 \\ 0.0489 \end{bmatrix} = \mathbf{r}_{yi}\mathbf{r}_{iy}$$

$$\mathbf{R}^2 = r_{y.x1}^2 + r_{y.x2}^2 + r_{y.x3}^2 = 0.6351^2 + 0.1466^2 + 0.0489^2.$$
$$= 0.4033 + 0.0215 + 0.0024 = 0.4272$$

Knowing that $R^2 = \text{SSModel}/\text{SST}$ we can also calculate SSModel from the R^2 and SST:

$$\text{SSModel} = R^2(\text{SST}) = 0.4033(130.9375) = 52.813.$$

This is actually a fairly simple way to accomplish ANOVA when the data are orthogonal—simply from the correlations between the orthogonal coding vectors and the dependent variable. With SST and SSE already known from the regression, all of the SS values can be derived from the R^2 column, and the other columns (*df*, MS, *F*, and *P*) follow. We will now examine the multiple regression model of ANOVA when the model is nonorthogonal.

9.3.3 Nonorthogonal AV2 Design (Unbalanced) and the General Linear Model

In the same year that Cohen and Cohen (1975) published their now classic work on the general linear models approach to ANOVA, Overall et al. (1975) published an important paper detailing how multiple regression can function as a general paradigm to effectively deal with the problem of nonorthogonal ANOVA designs, that is, factorial designs with an unequal number of observations in the cells. It is interesting to note that as late as the early 1970s, perhaps one of the best linear models books of the century, Winer's (1971) *Statistical Principles in Experimental Design, Second Edition*, was still using the awkward and ad-hoc adjustment factors approaches to dealing with non-orthogonal data. These other methods were, however, overtaken and soon eclipsed by the general linear models approach to dealing with unequal cell sizes.

We will use the same AV2 demonstration data from the previous section, but with several data points dropped out (rows 4, 8, 10, and 18) to create unequal cell sizes in the analysis. The code in the SAS program is no different for running the unbalanced data, but if we look carefully at the data statement, we see that the four groups vary in size. In the first block of data, corresponding to row 1 and column 1, there are four entries with the 1 1 subscripts (9, 8, 4, and 3). In the second block of data, corresponding to row 1 and column 2, there are three entries (7, 8, and 1). In the third block, row 2 and column 1 with the 2 1 subscripts, there are five entries (2, 4.5, 1, 3, and 2). Finally in the last block with the 2 2 subscripts (row 2 and column 2), there are five entries (2, 5, 1, 1, and 1), for a total sample size of $N = 17$ rather than the $N = 20$ for the orthogonal analysis in the last section. Everything in the SAS code other than the structure of the input data is the same for the nonorthogonal analysis as for the orthogonal analysis.

```
data MAV2unb;
  input R C verbal;
  datalines;
    1    1    9
    1    1    8
    1    1    4
    1    1    3
    1    2    7
    1    2    8
    1    2    1
    2    1    2
    2    1    4.5
    2    1    1
    2    1    3
    2    1    2
    2    2    2
    2    2    5
    2    2    1
    2    2    1
    2    2    1
  ;
run;

proc glm data=MAV2unb;
  class R C;
  model verbal = R C R*C;
  lsmeans R C R*C;
  title 'Two-way Unbalanced ANOVA';
run;
```

What impact does the unbalanced data structure have on the GLM analysis? Again, the SAS output is all contained on a single page. The ANOVA information summarizing the entire model, shown in the top third of the output, is not much different. The total *df* is 16 (for $N = 17$) as compared with the total *df* of 19 ($N = 20$) for the orthogonal analysis in the previous section. The *F*-value in the top section and the R^2 value in the middle section are each a little smaller (and the *P*-value is not significant), reflecting the smaller sample size, but the big difference is in the bottom third of the output, the detailed analysis of the row, column, and interaction effects.

The GLM Procedure

Dependent Variable: verbal

Source	DF	Sum of Squares	Mean Square	F Value	Pr > F
Model	3	50.8039216	16.9346405	2.99	0.0700
Error	13	73.6666667	5.6666667		
Corrected Total	16	124.4705882			

R-Square	Coeff Var	Root MSE	verbal Mean
0.408160	64.74895	2.380476	3.676471

Source	DF	Type I SS	Mean Square	F Value	Pr > F
R	1	49.41701681	49.41701681	8.72	0.0112
C	1	1.35865617	1.35865617	0.24	0.6325
R*C	1	0.02824859	0.02824859	0.00	0.9448

Source	DF	Type III SS	Mean Square	F Value	Pr > F
R	1	47.48587571	47.48587571	8.38	0.0125
C	1	1.38418079	1.38418079	0.24	0.6294
R*C	1	0.02824859	0.02824859	0.00	0.9448

In the orthogonal analysis, the "Type 1 SS" analysis, the first one shown in the lower third of the table, was identical to the "Type III SS" analysis below it, but here the Type I and Type III analyses are different from one another. When the data design is orthogonal, there is no correlation among the independent variables, no overlap in their effects upon the dependent variable. However, when they are nonorthogonal, the independent variable sources (row effects, column effects, and interaction effects, in this case) do overlap in their effects upon the dependent variable, as shown in Figure 9.1. The Type I sum of squares is like a stepwise regression, where the first term entered into the model (row effects in this case) includes all of the overlapping variance in its effects, the second term entered (column effects in this case) includes all except that which is common with rows, and the last term entered (RC interaction in this case) includes none of the overlapping variance, as shown in the lower left diagram of Figure 9.1. As shown in the Figure 9.1 Venn diagram, these three quantities are all additive and will sum to the total variation units (sums of squares) in the model. Notice in the table of SAS output for this model that the Type I sums of squares for rows, columns, and interaction all add up to the sum of squares value for the total.

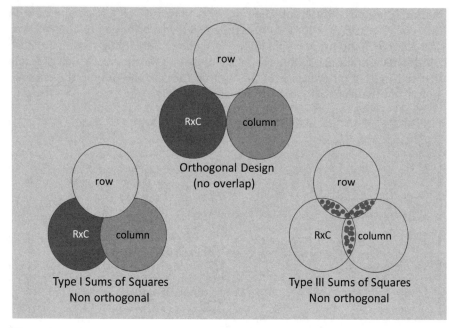

Figure 9.1   A contrast of relative variance distribution for Type I sums of squares and Type III sums of squares, as compared with orthogonal designs.

$$SSmodel = SSrows + SScolumns + SSinteraction.$$

$$50.8039 = 49.4170 + 1.3587 + .0282.$$

The Type III sum of squares, on the other hand, has a tendency to treat each term as though it were entered last in the model, as shown on the right of Figure 9.1, with the overlapping information not entering into any of the terms. Notice from the SAS output above, that the Type III sums of squares values fall short of the total sum of squares units in the model.

$$SSmodel = SSrows + SScolumns + SSinteraction.$$

$$50.8039 \neq 47.4859 + 1.3842 + .0282 = 48.8981.$$

The Type III sum of squares is the most conservative and the one usually used in published research. When the nonorthogonality is accidental, that is, not an inherent characteristic of the population being sampled, the Type III sum of squares corresponds most closely to what would be expected in the data if there had been no accidental imbalance. When the imbalance in the data is empirically real, there will sometimes be a good rationale for choosing another sum of squares model. There are actually four types of sums of squares available in SAS, but the two default types, and by far the most commonly used are the Type III and the Type I sums of squares. As shown in the previous section, Stata in the default condition only reports the Type III sums of squares. For a detailed explanation of the four types of sums of squares, what each means, and how to interpret each, see the SAS document "The four types of estimable functions" (SAS, 2009a, Chapter 15). See also Freund et al. (1991) and Milliken and Johnson (1984).

In the previous section, we used coding of the binary predictor variables (rows, columns, and interaction) to prepare an orthogonal ANOVA model to be analyzed with a multiple regression package (the **reg** command in Stata). We found that the regression output, though expressed in a somewhat different form, gave us precisely the same results as when we calculated the two-way ANOVA from sums of squares formulas or when we ran it on SAS's GLM package. We also found that the R_{xx} matrix in the regression process (the matrix of correlations among the predictors) was an identity matrix for orthogonal data. It will be instructive now to follow that same procedure in analyzing regression-coded data for the nonorthogonal model of this section. Table 9.9 shows the nonorthogonal form of the persuasion study data ready to be analyzed through multiple regression.

TABLE 9.9. Nonorthogonal Persuasion Data Ready to be Analyzed with a Multiple Regression Package to Accomplish a Two-Way Analysis of Variance, AV2

Verbal	Rows	Columns	RC
9	1	1	1
8	1	1	1
4	1	1	1
3	1	1	1
7	1	−1	−1
8	1	−1	−1
1	1	−1	−1
2	−1	1	−1
4.5	−1	1	−1
1	−1	1	−1
3	−1	1	−1
2	−1	1	−1
2	−1	−1	1
5	−1	−1	1
1	−1	−1	1
1	−1	−1	1
1	−1	−1	1

We can see that the coded weights here are not actually orthogonal. By definition, when weights are orthogonal, the sums of products and also the covariances and correlation coefficients will all be zero (Chapter 3, Section 3.9). For example, we see that this is true of the matrix of orthogonal coding weights, $\mathbf{W}_o$, from Table 9.6 in the previous section.

$$
\mathbf{W}_O'\mathbf{W}_O =
\begin{bmatrix}
1 & 1 & 1 & 1 & 1 & 1 & 1 & 1 & 1 & 1 & -1 & -1 & -1 & -1 & -1 \\
1 & 1 & 1 & 1 & 1 & -1 & -1 & -1 & -1 & -1 & 1 & 1 & 1 & 1 & 1 \\
1 & 1 & 1 & 1 & 1 & -1 & -1 & -1 & -1 & -1 & -1 & -1 & -1 & -1 & -1
\end{bmatrix}
\begin{bmatrix}
-1 & -1 & -1 & -1 & -1 \\
-1 & -1 & -1 & -1 & -1 \\
1 & 1 & 1 & 1 & 1
\end{bmatrix}
\begin{bmatrix}
1 & 1 & 1 \\
1 & 1 & 1 \\
1 & 1 & 1 \\
1 & 1 & 1 \\
1 & 1 & 1 \\
1 & -1 & -1 \\
1 & -1 & -1 \\
1 & -1 & -1 \\
1 & -1 & -1 \\
1 & -1 & -1 \\
-1 & 1 & -1 \\
-1 & 1 & -1 \\
-1 & 1 & -1 \\
-1 & 1 & -1 \\
-1 & 1 & -1 \\
-1 & -1 & 1 \\
-1 & -1 & 1 \\
-1 & -1 & 1 \\
-1 & -1 & 1 \\
-1 & -1 & 1
\end{bmatrix}
$$

$$
=
\begin{bmatrix}
20 & 0 & 0 \\
0 & 20 & 0 \\
0 & 0 & 20
\end{bmatrix}.
$$

However, it is not true of the coding weights, $\mathbf{W}_{no}$, for these data. They are not orthogonal, as shown by the nonzero sums of products.

$\mathbf{W}'_{no}\,\mathbf{W}_{no}$

$$= \begin{bmatrix} 1 & 1 & 1 & 1 & 1 & 1 & 1 & -1 & -1 & -1 & -1 & -1 & -1 & -1 & -1 & -1 & -1 \\ 1 & 1 & 1 & 1 & -1 & -1 & -1 & 1 & 1 & 1 & 1 & 1 & -1 & -1 & -1 & -1 & -1 \\ 1 & 1 & 1 & 1 & -1 & -1 & -1 & -1 & -1 & -1 & -1 & -1 & 1 & 1 & 1 & 1 & 1 \end{bmatrix}$$

$$\begin{bmatrix} 1 & 1 & 1 \\ 1 & 1 & 1 \\ 1 & 1 & 1 \\ 1 & 1 & 1 \\ 1 & -1 & -1 \\ 1 & -1 & -1 \\ 1 & -1 & -1 \\ -1 & 1 & -1 \\ -1 & 1 & -1 \\ -1 & 1 & -1 \\ -1 & 1 & -1 \\ -1 & 1 & -1 \\ -1 & -1 & 1 \\ -1 & -1 & 1 \\ -1 & -1 & 1 \\ -1 & -1 & 1 \\ -1 & -1 & 1 \end{bmatrix} = \begin{bmatrix} 17 & 1 & 1 \\ 1 & 17 & -3 \\ 1 & -3 & 17 \end{bmatrix}.$$

When the regression-coded data is run through the regression analysis of Stata, all of the ANOVA information for the entire model (as shown in the top several lines of Table 9.10) is absolutely consistent with the comparable statistics in the top third of the unbalanced ANOVA output from SAS's GLM. The detailed analysis of the row, column, and interaction predictors are also consistent with the Type III model from the SAS GLM analysis of the ANOVA model, even though the two sets of results are expressed in different formats. For example, the P-values for rows, columns, and interaction are the same for the SAS ANOVA results and the Stata regression results, but the ANOVA output gives F-ratios, while the regression gives t-ratios. The squares of the individual t-ratios should be equivalent to the corresponding F-ratio s from the ANOVAs and in fact they are. Presumably, it would be possible using a stepwise regression to obtain regression analysis results consistent with the Type I sum of squares model.

It is instructive to examine the partitioned correlation matrix in Table 9.11 of the regression-coded data from which the regression analysis is calculated. The orthogonal form of the data had an identity matrix, $\mathbf{I}$, for the $\mathbf{R}_{xx}$ matrix.

TABLE 9.10. Output from Stata Command "reg verbal rows columns RC" Applied to Table 9.9 Nonorthogonal Input Data to Accomplish a Two-Way Analysis of Variance with a Regression Package

```
. reg verbal rows columns rc

    Source |      SS       df       MS              Number of obs =      17
-----------+------------------------------          F(  3,    13) =    2.99
     Model | 50.8039216     3  16.9346405           Prob > F      =  0.0700
  Residual | 73.6666667    13  5.66666667           R-squared     =  0.4082
-----------+------------------------------          Adj R-squared =  0.2716
     Total | 124.470588    16  7.77941176           Root MSE      =  2.3805


    verbal |     Coef.   Std. Err.      t     P>|t|     [95% Conf. Interval]
-----------+----------------------------------------------------------------
      rows | 1.708333   .5901389     2.89   0.013     .4334158    2.983251
   columns |  .2916667   .5901389     0.49   0.629    -.9832509    1.566584
        rc |  .0416667   .5901389     0.07   0.945   -1.233251    1.316584
     _cons | 3.958333   .5901389     6.71   0.000     2.683416    5.233251
```

(StataCorp, 2009. Stata: Release 11. Statistical Software. College Station, TX: StataCorp LP.)

TABLE 9.11. Partitioned Correlation Matrix for the Unbalanced Data of Table 9.9, Showing Matrix R_{xx}, Vector r_{xy} and Vector r_{yx}

	Y	X1	X2	X3
Y: verbal	1	0.6301	0.1486	0.0397
X1: rows	0.6301	1	0.0704	0.0704
X2: columns	0.1486	0.0704	1	−0.1806
X3: RC	0.0397	0.0704	−0.1806	1

This R_{xx} matrix is not an identity matrix, of course, because the data are not orthogonal. The calculation of the multiple R^2 from the partitions of this matrix are shown below, using Equation 9.21.

$$R^2 = r_{yi} R_{ii}^{-1} r_{iy} \qquad (9.21)$$

$$R^2 = r_{yi} R_{ii}^{-1} r_{iy} = [0.6301 \quad 0.1486 \quad 0.0397] \begin{bmatrix} 1 & 0.0704 & 0.0704 \\ 0.0704 & 1 & -0.1806 \\ 0.0704 & -0.1806 & 1 \end{bmatrix}^{-1} \begin{bmatrix} 0.6301 \\ 0.1486 \\ 0.0397 \end{bmatrix}$$

$$= [0.6301 \quad 0.1486 \quad 0.0397] \begin{bmatrix} 1.0122 & -0.0870 & -0.0870 \\ -0.0870 & 1.0412 & 0.1942 \\ -0.0870 & 0.1942 & 1.0412 \end{bmatrix} \begin{bmatrix} 0.6301 \\ 0.1486 \\ 0.0397 \end{bmatrix} = 0.4082.$$

9.3.4 Other Designs Using Linear Contrasts

As we saw in Section 9.3.2, the AV2 ANOVA of the persuasion study data can be accomplished by first doing an AV1 one-way ANOVA analysis of the four means, and then following up the significant F-ratio with a set of linear contrasts that correspond to the AV2 analysis of the effects of rows, columns, and interaction. This is in fact, parallel to the reporting method of both SAS and also Stata of giving the significance test and R^2 of the overall model in the top section of the output, and then to give the tests of row, column, and interaction effects in the bottom section of the output.

This linear contrasts view of a two-way ANOVA is productive, in that it reminds us that the contrasts corresponding to AV2 are just one design of several possibilities for examining the relationships among the means of the four experimental groups. We will consider two other ways to follow up the significant model effects for the orthogonal data of Section 9.3.2.

One approach to using linear contrasts is to investigate hierarchical organization, somewhat like hierarchical cluster analysis but one dimensional. The four treatment groups in the persuasion study differ substantially in the impact of the four types of persuasive appeals. In Table 9.3, we saw that the test of the overall "model" level hypothesis (that various ways of persuading have differential effect) was supported with a P-value of 0.0171 and an R^2 of 0.4272. The means varied from an average rating of 2 on a 10-point scale for the lowest group up to an average rating of 6 for the highest group. Suppose we wish to now examine whether there is a clear grouping of the four persuasive approaches, and to determine how much variance that grouping accounts for. We put the means in ascending order, as shown in Table 9.12, and find the pair of means that are closest to one another. It is the mean for r2, c2 (social nonreinforcement combined with different or unfamiliar research assistant), and the mean for r2, c1 (social nonreinforcement combined with same or familiar research assistant). We see that different versus same does not have much impact when there is no social reinforcement, a difference between means of only 0.5. We create a contrast to test that difference and from it obtain a sum of squares of 0.6250 (Table 9.12) for which we show the

TABLE 9.12. Set of Linear Contrasts to Test the Hierarchical Organization of Persuasive Impact

	Social Nonreinforcement		Social Reinforcement			
	Different	Same	Different	Same		
	r2,c2	r2,c1	r1,c2	r1,c1	Contrast	SS(Contrast)
	2	2.5	5	6		
Closest pair	1	−1	0	0	−0.5	0.6250
Next closest pair	0	0	1	−1	−1	2.5000
Farthest groups	1	1	−1	−1	−6.5	52.8125

TABLE 9.13. Detailed Analysis of Hierarchical Organization of Persuasive Impact

Source	R^2	SS	df	MS	F	P
Closest pair	0.0048	0.6250	1	0.625	0.133	0.7198
Next closest pair	0.0191	2.5000	1	2.5	0.533	0.4758
Farthest groups	0.4033	52.8125	1	52.8125	11.267	0.0040

significance test in Table 9.13. The difference is nonsignificant and accounts for less that 1% of the variance.

We now go to the next smallest difference between means (or grouped means, treating the pair just tested now as a group) and find it to be a difference of one unit for the comparison between r1, c2 (social reinforcement with different assistant) and r1, c1 (social reinforcement with same assistant). We construct a set of weights for the contrast between those two means. We obtain a contrast sum of squares of 2.5, which is seen in Table 9.13 to not be significant and to account for less than 2% of the variance. Once we have done a test on two means, we will only deal with them as a group, which ensures that the contrasts will be orthogonal. We now have two groups, the r2,c2 versus r2,c1 group (social nonreinforcement) and the r1,c2 versus r1,c1 group (social reinforcement), and we create a contrast to test these two groups. This time, the sum of squares for the contrast is 52.8125, and we see in Table 9.13 that this difference is significant, with an F-ratio of 11.267 and a P-value of 0.004, and it accounts for over 40% of the variance (an R^2 of 0.4033).

We have seen that fully crossed factors (rows and columns and their interaction) can produce an orthogonal set of contrasts. Orthogonal contrasts can also be created hierarchically, as we can demonstrate by testing this set of hierarchical contrast weights to ensure that their sums of products are zero.

$$\mathbf{W}_h'\mathbf{W}_h = \begin{bmatrix} 1 & -1 & 0 & 0 \\ 0 & 0 & 1 & -1 \\ 1 & 1 & -1 & -1 \end{bmatrix} \begin{bmatrix} 1 & 0 & 1 \\ -1 & 0 & 1 \\ 0 & 1 & -1 \\ 0 & -1 & -1 \end{bmatrix} = \begin{bmatrix} 2 & 0 & 0 \\ 0 & 2 & 0 \\ 0 & 0 & 4 \end{bmatrix}.$$

Another hierarchical set of contrasts is possible with these four means. We have seen that the major factor accounting for the effectiveness of persuasion is social reinforcement and that the effects of familiarity are much smaller. Suppose that we now wish to assess and compare the relative impact of social reinforcement within the two familiarity groups, the high familiarity group that employed the same assistant and the low familiarity group that employed a

TABLE 9.14. Linear Contrasts to Test the Hypothesis that Social Reinforcement is Important with or Without Familiarity

	c1		c2			
	r1	r2	r1	r2		
	6	2.5	5	2	Contrast	SS(Contrast)
Same versus different Asst.	1	1	−1	−1	1.5	2.8125
Social impact within same Asst.	1	−1	0	0	3.5	30.6250
Social impact within different Asst.	0	0	1	−1	3	22.5000

TABLE 9.15. Detailed Analysis of Impact of Social Reinforcement with or without Familiarity

Source	R^2	SS	df	MS	F	P
Same versus different Asst.	0.0215	2.8125	1	2.8125	0.600	0.4499
Social impact within same Asst.	0.2339	30.6250	1	30.625	6.533	0.0211
Social impact within different Asst.	0.1718	22.5000	1	22.5	4.800	0.0436

different assistant. We create the contrast $(1, 1, -1, -1)$ for the two low familiarity groups against the two high familiarity groups, as shown in Table 9.14, and find a contrast sum of squares of 2.8125. In Table 9.15, we find that this contrast is not statistically significant and accounts for only a little over 2% of the variance. The other two contrasts that are orthogonal to this one are the two that correspond to what we wish to test—the impact of social reinforcement within each of the familiarity groups. We first test the impact of social reinforcement within the high familiarity group (same assistant) and find a contrast value of 30.625, with a statistically significant effect ($P = 0.0211$) that accounts for over 23% of the variance ($R^2 = 0.2339$).

Finally, we create a contrast to test the impact of social reinforcement within the low familiarity group (different assistant) and find another significant effect that accounts for over 17% of the variance. This set of contrasts is also orthogonal and uses up the three df, and thus constitutes a complete set of contrasts.

9.4 ANALYSIS OF COVARIANCE AND THE GENERAL LINEAR MODEL

ANCOVA can also be productively thought of as a nonorthogonal application of the general linear model. Consider the simple data set below in a screen capture from the SPSS data editor. It consists of three groups of three persons each with two scores for each person, one the dependent variable (dv), and one a covariate (cov). This could, for example, be scores in junior league fall basketball practice of a standard test of the number of free throws missed out of 10 tries. The coach is trying out three different methods of practicing free throws, with three players assigned to each of the treatments. Free throw misses were tested twice: a baseline measure as fall camp began (the covariate), and also after a week of practice (the dv). The coach would like to get an idea of which of the three groups is making more progress in training, corrected for initial starting place, so he does an analysis of covariance to test progress on the 1-week test, adjusted for their initial baseline performance.

	treatment	cov	dv
1	1	7.00	8.00
2	1	9.00	6.00
3	1	8.00	8.00
4	2	10.00	6.00
5	2	4.00	5.00
6	2	5.00	5.00
7	3	6.00	4.00
8	3	3.00	2.00
9	3	2.00	2.00
10			

(Reprinted courtesy of International Business Machines Corporation, © SPSS, Inc., an IBM Company.)

Analysis of covariance results are often compared with ordinary ANOVA results to get an idea of the difference the covariate makes in the analysis. For these simple data, we first examine the direct results (without the covariate) from the ANOVA summary table. Clearly, there are some major differences in performance under the three treatment methods, with an F-ratio of 16.444 and a p-value of 0.0037.

Analysis of Variance Summary Table					
Source	SS	df	MS	F	P
Groups	32.8889	2	16.4445	16.444	0.0037
Within	6.0000	6	1.0000		
Total	38.8889	8			

In looking at the means table, it can be seen that the players in the third treatment group have substantially fewer free throws missed. It can also be seen that the first treatment group has substantially more free throws missed than the other two groups. However, in examining the baseline scores, we also see that treatment group 3 had substantially fewer on the baseline measure and group one had substantially more on the baseline measure. We see the importance of adjusting performance for baseline level with an analysis of covariance.

Mean Missed Free Throws for the Three Treatment Groups

	Baseline	1 week
Treatment 1	8.00	7.33
Treatment 2	6.33	5.33
Treatment 3	3.67	2.67

The analysis of covariance summary table confirms our concern that the group differences might in large part be due to the baseline differences, in that we see considerable shrinkage in the effects, with an F-test ratio in the adjusted analysis of only 6.013, and a P-value of 0.0467, which is barely significant. Notice, by the way, that the df values for groups and for within do not sum to the total df. They are off by one. That is because the covariate uses up 1 degree of freedom, and for clarity it is usually not included in the table.

Analysis of Covariance Summary Table

Source	Adjusted SS	df	MS	F	P
Groups	11.3524	2	5.6762	6.013	0.0467
Within	4.7199	5	0.9440		
Total	38.8889	8			

These ANOVA and ANCOVA results were created with an SPSS analysis of the small data set shown in the SPSS editor screen capture above. Of the three packages, SPSS, SAS, and Stata, SPSS has the easiest learning curve and is most convenient for doing an analysis of covariance. However, there are interpretative advantages of each of the three, and there is something to be learned from comparing the output across the packages. The point and click instructions for analysis of covariance in SPSS are to select **Analyze** from the menu, then **General Linear Model**, then **Univariate**. When the small "Univariate" screen (shown) appears, "treatment" is moved into **Fixed Factors**, "cov" is moved into **Covariates**, and "dv" is moved into **Dependent Variable**.

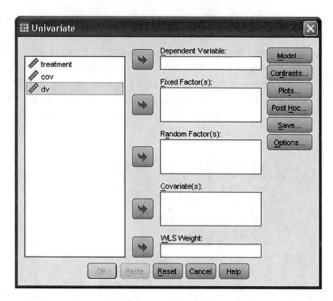

(Reprinted courtesy of International Business Machines Corporation, © SPSS, Inc., an IBM Company.)

You then click the **OK** button, and the following output appears. The three shaded lines give the statistics for treatment, error, and corrected total that make up the analysis of covariance table that was presented above.

SPSS Analysis of Covariance Output					
Tests of Between-Subjects Effects					
Dependent Variable:dv					
Source	Type III Sum of Squares	df	Mean Square	F	Sig.
Corrected Model	34.169[a]	3	11.390	12.066	.010
Intercept	12.061	1	12.061	12.777	.016
cov	1.280	1	1.280	1.356	.297
treatment	11.352	2	5.676	6.013	.047
Error	4.720	5	.944		
Total	274.000	9			
Corrected Total	38.889	8			
a. R Squared = .879 (Adjusted R Squared = .806)					

(Reprinted courtesy of International Business Machines Corporation, © SPSS, Inc., an IBM Company.)

The SAS code for this analysis is also straightforward, but a little more knowledge is required to be able to find what you want for the analysis of covariance table from the output. However, the presentation of results in SAS is also quite helpful in promoting an understanding of how ANCOVA fits within the structure of the general linear model. In the SAS code shown here, we have included both the analysis of covariance and also a direct ANOVA without the covariate.

```
data groups3data;
  input group cov dv;
  datalines;
  1   7   8
  1   9   6
  1   8   8
  2  10   6
  2   4   5
  2   5   5
  3   6   4
  3   3   2
  3   2   2
  ;
run;

proc glm data=groups3data;
  class group;
  model dv = group;
  lsmeans group;
  title 'One-way ANOVA';
run;

proc glm data=groups3data;
  class group;
  model dv = cov group;
  lsmeans group;
  title 'One-way ANCOVA';
run;
```

Notice in the SAS output that the analysis of covariance is handled within the general linear model format, so that one must understand the operation of the general linear model somewhat to know where to look for the SS, MS, and F-ratio adjusted for the covariate. The three lines of adjusted statistics that are needed for the analysis of covariance summary table are shaded on the output. The adjusted error term is the one that appears in the general model summary statistics in the top third of the output. The corrected total (corrected for the grand mean, that is) and its df from this section are also used in the analysis of covariance summary table. The treatment group SS, df, MS, and F-ratio adjusted for the covariate are found in the bottom third of the table where the detailed analyses for specific predictors are found.

```
                              SAS Analysis of Covariance Output
                          One-way ANCOVA                  10:19 Tuesday, May 3, 2011   2

                              The GLM Procedure
Dependent Variable: dv

                                        Sum of
     Source                 DF          Squares      Mean Square    F Value    Pr > F
     Model                  3        34.16903073     11.38967691     12.07     0.0100

     Error                  5         4.71985816      0.94397163

     Corrected Total        8        38.88888889


              R-Square       Coeff Var      Root MSE        dv Mean

              0.878632       19.00921       0.971582        5.111111


     Source                 DF        Type I SS      Mean Square    F Value    Pr > F
     cov                    1        22.81666667     22.81666667     24.17     0.0044
     group                  2        11.35236407      5.67618203      6.01     0.0467


     Source                 DF        Type III SS    Mean Square    F Value    Pr > F
     cov                    1         1.28014184      1.28014184      1.36     0.2967
     group                  2        11.35236407      5.67618203      6.01     0.0467
```

The rationale for the general linear models approach to analysis of covariance is clarified somewhat by considering how ANCOVA can be analyzed using a regression package. Table 9.16 shows how the three treatment groups can be expressed with orthogonal coding using two fixed effects variables, corresponding to the two *df* for treatment.

Table 9.17 shows the partitioned correlation matrix for a multiple regression analysis of these data, with matrix $\mathbf{R}_{xx}$ in the lower right, vector $\mathbf{r}_{yx}$ above it and $\mathbf{r}_{xy}$ to the left of it. Notice that because of the quantitative covariate, the $\mathbf{R}_{xx}$ matrix is not an identity matrix. That is, the covariate makes the multiple regression a nonorthogonal case. The analysis of covariance adjustment of terms is accomplished through the ordinary multivariate regression process for dealing with nonorthogonal models.

TABLE 9.16. Basketball Data for Analysis of Covariance in Multiple Regression Format

dv	cov	lc1	lc2
8	7	1	1
6	9	1	1
8	8	1	1
6	10	−1	1
5	4	−1	1
5	5	−1	1
4	6	0	−2
2	3	0	−2
2	2	0	−2

TABLE 9.17. Partitioned Correlation Matrix for the Analysis of Covariance Data of Table 9.16, Showing Matrix R_{xx}, Vector r_{xy} and Vector r_{yx}

	dv	Covariate	Lincont1	Lincont2
dv	1	0.7660	0.3928	0.8315
Covariate	0.7660	1	0.2635	0.6390
Lincont1	0.3928	0.2635	1	0.0000
Lincont2	0.8315	0.6390	0.0000	1

9.5 LINEAR CONTRASTS AND COMPLEX DESIGNS

In 1985, Sithichoke Waranusuntikule, a young officer in the Thai navy, completed his PhD in the applied social psychology program at Brigham Young University. His dissertation (Waranusuntikule, 1985) is particularly interesting because of the unusual statistical analysis design, a three-way ANOVA plus a control group. This unusual design can be used to illustrate the malleability and power of linear contrasts, even multivariate linear contrasts, in dealing with specifically targeted questions such as this.

In particular, Waranusuntikule wanted to test the efficacy of a method of persuasion, the "foot-in-the door technique," in contrast to a control group where no persuasive method was used. The hypothesis was that there would be more verbal and behavioral compliance with the use of the persuasive method. But in addition to the experimental versus control hypothesis, he also wanted to manipulate three aspects of how the technique was implemented: factor A (with or without material payment), factor B (with or without a social reinforcement manipulation), and factor C (levels of "familiarity with the approaching person," i.e., same assistant versus different assistant). These three binary manipulations were combined into eight combinations, in addition to the control group, to create nine experimental groups. The study was conducted with 140 female college students in a teacher's college in Thailand in the summer of 1984. Figure 9.1 shows the experimental design. In the actual dissertation, Waranusuntikule used a quantitative measure of verbal compliance and a binary measure of behavioral compliance ("show" versus "no-show" for planting trees). For this demonstration, however, we have constructed artificial data that is consistent with his actual results, but quantitative on both the verbal and the behavioral compliance measures (to enable us to demonstrate the analysis with a MANOVA using multivariate linear contrasts), and with a smaller number of subjects to make it manageable as a demonstration.[8]

---

[8] The two major differences between the artificial data, we will use for this analysis demonstration, and Waranusuntikule's actual data are (1) the construction of a *quantitative* behavioral measure consistent with the results of his *binary* behavioral measure, and (2) the number of subjects. For simplification, we have constructed data for only 45 subjects (five in each of the nine experimental groups) rather than the full 140 in his study. Since we did not have access to the original data, but only the tables of means and the analysis of variance summary table in the dissertation, it was necessary to reconstruct the data anyway, but we did it for a smaller sample size.

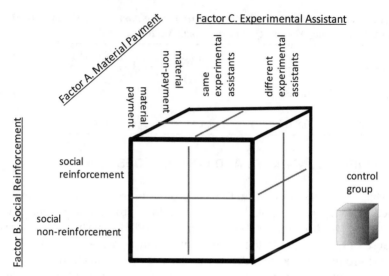

Figure 9.2 Research design for Waranusuntikule's (1985) dissertation, consisting of a three-way ANOVA, with a control group added.

We will first use a part of the Waranusuntikule data to demonstrate the use of an ordinary three-way MANOVA, a simple MAV3 design, completely randomized and each of the three factors having only two levels as shown in Figure 9.2. Factor A, material payment was a manipulation of whether or not they were given a pen for their participation in the study. Factor B, social reinforcement, is in many ways the most interesting manipulation in the study. Waranusuntikule describes it like this on page 32 of the dissertation:

> In the social reinforcement condition, the assistant thanked her and gave her compliments, smiled, and looked at the subject while in the social non-reinforcement condition she only wrote the subject's name on the pad.

The "compliments" were such things as, "thank you very much, you are so cooperative." The third factor to be manipulated was whether they were tested for compliance (verbal or behavioral) with the same persons who originally approached them about participating in the study, with whom they had begun to be acquainted, or whether it was a different experimenter's assistant. These three binary manipulations occurred in all eight possible combinations, and with equal numbers of subjects in each of the eight combinations.

We now demonstrate the analysis of the eight types of persuasion in a three-way MANOVA. The reconstructed data are shown in a SAS data statement below.

```
data MAV3data;
   input A B C verbal behavioral;
   datalines;
   1   1   1      6       11
   1   1   1      8.5      9
   1   1   1      5        9
   1   1   1      7        8
   1   1   1      6        8
   1   1   2      5        6
   1   1   2      7.5      5
   1   1   2      4        2
   1   1   2      6        3
   1   1   2      5        4
   1   2   1      7        3
   1   2   1      5        3
   1   2   1      5        6
   1   2   1      2        4
   1   2   1      1        4
   1   2   2      4        1
   1   2   2      7        2
   1   2   2      1        4
   1   2   2      0        2
   1   2   2      3        1
   2   1   1      9       10
   2   1   1      8        9
   2   1   1      4        6
   2   1   1      6        7
   2   1   1      3        8
   2   1   2      7        8
   2   1   2      8        7
   2   1   2      4        4
   2   1   2      1        5
   2   1   2      5        6
   2   2   1      2        1
   2   2   1      4.5      1
   2   2   1      1        4
   2   2   1      3        2
   2   2   1      2        2
   2   2   2      2        8
   2   2   2      5        6
   2   2   2      1        6
   2   2   2      1        5
   2   2   2      1        5
;
run;
```

The SAS code for accomplishing a three-way MANOVA of the data is given below. For simplicity, we have labeled the three factors as A, B, and C, each of which is listed in the class statement. The "class" statement indicates that each of these is a classification variable, categorical rather than quantitative. Each of the seven terms of the model (three main effect terms, three two-way interactions, and one three-way interaction) are listed in the model statement. Listing out all seven terms of the model as we have done is optional. The model statement could also have been written more simply as:

model verbal behavioral = A | B | C

The vertical bars instruct SAS to fill in the additional terms in the three-way model.

```
proc glm data=MAV3data;
  class A B C;
  model verbal behavioral = A B C A*B A*C B*C A*B*C;
  manova h=A;
  manova h=B;
  manova h=C;
  manova h=A*B;
  manova h=A*C;
  manova h=B*C;
  manova h=A*B*C;
  lsmeans A B C A*B A*C B*C A*B*C;
  title 'Three-way MANOVA of Sithi's Data';
  run;
```

To obtain the multivariate tests, a manova instruction must be written for every term that is to be tested, followed by an "h=" and the name of the term. In the **manova** instruction, it is usual to enter both an "h=" instruction for naming the hypothesis matrix, and also an "e=" instruction for identifying the proper error term. In a simple, completely randomized design like this one, since the error matrix is the error term in the holistic model, "e=" need not be specified. If it were a more complex model, such as one with random effects variables, as well as fixed effects ones, the manova statement could be something like this—**manova h=A e=A*B**—indicating that the two-way interaction AB is the proper error term for factor A.

We have also asked for least-squares means with the statement **lsmeans,** followed by a list of terms for which the means should be given, including main effects, two-way interactions, and the three-way interaction. Least squares means are regressed means that take into account such things as imbalance in the model or covariates. When the design is a simple balanced one as this is, there is no difference between least squares means and ordinary arithmetic means.

Twelve pages of GLM output is generated by the MAV3 program. We will examine three of these to get an idea of the results of Waranusuntikule's study, at least that half of the study dealing with the relative effectiveness of the three dimensions of persuasion that were manipulated—payment, social reinforcement, and familiarity of the assistant. The results are probably shown best in the screen capture below from page two, the univariate results on the dependent variable measure of verbal compliance.

Three-way MANOVA of Waranusuntikule Data   02:38 Wednesday, May 4, 2011   2

The GLM Procedure

Dependent Variable: verbal

Source	DF	Sum of Squares	Mean Square	F Value	Pr > F
Model	7	99.8437500	14.2633929	3.19	0.0112
Error	32	143.0000000	4.4687500		
Corrected Total	39	242.8437500			

R-Square	Coeff Var	Root MSE	verbal Mean
0.411144	49.01894	2.113942	4.312500

Source	DF	Type I SS	Mean Square	F Value	Pr > F
A	1	7.65625000	7.65625000	1.71	0.1999
B	1	82.65625000	82.65625000	18.50	0.0001
C	1	7.65625000	7.65625000	1.71	0.1999
A*B	1	1.40625000	1.40625000	0.31	0.5787
A*C	1	0.15625000	0.15625000	0.03	0.8528
B*C	1	0.15625000	0.15625000	0.03	0.8528
A*B*C	1	0.15625000	0.15625000	0.03	0.8528

This is the standard GLM screen that by now is becoming familiar, with overall model statistics at the top, R^2 for the model in the middle and the tests of individual effects at the bottom. Since the Type I and Type III sums of squares are equivalent with balanced data like this, we have left off the Type III data at the bottom of the page. Obviously, the strongest effect here by far is the effect of social reinforcement, with an SS value of 82.656, an F-ratio of 18.50, and a p-value of 0.0001. It will not be shown here, but a similar pattern is found for the second dependent variable, behavioral compliance. Figure 9.3 gives a bar chart of the R^2 values for each of the seven terms of the MAV3 model, and we see that social reinforcement is very high in R^2 for both dependent variables, accounting for about one-third of the variance in each. The AC and BC interactions are also fairly strong, but only on the behavioral compliance dependent variable. Factor C, familiar versus unfamiliar assistant reaches about 5% R^2 on behavioral compliance, but factor A, material payment, is quite small on both.

To show the contrast in outcomes, multivariate statistics for both a weak effect and also a strong effect are now shown. The weak effect is for factor A, material payment.

```
                      Three-way MANOVA of Waranusuntikule Data                    4
                                                       02:38 Wednesday, May 4, 2011

                                 The GLM Procedure
                          Multivariate Analysis of Variance

             Characteristic Roots and Vectors of: E Inverse * H, where
                            H = Type III SSCP Matrix for A
                            E = Error SSCP Matrix

            Characteristic                    Characteristic Vector  V'EV=1
                  Root       Percent               verbal         behavioral

              0.18168605     100.00            -0.05930803        0.11052859
              0.00000000       0.00             0.06123724        0.07144345

    MANOVA Test Criteria and Exact F Statistics for the Hypothesis of No Overall A Effect
                            H = Type III SSCP Matrix for A
                            E = Error SSCP Matrix

                       S=1       M=0       N=14.5

       Statistic                    Value     F Value   Num DF   Den DF   Pr > F

       Wilks' Lambda             0.84624846     2.82       2        31     0.0752
       Pillai's Trace            0.15375154     2.82       2        31     0.0752
       Hotelling-Lawley Trace    0.18168605     2.82       2        31     0.0752
       Roy's Greatest Root       0.18168605     2.82       2        31     0.0752
```

Even though this is not a particularly strong effect, it almost reaches statistical significance ($P = 0.0752$). Remember that Wilks' lambda has an inverse relationship—the smaller it is, the more significant—so a lambda of 0.8462 indicates a weak relationship. You may remember from Chapter 8 that one minus Wilks' lambda in the one-way case is essentially equivalent to a multivariate R^2 (Rencher, 2002, 350). An F approximation value of 2.82 is also indicative of a weak relationship.

By contrast, the multivariate effects of factor B, social reinforcement, are very strong, with a Wilk's lambda of 0.3308, an F approximation value of 31.35, and $P < 0.0001$. You may notice that the P-values for all four of the multivari-

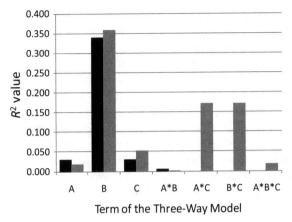

Figure 9.3 R^2 value for each of the seven terms of the three-way MANOVA model, with the bar for dv1, the verbal compliance measure on the left (black), and the bar for dv2, the behavioral compliance measure on the right (gray).

ate statistics are identical, as are the F-values. That is because there are only two multivariate means (one degree of freedom) being compared in each of these analyses. Whenever there are only two levels being compared, the four multivariate tests all give the same answer, and they are essentially like Hotelling's T^2 statistic. It is when there are three or more multivariate means that the four multivariate statistics come into play, and one will be more powerful under one configuration of multivariate means, and another more powerful in a different configuration. The more multivariate means involved in a comparison, the greater the number of possible patterns.

```
                    Three-way MANOVA of Waranusuntikule Data                    5
                                                      02:38 Wednesday, May 4, 2011

                              The GLM Procedure
                        Multivariate Analysis of Variance

            Characteristic Roots and Vectors of: E Inverse * H, where
                          H = Type III SSCP Matrix for B
                          E = Error SSCP Matrix

               Characteristic              Characteristic Vector  V'EV=1
                       Root      Percent          verbal        behavioral

                2.02277132       100.00         0.03070165      0.11122617
                0.00000000         0.00        -0.07952893      0.07035251


   MANOVA Test Criteria and Exact F Statistics for the Hypothesis of No Overall B Effect
                          H = Type III SSCP Matrix for B
                          E = Error SSCP Matrix

                          S=1      M=0      N=14.5

       Statistic                    Value    F Value   Num DF   Den DF   Pr > F

       Wilks' Lambda              0.33082225   31.35      2        31     <.0001
       Pillai's Trace             0.66917775   31.35      2        31     <.0001
       Hotelling-Lawley Trace     2.02277132   31.35      2        31     <.0001
       Roy's Greatest Root        2.02277132   31.35      2        31     <.0001
```

Table 9.18 summarizes the multivariate results for each of the seven of Waranusuntikule's main and interaction effects.[9]

Precisely the same results would be obtained were we to now carry out the MAV3 analysis using linear contrasts. That means that fully crossed designs like MAV3 for a particular data set are only one of a variety of possible analyses that could be done if the contrasts method is understood.

However, our motivation for doing the analysis with multivariate linear contrasts is because Waranusuntikule wanted to answer two questions within the same analysis: (1) does persuasion have an effect in comparison with a control group (nonpersuasion), and (2) which of three factors of persuasion most affect its effectiveness? The linear contrasts approach as shown below in SAS's GLM package offers that flexibility.

```
data sithi;
   input group verbal behavioral @@;
datalines;
1 6   11      1 8.5 9     1 5  9     1 7   8     1 6 8
2 5   6       2 7.5 5     2 4  2     2 6   3     2 5 4
3 7   3       3 5   3     3 5  6     3 2   4     3 1 4
4 4   1       4 7   2     4 1  4     4 0   2     4 3 1
5 9   10      5 8   9     5 4  6     5 6   7     5 3 8
6 7   8       6 8   7     6 4  4     6 1   5     6 5 6
7 2   1       7 4.5 1     7 1  4     7 3   2     7 2 2
8 2   8       8 5   6     8 1  6     8 1   5     8 1 5
9 2.3 6       9 2.1 5     9 0.5 2    9 2.35 3    9 2 4
;
run;

proc print; run;

proc glm data=Sithi;
   class group;
   model verbal behavioral=group;
   contrast 'A' group       1 1 1 1 -1 -1 -1 -1 0;
   contrast 'B' group       1 1 -1 -1 1 1 -1 -1 0;
   contrast 'C' group       1 -1 1 -1 1 -1 1 -1 0;
   contrast 'AB' group      1 1 -1 -1 -1 -1 1 1 0;
   contrast 'AC' group      1 -1 1 -1 -1 1 -1 1 0;
   contrast 'BC' group      1 -1 -1 1 1 -1 -1 1 0;
   contrast 'ABC' group     1 -1 -1 1 -1 1 1 -1 0;
   contrast 'control' group 1 1 1 1 1 1 1 1 -8;
   MANOVA h=group;
   lsmeans group;
   run;
```

[9] Since each of these multivariate tests only has one degree of freedom in the numerator, only one of the four multivariate tests is needed to quantify the relationship. The four multivariate tests will differ from one another only when there are three or more means.

TABLE 9.18. Wilks' Lambdas and *P*-Values for the Seven Multivariate Tests of the MAV3 MANOVA of the Waranusuntikule's (1985) Data

Term of the MAV3 Model	Lambda	*P*-Value
A: Material payment	0.84625	0.0752
B: Social reinforcement	0.33082	<0.0001
C: Same/different assistant	0.78211	0.0222
A*B	0.97548	0.6806
A*C	0.53598	<0.0001
B*C	0.53598	<0.0001
A*B*C	0.91368	0.2468

Besides demonstrating the construction of multivariate contrasts in GLM, this code also contains a more compact way of entering data for a long, strung-out data set like this one. Notice the two characters "@@" at the end of the input line of the data statement. This indicates to SAS to keep reading that same line with the instruction to enter the three things (group, verbal, and behavior) until there are no more of them on that line. This enables us to put five observation units on each row of data. Notice that there are nine rows of data and five groups of three data points on each line. The first row contains all five replications of data for the first group, the second row contains all five for the second group, and so forth.

The first three lines of PROC GLM look pretty much like any other MAV1 model of MANOVA. The 4th through 11th lines contain the weights that define the eight sets of contrasts corresponding to the MAV3 one-way, two-way, and three-way terms, plus the control group tested against all eight of them combined, following the principles outlined earlier in the chapter. Each contrast is identified with its label in single quotes, "A," "B," "ABC" interaction, and so forth. The MANOVA statement must come *after* the contrasts are specified or the multivariate contrast tests will not be created.

Table 9.19 gives the univariate results for the first dependent variable, verbal compliance. It contains the df, SS, MS, F, and P-values for all eight of the significance tests that have been defined by contrasts, the seven MAV3 terms, plus the test of the control group. We see here that the control group test is statistically significant, with a P-value of 0.0140 (and it is also backed up by a significant corresponding multivariate test). Waranusuntikule can reject the null hypothesis and conclude that the "foot-in-the-door" persuasion technique is having an effect. In fact, the only two significant effects on this first dependent variable are the control group test and factor B, social reinforcement.

Figure 9.4 shows the R^2 value for each of the eight terms of the expanded linear contrast MAV3 model. The results are seen to be identical to those from the actual MAV3 analysis (reported in Fig. 9.3), with the exception that this figure also shows the significant effect for the control group test, with about

TABLE 9.19. Univariate Analysis of Variance Summary Table of the First Dependent Variable, Verbal Compliance, for the Eight Linear Contrasts of the MAV3 Model Plus Control Group for Waranusuntikule's (1985) Data, Output Table from the SAS Enterprise Guide

The GLM Procedure

Dependent Variable: verbal

Source	DF	Sum of Squares	Mean Square	F Value	Pr > F
Model	8	126.7944444	15.8493056	3.93	0.0020
Error	36	145.3600000	4.0377778		
Corrected Total	44	272.1544444			

R-Square	Coeff Var	Root MSE	verbal Mean
0.465892	49.75186	2.009422	4.038889

Source	DF	Type I SS	Mean Square	F Value	Pr > F
group	8	126.7944444	15.8493056	3.93	0.0020

Source	DF	Type III SS	Mean Square	F Value	Pr > F
group	8	126.7944444	15.8493056	3.93	0.0020

Contrast	DF	Contrast SS	Mean Square	F Value	Pr > F
A	1	7.65625000	7.65625000	1.90	0.1770
B	1	82.65625000	82.65625000	20.47	<.0001
C	1	7.65625000	7.65625000	1.90	0.1770
AB	1	1.40625000	1.40625000	0.35	0.5588
AC	1	0.15625000	0.15625000	0.04	0.8452
BC	1	0.15625000	0.15625000	0.04	0.8452
ABC	1	0.15625000	0.15625000	0.04	0.8452
control	1	26.95069444	26.95069444	6.67	0.0140

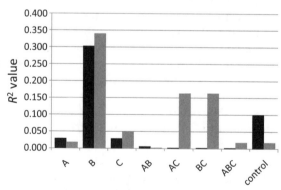

Term of the Multivariate Contrasts Model

Figure 9.4 R^2 value for each of the eight terms of the expanded linear contrast MANOVA model, with terms for each of the seven terms of the MAV3 model, plus one term for control group versus the other eight groups. The bar for dv1, the verbal compliance measure, is on the left, and the bar for dv2, the behavioral compliance measure, is on the right.

TABLE 9.20. The Four Multivariate Significance Tests for Contrast B, the Test of Main Effects for the Social Reinforcement Independent Variable for Waranusuntikule's (1985) Data, Output from the SAS Enterprise Guide

MANOVA Test Criteria and Exact F Statistics for the Hypothesis of No Overall B Effect H = Contrast SSCP Matrix for B E = Error SSCP Matrix S=1 M=0 N=16.5					
Statistic	Value	F Value	Num DF	Den DF	Pr > F
Wilks' Lambda	0.36172954	30.88	2	35	<.0001
Pillai's Trace	0.63827046	30.88	2	35	<.0001
Hotelling-Lawley Trace	1.76449636	30.88	2	35	<.0001
Roy's Greatest Root	1.76449636	30.88	2	35	<.0001

TABLE 9.21. The Four Multivariate Significance Tests for the Overall Control Group Contrast for Waranusuntikule's (1985) Data, Output from the SAS Enterprise Guide

MANOVA Test Criteria and Exact F Statistics for the Hypothesis of No Overall control Effect H = Contrast SSCP Matrix for control E = Error SSCP Matrix S=1 M=0 N=16.5					
Statistic	Value	F Value	Num DF	Den DF	Pr > F
Wilks' Lambda	0.81690321	3.92	2	35	0.0290
Pillai's Trace	0.18309679	3.92	2	35	0.0290
Hotelling-Lawley Trace	0.22413522	3.92	2	35	0.0290
Roy's Greatest Root	0.22413522	3.92	2	35	0.0290

10% of the variance accounted for on the verbal compliance dependent variable.

Table 9.20 reports the four multivariate statistics tests for factor B, social reinforcement. The results are almost identical to those found in the actual MAV3 analysis. They would be exactly the same, except that the error matrix for this analysis includes the control group data in addition to the eight groups that constitute the MAV3 data set.

Table 9.21 contains the four multivariate statistical tests for the test of the control group against the other eight groups. The results are not so strong as those for factor B, the social reinforcement factor, but they are statistically significant, $P = 0.0290$.

Figure 9.5a shows the pattern of the nine treatment groups (the eight for the MAV3 combinations plus the control group) within the multivariate space of the verbal compliance dependent variable (horizontal axis) and the behavioral compliance dependent variable (vertical axis). There is a clear separation between the two social reinforcement groups, with the four means for social nonreinforcement in the lower left (low on both verbal compliance and behavioral compliance), and the four means for social reinforcement located in the upper right (high on both verbal compliance and also behavioral compliance). This clearly expresses the strong statistical effects of social reinforcement in the multivariate and univariate significance tests.

Figure 9.5 (parts a, b, c, d, and e) demonstrates principle number eight in the list of nine "Principles and Powers of Multivariate Analysis of Variance" given at the beginning of Chapter 8. This is the principle that a multivariate plot of three-way means gives much more information than what is reflected in a three-way interaction such as ABC. It in fact contains the combined information of all the lower order interactions and simple effects. Understanding the three-way pattern has within it the information of the two-way and one-way patterns that are part of it—in this case, the AB, AC, and BC patterns, and the patterns for the main effects of A, B, and C. To illustrate, we will examine several of the subpatterns that are contained within this simple graph of eight bivariate means.

Figure 9.5b focuses on the two-way interaction between factor A and factor B at only one level of factor C, the one for "different assistant." Here we see the four means for all combinations of factor A (material payment, PMT, and nonpayment, nonPMT) and factor B (social reinforcement and social nonreinforcement), but only at the second level of C, that of "different assistant."

A parallelogram would indicate zero interaction effects, even though this diverges somewhat. The two means on the right are for the social reinforcement manipulation, and the two on the left are for the social nonreinforcement manipulation. The two at the top are for the nonpayment manipulation and the two more toward the bottom are for payment. This set of two-way means could be summarized with three linear contrasts—one for social reinforcement effects, one for payment effects, and one for the interaction between the two. The three contrasts would have "1" and "–1" appropriately placed for each of the four means involving the "different assistant" condition, but "0" entries for the four means that involve the "same assistant" condition. The shape roughly approximating a parallelogram indicates that this one-way and two-way relationship is primarily defined by the main effects A and B, with very little AB interaction effect present. A reasonably compelling rationale could be identified for this pattern. It makes some intuitive sense that a relative stranger who is kind and reinforcing would elicit some verbal compliance, but it would perhaps have little impact upon behavioral compliance (as indicated by the arrows from nonreinforcement to reinforcement going primarily

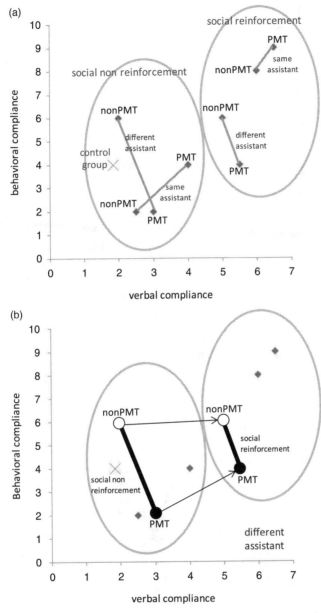

Figure 9.5 (a) Bivariate scatter plot of the eight three-way means and the control group mean, including all eight combinations of factor A, material payment (PMT or nonPMT); factor B, social reinforcement; and factor C, same versus different experimental assistant. (b) Same bivariate scatter plot as Figure 9.5a, but showing the AB interaction (material payment by social reinforcement) nested within the C_2 condition of "different assistant."

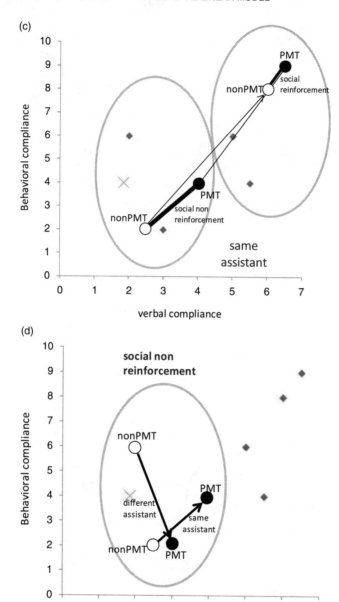

Figure 9.5 (c) Same bivariate scatter plot as Figure 9.5a, but showing the AB interaction (material payment by social reinforcement) nested within the C_1 condition of "same assistant." (d) Same bivariate scatter plot as Figure 9.5a, but showing the AC interaction (material payment by same/different assistant) nested within the B2 condition of "social nonreinforcement."

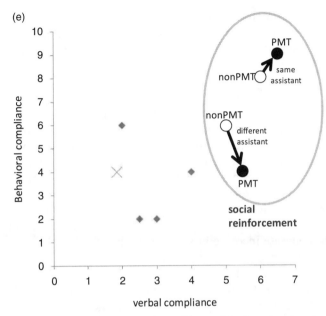

Figure 9.5   (e) Same bivariate scatter plot as Figure 9.5a, but showing the AC interaction (material payment by same/different assistant) nested within the B1 condition of "same assistant."

horizontally to the right, with little vertical direction). On the other hand, the gift of the pen from the relative stranger would actually lead to less behavioral compliance than the no-gift condition (both PMT conditions are much lower on behavioral compliance than the corresponding non-PMT conditions).

Figure 9.5c articulates the other side of this "AB interaction within a level of C," the one within the "same assistant" condition. When there is at least some acquaintance with the experimenter's assistant, both social reinforcement and also the "material payment" gift lead to increased compliance, both verbal compliance and also behavioral compliance (as indicated by the arrows at a nearly 45° angle). Putting these two nested two-way interactions together, with three df each, and adding the overall test of the four "same assistant" means against the four "different assistant" means, creates an orthogonal set of seven contrasts that rivals, and is perhaps competitive with, the MAV3 set of linear contrasts. In other words, a three-way MANOVA is only one of a number of patterned orthogonal linear contrasts that could be used to explain the 7 df implicit within the eight means. The multivariate graph (which is only a bivariate graph in this case) can be used to identify the more compelling story in the data.

Another possible set of seven orthogonal contrasts is contained in the patterns of Figures 9.5d and 9.5e. Here we are examining the AC interaction, nested within either B1 social reinforcement or B2 social nonreinforcement. Figure

9.5d, the one showing the AC interaction within nonreinforcement is a classic example of a strong two-way interaction within a multivariate space, with the two lines completely crossing one another. In other words, material payment (PMT) or nonpayment (nonPMT) has a completely different meaning, depending on whether it is the same assistant condition or the different assistant condition.

All four of these means are over on the left side (low verbal compliance), because all four of these means are within the social nonreinforcement condition. We could summarize. When the respondent is not receiving social kindness from the experimenter's assistant, but it is a familiar assistant, then the material payment of a pen helps the relationship somewhat. This is true both for verbal compliance and also for behavioral compliance. On the other hand, when social nonreinforcement is combined with the condition of a different assistant, then the payment of a pen leads to much lower behavioral compliance and a moderate increase in verbal compliance.

This interactive pattern, although a little surprising and strange, is consistent both in the social non-reinforcement condition on the left side of the space (Fig. 9.5d) and also on the right hand side of the space in Figure 9.5e, the social reinforcement condition. The effect is a small one, but payment of the pen helps when the same assistant condition is paired with social reinforcement (in the upper right of the graph), but again, when there is a different assistant, the payment hurts behavioral compliance, but helps slightly on verbal compliance. In other words, with a relative stranger, the gift of a pen induces them to talk a little more compliantly but behave less compliantly.

We have examined now two models that compete with the MAV3 fully crossed model as explanations for the bivariate patterns in Figure 9.5a. There is a third possible nested two-way interaction scheme, the two BC interactions nested within factor A. There are also three possible single effects nested within two-way interactions (A within BC, B within AC, and C within AB), for a total of seven complete orthogonal models for a small data set of only eight means. Linear contrasts open the way to a variety of ways of making sense of such data. Some of them might provide better insights into the data than the usual fully crossed designs.

9.6 REGRESSING CATEGORICAL VARIABLES

We have seen that the predictor variables in regression can either be "class" variables or quantitative variables. That is, they can be categorical variables, as in the GLM approach to ANOVA, or they can be random variables as in most ordinary least squares regression applications (the demonstration of Section 9.2). These two models—the fixed-effects regression of GLM and the common random-effects regression—will cover a great many applications. However, it is often the case that we wish to predict binary or categorical outcomes, such as whether a person contracts a disease, whether students pass or fail a class, or whether persons choose one or another of several occupations. There is a need to go beyond the *general linear model* described in Sections 9.3–9.7 to expand the general linear model to also deal with binary or categorical dependent variables. This next step is *generalized linear models*, a generalization of the linear model to deal with dependent variables of many kinds (McCullagh and Nelder, 1989; Agresti, 2002; Long and Freese, 2006). Darlington (1990) gives a particularly accessible treatment of these methods.

In this book, we deal with three methods that can be used for predicting binary or categorical dependent variables: discriminant analysis (in chapter seven) and two generalized linear models in this chapter—log-linear analysis, and logistic regression. The most general of these is logistic regression. It can be used with any combination of numerical and categorical independent variables, as long as the dependent variable is binary or categorical. Discriminant analysis applies to data with independent variables that are all numerical. Log-linear analysis requires that the independent variables all be categorical. In that sense, log-linear analysis can be thought of as a special case of logistic regression, in the same way that ANOVA can be thought of as a special case of ordinary least squares regression.

9.6.1 Log-Linear Analysis

Both log-linear analysis and also logistic regression are broad and varied, and there is an almost bewildering array of possible models from which to choose. To give just a brief introduction at the end of this regression chapter, perhaps the best thing to do would be to show just one clear and simple example of how it can be useful, and to compare and contrast it with Pearson's chi-square, which is familiar and generally well understood.

Let us return to the example that was given in Chapter 5 about the relationship between parent dyadic satisfaction scores and child hope/optimism. Suppose we have a sample of 264 families with scores of the parents on a nationally normed parent dyadic scale, and scores of the children on a nationally normed hope/optimism scale, and that the distribution of our sample with respect to the national norms is as shown.

hypothetical data		Parent Dyadic Scale		
		lowest tertile	middle tertile	highest tertile
child optimism	upper half	6	8	24
	lower half	26	58	142

We first analyze this table with a chi-square analysis and then explain how the likelihood ratio of log-linear analysis can be calculated from the Pearson chi-square calculations. The first computational step is to obtain row and column marginal sums of frequencies and the grand sum of the frequencies.

	low	mid	high	column sums
upper	6	8	24	38
lower	26	58	142	226
row sums	32	66	166	264

Each expected cell frequency E_{rc} is obtained by multiplying each corresponding row sum T_r by the corresponding column sum T_c, and dividing the product by the grand sum T, as shown in Equation 9.22.

$$E_{rc} = \frac{T_r T_c}{T}.\qquad(9.22)$$

The formula for Pearson's chi-square is given in Equation 9.23.

$$\chi^2 = \sum_r \sum_c \frac{(O_{rc} - E_{rc})^2}{E_{rc}}.\qquad(9.23)$$

The E_{rc} values are found by Equation 9.22.

$$E_{rc} = \begin{bmatrix} 4.606 & 9.500 & 23.894 \\ 27.390 & 56.500 & 142.106 \end{bmatrix}.$$

The matrix of E_{rc} values is subtracted from the matrix of O_{rc} observed scores.

$$O_{rc} - E_{rc} = \begin{bmatrix} 6 & 8 & 24 \\ 26 & 58 & 142 \end{bmatrix} - \begin{bmatrix} 4.606 & 9.500 & 23.894 \\ 27.390 & 56.500 & 142.106 \end{bmatrix}$$
$$= \begin{bmatrix} 1.394 & -1.500 & 0.106 \\ -1.394 & 1.500 & -0.106 \end{bmatrix}.$$

Each element in the matrix of $O_{rc} - E_{rc}$ is squared.

$$(O_{rc} - E_{rc})^2 = \begin{bmatrix} (1.394)^2 & (-1.500)^2 & (0.106)^2 \\ (-1.394)^2 & (1.500)^2 & (-0.106)^2 \end{bmatrix} = \begin{bmatrix} 1.943 & 2.250 & 0.011 \\ 1.943 & 2.250 & 0.011 \end{bmatrix}.$$

These squared elements are now entered into Equation 9.23. Each is divided by its respective E_{rc} value, and the quotients are all summed to obtain the value of chi-square.

$$\chi^2 = \sum_r \sum_c \frac{(O_{rc} - E_{rc})^2}{E_{rc}} = \frac{1.943}{4.606} + \frac{2.250}{9.500} + \frac{0.011}{23.894} + \frac{1.943}{27.394} + \frac{2.250}{56.500} + \frac{0.011}{142.106}$$

$$= 0.4219 + 0.2368 + 0.0005 + 0.0709 + 0.0398 + 0.0001$$

$$= 0.7700$$

The df for this contingency table chi-square is $(R-1)(C-1) = (2-1)$ $(3-1) = 2$. We look up the critical chi square ratio for 2 df and find it to be 5.99. We fall far short of that value and fail to reject the null hypothesis. There is no evidence for a contingent relationship between parent dyadic score and child hope/optimism score.

One version of the calculating equations for log-linear analysis is quite similar to the chi-square calculations, as shown in Equation 9.24. There are several reasonably accessible demonstrations within the published literature using these simple equations (Howell, 2012; Brown and Hendrix, 2005; Tabachnick and Fidell, 2007).

$$D^2 = 2 \sum_r \sum_c O_{rc} \ln\left(\frac{O_{rc}}{E_{rc}}\right). \tag{9.24}$$

This formula employs the same O_{rc} and E_{rc} matrices, but we sum the product of the observed frequency times the natural logarithm of the ratio of the observed frequency to the expected frequency, and then multiply that sum by 2. We calculate the D^2 likelihood ratio test for these data.

$$D_{rc}^2 = 2 \sum_r \sum_c O_{rc} \ln\left(\frac{O_{rc}}{E_{rc}}\right)$$

$$= (2)\left[6\ln\left(\frac{6}{4.606}\right) + 8\ln\left(\frac{8}{9.500}\right) + 24\ln\left(\frac{24}{23.894}\right) + 26\ln\left(\frac{26}{27.394}\right)\right.$$

$$\left. + 58\ln\left(\frac{58}{56.500}\right) + 142\ln\left(\frac{142}{142.106}\right)\right]$$

$$= (2)(1.5863 - 1.3748 + 0.1063 - 1.3579 + 1.5197 - 0.1060) = (2)(0.3737)$$

$$= 0.7473.$$

The df are the same for this likelihood ratio test, and it is also compared to a chi-square table to determine significance. On this test also, the null hypothesis cannot be rejected.

The chi-square analysis and also the likelihood ratio analysis can be run on SAS's PROC FREQ. We check ourselves by running this analysis with the SAS code given below.

```
data conting;
    input hope pds Fcount;
    datalines;
        1    1    6
        1    2    8
        1    3    24
        2    1    26
        2    2    58
        2    3    142
    ;
run;

proc freq data=conting;
    weight Fcount;
    tables hope*pds/chisq;
run;
```

The SAS output includes the contingency table in several forms (frequencies, percents of total, percents of row totals, and percents of column totals), and gives six analytical coefficients, the first two of which are the Pearson chi-square value and the likelihood ratio chi-square. Both agree with the values that we obtained.

```
                    The SAS System              14:20 Wednesday, May 4,

                    The FREQ Procedure

                Table of hope by pds

    hope          pds

    Frequency|
    Percent  |
    Row Pct  |
    Col Pct  |      1|      2|      3|  Total
    ---------+-------+-------+-------+
         1   |    6  |    8  |   24  |    38
             |  2.27 |  3.03 |  9.09 | 14.39
             | 15.79 | 21.05 | 63.16 |
             | 18.75 | 12.12 | 14.46 |
    ---------+-------+-------+-------+
         2   |   26  |   58  |  142  |   226
             |  9.85 | 21.97 | 53.79 | 85.61
             | 11.50 | 25.66 | 62.83 |
             | 81.25 | 87.88 | 85.54 |
    ---------+-------+-------+-------+
    Total        32      66     166      264
              12.12   25.00   62.88  100.00


        Statistics for Table of hope by pds

    Statistic                    DF     Value      Prob
    --------------------------------------------------------
    Chi-Square                    2     0.7700    0.6805
    Likelihood Ratio Chi-Square   2     0.7473    0.6882
    Mantel-Haenszel Chi-Square    1     0.1032    0.7481
    Phi Coefficient                     0.0540
    Contingency Coefficient             0.0539
    Cramer's V                          0.0540

            Sample Size = 264
```

The likelihood ratio we just completed is the one that tests the interaction between the rows of our two-way table (child optimism/hope) and the columns of our two-way table (parent dyadic score). There are three other likelihood ratio statistics that can be calculated with various versions of this same formula, one for rows, one for columns, and one for the total two-way matrix, Equations 9.25–9.27, respectively.

$$D_r^2 = 2\sum_r O_r \ln\left(\frac{O_r}{E_r}\right). \tag{9.25}$$

$$D_c^2 = 2\sum_c O_c \ln\left(\frac{O_c}{E_c}\right). \tag{9.26}$$

$$D_{total}^2 = 2\sum_r \sum_c O_{rc} \ln\left(\frac{O_{rc}}{E_{rc}}\right). \tag{9.27}$$

Equation 9.27 for the total likelihood ratio looks very much like Equation 9.24, the one for the row by column interaction. The difference is in how the expected value is calculated. The expected values of the cell frequencies for the interaction test are given in Equation 9.22. The expected values of the cell frequencies for the total matrix are given in Equation 9.28 and are seen to be just the total frequency in the matrix divided by the number of cells (RC), which for these data comes out to 44. That is, all six cells have that same expected value of 44.

$$E_{total} = \frac{T}{RC} = \frac{264}{(2)(3)} = 44. \tag{9.28}$$

The likelihood test for rows tests the null hypothesis that the rows are equal in their frequencies, which in the hypothetical study we described above would be the null hypothesis that the children in our sample have the same distribution of hope/optimism scores as the national sample on which the norms were based. The likelihood test for columns likewise tests the null hypothesis that our sample of couples have the same distribution of dyadic couple satisfaction scores as the national sample. The likelihood ratio test for the entire matrix tests the holistic null hypothesis that the distribution of the entire table in our sample does not differ from the distribution in the national sample on which the norms are based. The interesting thing about likelihood ratio statistics is that they are additive as shown in Equation 9.29. That is, the likelihood ratio for rows, and that for columns, and that for the R by C interaction, will all sum to the likelihood ratio for the entire matrix.

$$D_{total}^2 = D_{rows}^2 + D_{columns}^2 + D_{rc}^2. \tag{9.29}$$

That is how we were able to say at the end of Chapter 5, Section 5.5, that child hope scores accounted for 57.2% of the information in the contingency table, Table 5.2, parent dyadic scores accounted for 42.1% of the information, and the interaction between the two accounted for only 0.6% of the information. We demonstrate the calculation of the likelihood ratio for rows (child hope/optimism) for our fictitious data using Equation 9.25. As we can see from the numerical values in carrying out this formula, the null hypothesis holds that the two rows will have equal frequencies.

$$D_{\text{rows}}^2 = 2\sum_r O_r \ln\left(\frac{O_r}{E_r}\right) = (2)\left[38\ln\left(\frac{38}{132}\right) + 226\ln\left(\frac{226}{132}\right)\right] = 148.4190.$$

The *df* value for the rows likelihood test is $R - 1$, which is 1. Likewise, the *df* value for the columns likelihood test is $C - 1$, which is 2. The *df* value for the total table likelihood test is $RC - 1$, which is 5. The *df* for the likelihood tests are also additive. The total *df* of 5 for the entire table is the sum of the rows df (1) plus the columns *df* (2), plus the R by C interaction *df* (2).

In a similar manner, the likelihood ratio for columns and the likelihood ratio for the total matrix can be calculated using Equations 9.26 and 9.27.

$$D_{\text{columns}}^2 = 2\sum_c O_c \ln\left(\frac{O_c}{E_c}\right) = 107.9876.$$

$$D_{\text{total}}^2 = 2\sum_r\sum_c O_{rc} \ln\left(\frac{O_{rc}}{E_{rc}}\right) = 257.1539.$$

Using Equation 9.30, we can validate that the likelihood ratios for this contingency matrix are indeed additive. The three individual likelihood ratios sum to the likelihood ratio for the entire table.

$$D_{\text{total}}^2 = D_{\text{rows}}^2 + D_{\text{columns}}^2 + D_{\text{rc}}^2 = 148.4190 + 107.9876 + 0.7473 = 257.1539.$$

The percents of table information attributed to rows, columns, and R by C interaction can also be calculated.

$$\text{percent for rows} = \frac{D_{\text{rows}}^2}{D_{\text{total}}^2} = (100)\frac{148.4190}{257.1539} = 41.99\%.$$

$$\text{percent for columns} = \frac{D_{\text{columns}}^2}{D_{\text{total}}^2} = (100)\frac{107.9876}{257.1539} = 57.72\%$$

$$\text{percent for } R \text{ by } C = \frac{D_{\text{rc}}^2}{D_{\text{total}}^2} = (100)\frac{0.7473}{257.1539} = 0.29\%.$$

We see that these hypothetical data were constructed to have a very similar distribution of information to the three sources (child hope, couple dyadic score, and interaction) as the actual results from the Flourishing Families Study in Chapter 5.

9.6.2   Logistic Regression

Logistic regression has been often used in medicine, epidemiology, and biostatistics (Lemmeshow and Hosmer, 2005; Vittinghoff et al., 2005). It has much to recommend it (Hosmer and Lemeshow, 2000). As discussed above, it can apply to a wide variety of research situations, and, compared with some of its competitors like discriminant analysis, it is relatively free of restrictive assumptions. One of the features of logistic regression that makes it particularly useful in the biomedical fields is its use of odds ratios, which make it possible to make statements such as "for every additional 0.2 percent of some substance found in the blood, the odds of a heart attack are increased by 2.7." We also found odds ratios to be of use in our analysis of child hope scores in the Flourishing Families data set, as shown in Figure 5.27 of Chapter 5, an "odds ratio scatter plot." We will return to that data example to give a brief explanation of the rationale of logistic regression.

Table 9.17 is taken from the Flourishing Families data set discussed in Chapter 5, Section 5.5. Of the 132 families analyzed for the results presented in that section, only 16 had a child with a hope/optimism score above the midpoint (3.00) on the scale. Seven of the 13 cluster groups of parents had zero children with a score above 3.00. The other six groups are shown in this table. Group 8 has 10 families in it, with only one family having a child with a hope score at or above 3.00. The odds of a child with a hope score of three or more in that group is therefore "1 to 9," which comes out to a numerical odds score of 0.111. This was the lowest odds of those groups that had any such children in their group at all, so it was taken to be the comparison couple. The odds ratios for the other cluster groups are therefore calculated in comparison to group 8. Three of the parent groups (2, 4, and 6) have odds of 0.167. The ratio of 0.167 to 0.111 is approximately 1.5, so these three groups are assigned an odds ratio of 1.5. Group 4 has four children at or above three and 23 not, which is odds of 0.174. The odds ratio for that group is therefore 1.565. The highest odds are in group 1, with "3 to 11" having high optimism scores, so their odds are 0.273. Their odds ratio score is therefore 0.273 divided by 0.111, or 2.455.

Logistic regression uses a linking function to convert odds ratios into values that can be analyzed using the mathematics of regression, by taking logarithms of the odds ratios. SAS reports both the odds ratios and also the log odds ratios (the regression coefficients) in their output. Stata uses two programs for this, **logit**, and **logistic**. We use the data in Table 9.22 to demonstrate the output from SAS for a logistic regression using PROC LOGISTIC (SAS Institute, Inc., 2009b, Chapter 51). The SAS code for running logistic is quite similar to the other regression analyses in SAS. In the data statement, we read in three variables. The first, cH, stands for "child hope," and indicates the binary category of "0" for high hope and "1" for low hope. The next column, pSat, stands

TABLE 9.22. Number of Children in Each of Six Cluster Groups of Parents Having and Not Having a Hope Score of Three or More, Including Odds and Odds Ratios

	Group 1	Group 2	Group 3	Group 4	Group 6	Group 8
Child hope Score ≥ 3.0	3	4	3	4	1	1
Child hope Score > 3.0	11	24	18	23	6	9
Odds of a high hope child	0.273	0.167	0.167	0.174	0.167	0.111
Odds ratio	2.455	1.500	1.500	1.565	1.500	1.000

for "parent satisfaction" meaning the marriage satisfaction group that each couple belonged to. Six groups are listed here, groups 1, 2, 3, 4, 6, and 8. The third column, Fcount, indicates how many families are in each of those cross-classified categories. In other words, this input recreates the 2×6 table of data shown at the top of Table 9.22. In the PROC LOGISTIC code, the first line indicates the identity of the data set to be analyzed. The second line indicates that parent marriage satisfaction group, pSat, is a categorical or classification variable. The next line tells SAS that the variable named Fcount is in fact the weights that are to be used to create the 2×6 table frequencies. The model is specified with child hope as the dependent variable, and the pSat group of the parents as the independent variable.

```
data hope;
  input cH pSat Fcount;
  datalines;
        0        1        3
        0        2        4
        0        3        3
        0        4        4
        0        6        1
        0        8        1
        1        1        11
        1        2        24
        1        3        18
        1        4        23
        1        6        6
        1        8        9
;
run;

proc logistic data = hope;
  class pSat;
  weight Fcount;
  model cH = pSat;
run;
```

The SAS output from this PROC LOGISTIC run has much the look and organization as the other regression programs in SAS. It has the overall model information at the top and the detailed information about specific predictors at the bottom. This screen capture shows the two types of detailed information produced by PROC LOGISTIC, with the logit information on top (the regression coefficients from the log odds ratios) and the Odds Ratio information below for interpretation. Clearly, with the small sample sizes in this demonstration data set, there are no statistically significant relationships in the data.

```
               Analysis of Maximum Likelihood Estimates

                                 Standard       Wald
    Parameter       DF  Estimate    Error  Chi-Square   Pr > ChiSq

    Intercept        1   -1.7702   0.3196    30.6865      <.0001
    pSat    1        1    0.4709   0.6204     0.5760       0.4479
    pSat    2        1   -0.0216   0.5446     0.0016       0.9684
    pSat    3        1   -0.0216   0.6011     0.0013       0.9713
    pSat    4        1    0.0210   0.5457     0.0015       0.9694
    pSat    6        1   -0.0216   0.9380     0.0005       0.9816


                      Odds Ratio Estimates

                         Point           95% Wald
        Effect         Estimate     Confidence Limits

        pSat 1 vs 8      2.455      0.216      27.841
        pSat 2 vs 8      1.500      0.147      15.284
        pSat 3 vs 8      1.500      0.136      16.542
        pSat 4 vs 8      1.565      0.153      15.973
        pSat 6 vs 8      1.500      0.078      28.890
```

See Kleinbaum and Klein (2002) for an accessible treatment of logistic regression, and Gould (2000) for useful information on the interpretation of logistic regression.

9.7 SUMMARY AND CONCLUSIONS

Multiple regression has now been used for many years to predict scores on one variable from scores on several predictor variables. It has been a staple in the data analysis arsenal of a number of disciplines, but it in only in the past half-century that it has become abundantly clear that it is much more than just one more of the statistical methods. It has turned out to be one of the most productive of strategies in the sense of opening up new avenues for development in virtually every area of quantitative work.

STUDY QUESTIONS

A. Essay Questions

1. Do a little research in the library or on the Internet and find out how the least squares method came about, and in what ways it has been used.

2. Investigate the history of the *general linear models* approach to ANOVA and write an essay summarizing that history and the major figures in its development.

3. Investigate the history of the *generalized* linear models approach to regression analysis of categorical data and write an essay summarizing that history and the major figures in its development.

4. In the Stata output to the simple regression data in Section 9.2, explain why the confidence interval will have zero within the interval (with both a positive limit and also a negative limit to the interval) when the *t*-test is not significant. What would happen to the confidence interval when the *t*-test is significant?

B. Computational Questions

1. For the simple data below, do a multiple regression analysis on a spreadsheet, following the steps outlined in the chapter.

Y	X_1	X_2
12	8	9
16	6	3
10	5	8
8	5	5
10	4	7
4	2	4

2. Shown below are the data from the AV2 demonstration in Section 9.3.1, but with a second dependent variable added, behavioral compliance, to make the data set multivariate. Expand Equations 9.8–9.12 of Section 9.3.1 from univariate formulas into multivariate formulas. That is, change them from *vectors* of means and totals into *matrices* of means and totals (like the MANOVA formulas from Chapter 8, Section 8.3). Use these formulas on a spreadsheet to calculate the MAV2 multivariate two-way ANOVA on these data. Complete the univariate ANOVA summary tables, and calculate the Wilks' lambda values for rows, columns, and RxC interaction. Look up the critical *F*-ratios from Table C in the Appendix and the critical Wilks' lambda values from Table F of the Appendix to determine whether each statistical test reached significance.

sr	asst	verbal	behav
1	1	9	10
1	1	8	9
1	1	4	6
1	1	6	7
1	1	3	8
1	2	7	8
1	2	8	7
1	2	4	4
1	2	1	5
1	2	5	6
2	1	2	1
2	1	4.5	1
2	1	1	4
2	1	3	2
2	1	2	2
2	2	2	8
2	2	5	6
2	2	1	6
2	2	1	5
2	2	1	5

C. Data Analysis Questions

1. Use Stata, SPSS, or SAS to run a multiple regression on the small example data matrix in Section 9.2 of the chapter. Explain the meaning of the output.

2. Use Stata, SPSS, or SAS to run a multiple regression on the small example data matrix in question b1 above. Explain the meaning of the output.

3. Find a data matrix with three or more variables and several observations and run a multiple regression on it. Explain the meaning of the output.

4. Use Metrika to create a 3D graphic of the regression plane for the small data set in Section 9.2, and/or the small data set in question b1 above. (Metrika script files for this question are provided in the online answers to questions.)

5. Use the root MSE to show the extent of the 90% confidence interval around the regression plane of question 4.

6. Paste the data of question b2 above (the AV2 demonstration from Section 9.3.1) into Stata. Use the anova command in Stata to accomplish the two-way analysis of variance on this data (the "verbal" dependent variable only) and compare the results to those reported from SAS output in Section 9.3.1. The Stata command is:

anova verbal sr asst sr#asst

7. Use the same data set of question 6, but drop out rows 4, 8, 10, and 18 as we did for the demonstration in Section 9.4 to create a factorial unbalanced data set. Run this analysis in Stata (the command is exactly the same as in question 6) and compare the Stata output with that reported from SAS in Section 9.4. Notice that whereas SAS reports both the Type I and also the Type III sums of squares, Stata reports only one. Which one does Stata report in its default condition? Why?

8. For the Waranusuntikule data of Section 9.6 use Metrika or another graphing method to create a starplot (similar to Fig. 5.5 of chapter five) showing the high percentage of variance in the raw data accounted for by the three-way means.

9. Use one of the statistical packages to run the MAV2 two-way MANOVA on the data of question b2 above, and compare the results from the statistical package to the results of your spreadsheet analysis of the data.

10. Alter the SAS multivariate linear contrasts program given in Section 9.5 to calculate significance levels and variance accounted for in the analysis plan suggested in Figures 9.3b and 9.3c and the accompanying text.

REFERENCES

Agresti, A. 2002. *Categorical Data Analysis, Second Edition*. New York: John Wiley & Sons, Inc.

Brown, B. L., and Hendrix, K. A. 2005. Tests of independence. In Everitt, B., and Howell, D. C. (Eds.), *Encyclopedia of Behavioral Statistics*. New York: John Wiley & Sons, Inc.

Bryce, G. R. 1970. A unified method for the analysis of unbalanced designs (Unpublished master's thesis). Brigham Young University.

Bryce, G. R., and Carter, M. W. 1974. MAD—The analysis of variance in unbalanced designs—A software package. In Bruckman, G., Ferschi, F., and Schmetterer, L. (Eds.), *COMPSTAT 1974: Proceedings in Computational Statistics*. New York: Springer-Verlag.

Cohen, J. 1968. Multiple regression as a general data-analytic system. *Psychological Bulletin, 70*, 426–443.

Cohen, J., and Cohen, P. 1975. *Applied Multiple Regression/Correlation Analysis for the Behavioral Sciences*. Hillsdale, NJ: Lawrence Erlbaum.

Cohen, J., and Cohen, P. 1983. *Applied Multiple Regression/Correlation Analysis for the Behavioral Sciences, Second Edition*. Hillsdale, NJ: Lawrence Erlbaum.

Darlington, R. B. 1990. *Regression and Linear Models*. New York: McGraw-Hill Series in Psychology.

Francis, I. 1973. A comparison of several analysis of variance programs. *Journal of the American Statistical Association, 68*, 860–871.

Freund, R. J., Littell, R. C., and Spector, P. C. 1991. *SAS System for Linear Models*. Cary, NC: SAS Institute Inc.

Gould, W. W. 2000. sg124: Interpreting logistic regression in all its forms. *Stata Technical Bulletin*, *53*, 19–29. Reprinted in *Stata Technical Bulletin Reprints, Vol. 9*. College Station, TX: Stata Press; pp. 257–270.

Hosmer, D. W. Jr., and Lemeshow, S. 2000. *Applied Logistic Regression, Second Edition*. New York: John Wiley & Sons, Inc.

Howell, D. C. 2012. *Statistical Methods for Psychology, Eighth Edition*. Pacific Grove, CA: Duxbury.

Kleinbaum, D. G., and Klein, M. 2002. *Logistic Regression: A Self-Learning Text, Second Edition*. New York: Springer.

Kutner, M. H., Neter, J., Nachsheim, C., and Wasserman, W. 2005. *Applied Linear Statistical Models, Fifth Edition*. New York: McGraw-Hill.

Lawson, J. 1971. The validity of a unified method for the analysis of unbalanced designs (Unpublished master's thesis). Brigham Young University.

Legendre, A. M. 1805. *Nouvelles Methodes pour la Determination des Orbites des Cometes*. Paris: Courcier.

Lemmeshow, S., and Hosmer, D. W. 2005. Logistic regression. In Armitage, P., and Colton, T. (Eds.), *Encyclopedia of Biostatistics, Vol. 2*. Chichester, UK: Wiley; pp. 2870–2880.

Long, J. S., and Freese, J. 2006. *Regression Models for Categorical Dependent Variables Using Stata, Second Edition*. College Station, TX: Stata Press.

McCullagh, P., and Nelder, J. A. 1989. *Generalized Linear Models, Second Edition*. London: Chapman & Hall.

Milliken, G. A., and Johnson, D. E. 1984. *Analysis of Messy Data, Volume I: Designed Experiments*. Belmont, CA: Lifetime Learning Publications.

Neter, J., and Wasserman, W. 1974. *Applied Linear Statistical Models*. Homewood, IL: Richard D. Irwin, Inc.

Overall, J. E., Spiegel, D. K., and Cohen, J. 1975. Equivalence of orthogonal and non-orthogonal analysis of variance. *Psychological Bulletin*, *82*, 182–186.

Pedhazur, E. J. 1982. *Multiple Regression in Behavioral Research: Explanation and Prediction, Second Edition*. New York: Holt, Rinehart, and Winston.

Pedhazur, E. J. 1997. *Multiple Regression in Behavioral Research: Explanation and Prediction, Third Edition*. Fort Worth, TX: Harcourt-Brace.

Plackett, R. L. 1972. Studies in the history of probability and statistics. XXIX. The discovery of the method of least squares. *Biometrika*, *59*(2), 239–251.

Rao, C. R., Toutenburg, H., Fieger, A., Heumann, C., Nitter, T., and Scheid, S. 1999. *Linear Models: Least Squares and Alternatives*. New York: Springer.

Rencher, A. C. 2002. *Methods of Multivariate Analysis*. New York: Wiley.

Rencher, A. C., and Schaalje, G. B. 2008. *Linear Models in Statistics, Second Edition*. New York: Wiley.

SAS Institute Inc 2009a. *Chapter 15: The Four Types of Estimable Functions. SAS/STAT ® 9.2 User's Guide, Second Edition*. Cary, NC: SAS Institute Inc.

SAS Institute Inc 2009b. *Chapter 51: The LOGISTIC procedure. SAS/STAT ® 9.2 User's Guide, Second Edition*. Cary, NC: SAS Institute Inc.

Seal, H. L. 1967. Studies in the history of probability and statistics. XV. The historical development of the Gauss linear model. *Biometrika*, *59*(2), 239–251.

Searle, S. R. 1971. *Linear Models*. New York: Wiley.

Searle, S. R., Speed, F. M., and Henderson, H. V. 1981. Some computational an dmodel equivalencies in analysis of variance of unequal-subclass-numbers data. *The American Statistician*, *35*, 16–33.

Speed, F. M. 1969. *A New Approach to the Analysis of Linear Models*. Technical report, National Aeronautics and space Administration, Houston, TX; a NASA Technical memo, MASA TM X-58030.

Speed, F. M., Hocking, R. R., and Hackney, O. P. 1978. Methods of analysis of linear models with unbalanced data. *Journal of the American Statistical Association*, *73*, 105–112.

Stigler, S. M. 1981. Gauss and the invention of least squares. *The Annals of Statistics*, *9*(3), 465–474.

Tabachnick, B. G., and Fidell, L. S. 2007. *Using Multivariate Statistics, Fifth Edition*. Boston: Pearson.

Vittinghoff, E. D., Glidden, D. V., Shiboski, S. C., and McCulloch, C. E. 2005. *Regression Methods in Biostatistics: Linear Logistic, Survival, and Repeated Measures Models*. New York: Springer.

Waranusuntikule, S. 1985. Foot-in-the-door technique: Mediating effects of material payment, social reinforcement, and familiarity (Unpublished doctoral dissertation). Brigham Young University.

West, S. G., Cohen, P., and Aiken, L. 2003. *Applied Multiple Regression/Corelation Analysis for the Behavioral Sciences, Third Edition*. Hillsdale, NJ: Lawrence Erlbaum.

Winer, B. J. 1971. *Statistical Principles in Experimental Design, Second Edition*. New York: McGraw-Hill.

APPENDICES: STATISTICAL TABLES

Multivariate Analysis for the Biobehavioral and Social Sciences: A Graphical Approach,
First Edition. Bruce L. Brown, Suzanne B. Hendrix, Dawson W. Hedges, Timothy B. Smith.
© 2012 John Wiley & Sons, Inc. Published 2012 by John Wiley & Sons, Inc.

TABLE A. Areas under the Standard Normal Distribution

z	area between mean and z	z	area between mean and z	z	area between mean and z	z	area between mean and z	z	area between mean and z	z	area between mean and z	z	area between mean and z
0.00	.0000	0.40	.1554	0.80	.2881	1.20	.3849	1.60	.4452	2.00	.4772	2.40	.4918
0.01	.0040	0.41	.1591	0.81	.2910	1.21	.3869	1.61	.4463	2.01	.4778	2.41	.4920
0.02	.0080	0.42	.1628	0.82	.2939	1.22	.3888	1.62	.4474	2.02	.4783	2.42	.4922
0.03	.0120	0.43	.1664	0.83	.2967	1.23	.3907	1.63	.4484	2.03	.4788	2.43	.4925
0.04	.0160	0.44	.1700	0.84	.2995	1.24	.3925	1.64	.4495	2.04	.4793	2.44	.4927
0.05	.0199	0.45	.1736	0.85	.3023	1.25	.3944	1.65	.4505	2.05	.4798	2.45	.4929
0.06	.0239	0.46	.1772	0.86	.3051	1.26	.3962	1.66	.4515	2.06	.4803	2.46	.4931
0.07	.0279	0.47	.1808	0.87	.3078	1.27	.3980	1.67	.4525	2.07	.4808	2.47	.4932
0.08	.0319	0.48	.1844	0.88	.3106	1.28	.3997	1.68	.4535	2.08	.4812	2.48	.4934
0.09	.0359	0.49	.1879	0.89	.3133	1.29	.4015	1.69	.4545	2.09	.4817	2.49	.4936
0.10	.0398	0.50	.1915	0.90	.3159	1.30	.4032	1.70	.4554	2.10	.4821	2.50	.4938
0.11	.0438	0.51	.1950	0.91	.3186	1.31	.4049	1.71	.4564	2.11	.4826	2.55	.4946
0.12	.0478	0.52	.1985	0.92	.3212	1.32	.4066	1.72	.4573	2.12	.4830	2.60	.4953
0.13	.0517	0.53	.2019	0.93	.3238	1.33	.4082	1.73	.4582	2.13	.4834	2.65	.4960
0.14	.0557	0.54	.2054	0.94	.3264	1.34	.4099	1.74	.4591	2.14	.4838	2.70	.4965

z	Area	z	Area	z	Area	z	Area	z	Area	z	Area	z	Area
0.15	.0596	0.55	.2088	0.95	.3289	1.35	.4115	1.75	.4599	2.15	.4842	2.75	.4970
0.16	.0636	0.56	.2123	0.96	.3315	1.36	.4131	1.76	.4608	2.16	.4846	2.80	.4974
0.17	.0675	0.57	.2157	0.97	.3340	1.37	.4147	1.77	.4616	2.17	.4850	2.85	.4978
0.18	.0714	0.58	.2190	0.98	.3365	1.38	.4162	1.78	.4625	2.18	.4854	2.90	.4981
0.19	.0753	0.59	.2224	0.99	.3389	1.39	.4177	1.79	.4633	2.19	.4857	2.95	.4984
0.20	.0793	0.60	.2257	1.00	.3413	1.40	.4192	1.80	.4641	2.20	.4861	3.00	.49865
0.21	.0832	0.61	.2291	1.01	.3438	1.41	.4207	1.81	.4649	2.21	.4864	3.05	.49886
0.22	.0871	0.62	.2324	1.02	.3461	1.42	.4222	1.82	.4656	2.22	.4868	3.10	.49903
0.23	.0910	0.63	.2357	1.03	.3485	1.43	.4236	1.83	.4664	2.23	.4871	3.15	.49918
0.24	.0948	0.64	.2389	1.04	.3508	1.44	.4251	1.84	.4671	2.24	.4875	3.20	.49931
0.25	.0987	0.65	.2422	1.05	.3531	1.45	.4265	1.85	.4678	2.25	.4878	3.25	.49942
0.26	.1026	0.66	.2454	1.06	.3554	1.46	.4279	1.86	.4686	2.26	.4881	3.30	.49952
0.27	.1064	0.67	.2486	1.07	.3577	1.47	.4292	1.87	.4693	2.27	.4884	3.35	.49960
0.28	.1103	0.68	.2517	1.08	.3599	1.48	.4306	1.88	.4699	2.28	.4887	3.40	.49966
0.29	.1141	0.69	.2549	1.09	.3621	1.49	.4319	1.89	.4706	2.29	.4890	3.45	.49972
0.30	.1179	0.70	.2580	1.10	.3643	1.50	.4332	1.90	.4713	2.30	.4893	3.50	.49977
0.31	.1217	0.71	.2611	1.11	.3665	1.51	.4345	1.91	.4719	2.31	.4896	3.55	.49981
0.32	.1255	0.72	.2642	1.12	.3686	1.52	.4357	1.92	.4726	2.32	.4898	3.60	.49984
0.33	.1293	0.73	.2673	1.13	.3708	1.53	.4370	1.93	.4732	2.33	.4901	3.65	.49987
0.34	.1331	0.74	.2704	1.14	.3729	1.54	.4382	1.94	.4738	2.34	.4904	3.70	.49989
0.35	.1368	0.75	.2734	1.15	.3749	1.55	.4394	1.95	.4744	2.35	.4906	3.75	.49991
0.36	.1406	0.76	.2764	1.16	.3770	1.56	.4406	1.96	.4750	2.36	.4909	3.80	.49993
0.37	.1443	0.77	.2794	1.17	.3790	1.57	.4418	1.97	.4756	2.37	.4911	3.85	.49994
0.38	.1480	0.78	.2823	1.18	.3810	1.58	.4429	1.98	.4761	2.38	.4913	3.90	.49995
0.39	.1517	0.79	.2852	1.19	.3830	1.59	.4441	1.99	.4767	2.39	.4916		

Source: The entries in this table were computed by the authors.

445

TABLE B. Critical Values of Student's *t* Distribution

	Level of significance for a one-tailed test					
	.10	.05	.025	.01	.005	.0005
	Level of significance for a two-tailed test					
df	.20	.10	.05	.02	.01	.001
1	3.078	6.314	12.706	31.821	63.657	636.619
2	1.886	2.920	4.303	6.965	9.925	31.599
3	1.638	2.353	3.182	4.541	5.841	12.924
4	1.533	2.132	2.776	3.747	4.604	8.610
5	1.476	2.015	2.571	3.365	4.032	6.869
6	1.440	1.943	2.447	3.143	3.707	5.959
7	1.415	1.895	2.365	2.998	3.499	5.408
8	1.397	1.860	2.306	2.896	3.355	5.041
9	1.383	1.833	2.262	2.821	3.250	4.781
10	1.372	1.812	2.228	2.764	3.169	4.587
11	1.363	1.796	2.201	2.718	3.106	4.437
12	1.356	1.782	2.179	2.681	3.055	4.318
13	1.350	1.771	2.160	2.650	3.012	4.221
14	1.345	1.761	2.145	2.624	2.977	4.140
15	1.341	1.753	2.131	2.602	2.947	4.073
16	1.337	1.746	2.120	2.583	2.921	4.015
17	1.333	1.740	2.110	2.567	2.898	3.965
18	1.330	1.734	2.101	2.552	2.878	3.922
19	1.328	1.729	2.093	2.539	2.861	3.883
20	1.325	1.725	2.086	2.528	2.845	3.850
21	1.323	1.721	2.080	2.518	2.831	3.819
22	1.321	1.717	2.074	2.508	2.819	3.792
23	1.319	1.714	2.069	2.500	2.807	3.768
24	1.318	1.711	2.064	2.492	2.797	3.745
25	1.316	1.708	2.060	2.485	2.787	3.725
26	1.315	1.706	2.056	2.479	2.779	3.707
27	1.314	1.703	2.052	2.473	2.771	3.690
28	1.313	1.701	2.048	2.467	2.763	3.674
29	1.311	1.699	2.045	2.462	2.756	3.659
30	1.310	1.697	2.042	2.457	2.750	3.646
40	1.303	1.684	2.021	2.423	2.704	3.551
60	1.296	1.671	2.000	2.390	2.660	3.460
100	1.290	1.660	1.984	2.364	2.626	3.390
160	1.287	1.654	1.975	2.350	2.607	3.352
∞	1.282	1.645	1.960	2.326	2.576	3.291

Source: The entries in this table were computed by the authors.

TABLE C. Critical Values of the F Distribution

Degrees of freedom for denominator	α	Degrees of freedom for numerator															
		1	2	3	4	5	6	7	8	9	10	15	20	25	30	40	50
1	.05	161	199	216	225	230	234	237	239	241	242	246	248	249	250	251	252
	.01	4052	4999	5403	5625	5764	5859	5928	5981	6022	6056	6157	6209	6240	6261	6287	6303
2	.05	18.51	19.00	19.16	19.25	19.30	19.33	19.35	19.37	19.38	19.40	19.43	19.45	19.46	19.46	19.47	19.48
	.01	98.50	99.00	99.17	99.25	99.30	99.33	99.36	99.37	99.39	99.40	99.43	99.45	99.46	99.47	99.47	99.48
3	.05	10.13	9.55	9.28	9.12	9.01	8.94	8.89	8.85	8.81	8.79	8.70	8.66	8.63	8.62	8.59	8.58
	.01	34.12	30.82	29.46	28.71	28.24	27.91	27.67	27.49	27.35	27.23	26.87	26.69	26.58	26.50	26.41	26.35
4	.05	7.71	6.94	6.59	6.39	6.26	6.16	6.09	6.04	6.00	5.96	5.86	5.80	5.77	5.75	5.72	5.70
	.01	21.20	18.00	16.69	15.98	15.52	15.21	14.98	14.80	14.66	14.55	14.20	14.02	13.91	13.84	13.75	13.69
5	.05	6.61	5.79	5.41	5.19	5.05	4.95	4.88	4.82	4.77	4.74	4.62	4.56	4.52	4.50	4.46	4.44
	.01	16.26	13.27	12.06	11.39	10.97	10.67	10.46	10.29	10.16	10.05	9.72	9.55	9.45	9.38	9.29	9.24
6	.05	5.99	5.14	4.76	4.53	4.39	4.28	4.21	4.15	4.10	4.06	3.94	3.87	3.83	3.81	3.77	3.75
	.01	13.75	10.92	9.78	9.15	8.75	8.47	8.26	8.10	7.98	7.87	7.56	7.40	7.30	7.23	7.14	7.09
7	.05	5.59	4.74	4.35	4.12	3.97	3.87	3.79	3.73	3.68	3.64	3.51	3.44	3.40	3.38	3.34	3.32
	.01	12.25	9.55	8.45	7.85	7.46	7.19	6.99	6.84	6.72	6.62	6.31	6.16	6.06	5.99	5.91	5.86
8	.05	5.32	4.46	4.07	3.84	3.69	3.58	3.50	3.44	3.39	3.35	3.22	3.15	3.11	3.08	3.04	3.02
	.01	11.26	8.65	7.59	7.01	6.63	6.37	6.18	6.03	5.91	5.81	5.52	5.36	5.26	5.20	5.12	5.07
9	.05	5.12	4.26	3.86	3.63	3.48	3.37	3.29	3.23	3.18	3.14	3.01	2.94	2.89	2.86	2.83	2.80
	.01	10.56	8.02	6.99	6.42	6.06	5.80	5.61	5.47	5.35	5.26	4.96	4.81	4.71	4.65	4.57	4.52
10	.05	4.96	4.10	3.71	3.48	3.33	3.22	3.14	3.07	3.02	2.98	2.85	2.77	2.73	2.70	2.66	2.64
	.01	10.04	7.56	6.55	5.99	5.64	5.39	5.20	5.06	4.94	4.85	4.56	4.41	4.31	4.25	4.17	4.12
11	.05	4.84	3.98	3.59	3.36	3.20	3.09	3.01	2.95	2.90	2.85	2.72	2.65	2.60	2.57	2.53	2.51
	.01	9.65	7.21	6.22	5.67	5.32	5.07	4.89	4.74	4.63	4.54	4.25	4.10	4.01	3.94	3.86	3.81
12	.05	4.75	3.89	3.49	3.26	3.11	3.00	2.91	2.85	2.80	2.75	2.62	2.54	2.50	2.47	2.43	2.40
	.01	9.33	6.93	5.95	5.41	5.06	4.82	4.64	4.50	4.39	4.30	4.01	3.86	3.76	3.70	3.62	3.57

(Continued)

TABLE C. (*Continued*)

Degrees of freedom for denominator	α	1	2	3	4	5	6	7	8	9	10	15	20	25	30	40	50
13	.05	4.67	3.81	3.41	3.18	3.03	2.92	2.83	2.77	2.71	2.67	2.53	2.46	2.41	2.38	2.34	2.31
	.01	9.07	6.70	5.74	5.21	4.86	4.62	4.44	4.30	4.19	4.10	3.82	3.66	3.57	3.51	3.43	3.38
14	.05	4.60	3.74	3.34	3.11	2.96	2.85	2.76	2.70	2.65	2.60	2.46	2.39	2.34	2.31	2.27	2.24
	.01	8.86	6.51	5.56	5.04	4.69	4.46	4.28	4.14	4.03	3.94	3.66	3.51	3.41	3.35	3.27	3.22
15	.05	4.54	3.68	3.29	3.06	2.90	2.79	2.71	2.64	2.59	2.54	2.40	2.33	2.28	2.25	2.20	2.18
	.01	8.68	6.36	5.42	4.89	4.56	4.32	4.14	4.00	3.89	3.80	3.52	3.37	3.28	3.21	3.13	3.08
16	.05	4.49	3.63	3.24	3.01	2.85	2.74	2.66	2.59	2.54	2.49	2.35	2.28	2.23	2.19	2.15	2.12
	.01	8.53	6.23	5.29	4.77	4.44	4.20	4.03	3.89	3.78	3.69	3.41	3.26	3.16	3.10	3.02	2.97
17	.05	4.45	3.59	3.20	2.96	2.81	2.70	2.61	2.55	2.49	2.45	2.31	2.23	2.18	2.15	2.10	2.08
	.01	8.40	6.11	5.18	4.67	4.34	4.10	3.93	3.79	3.68	3.59	3.31	3.16	3.07	3.00	2.92	2.87
18	.05	4.41	3.55	3.16	2.93	2.77	2.66	2.58	2.51	2.46	2.41	2.27	2.19	2.14	2.11	2.06	2.04
	.01	8.29	6.01	5.09	4.58	4.25	4.01	3.84	3.71	3.60	3.51	3.23	3.08	2.98	2.92	2.84	2.78
19	.05	4.38	3.52	3.13	2.90	2.74	2.63	2.54	2.48	2.42	2.38	2.23	2.16	2.11	2.07	2.03	2.00
	.01	8.18	5.93	5.01	4.50	4.17	3.94	3.77	3.63	3.52	3.43	3.15	3.00	2.91	2.84	2.76	2.71
20	.05	4.35	3.49	3.10	2.87	2.71	2.60	2.51	2.45	2.39	2.35	2.20	2.12	2.07	2.04	1.99	1.97
	.01	8.10	5.85	4.94	4.43	4.10	3.87	3.70	3.56	3.46	3.37	3.09	2.94	2.84	2.78	2.69	2.64
21	.05	4.32	3.47	3.07	2.84	2.68	2.57	2.49	2.42	2.37	2.32	2.18	2.10	2.05	2.01	1.96	1.94
	.01	8.02	5.78	4.87	4.37	4.04	3.81	3.64	3.51	3.40	3.31	3.03	2.88	2.79	2.72	2.64	2.58
22	.05	4.30	3.44	3.05	2.82	2.66	2.55	2.46	2.40	2.34	2.30	2.15	2.07	2.02	1.98	1.94	1.91
	.01	7.95	5.72	4.82	4.31	3.99	3.76	3.59	3.45	3.35	3.26	2.98	2.83	2.73	2.67	2.58	2.53
23	.05	4.28	3.42	3.03	2.80	2.64	2.53	2.44	2.37	2.32	2.27	2.13	2.05	2.00	1.96	1.91	1.88
	.01	7.88	5.66	4.76	4.26	3.94	3.71	3.54	3.41	3.30	3.21	2.93	2.78	2.69	2.62	2.54	2.48
24	.05	4.26	3.40	3.01	2.78	2.62	2.51	2.42	2.36	2.30	2.25	2.11	2.03	1.97	1.94	1.89	1.86
	.01	7.82	5.61	4.72	4.22	3.90	3.67	3.50	3.36	3.26	3.17	2.89	2.74	2.64	2.58	2.49	2.44
25	.05	4.24	3.39	2.99	2.76	2.60	2.49	2.40	2.34	2.28	2.24	2.09	2.01	1.96	1.92	1.87	1.84
	.01	7.77	5.57	4.68	4.18	3.85	3.63	3.46	3.32	3.22	3.13	2.85	2.70	2.60	2.54	2.45	2.40

Degrees of freedom for numerator

26	.05	4.23	3.37	2.98	2.74	2.59	2.47	2.39	2.32	2.27	2.22	2.07	1.99	1.94	1.90	1.85	1.82
	.01	7.72	5.53	4.64	4.14	3.82	3.59	3.42	3.29	3.18	3.09	2.81	2.66	2.57	2.50	2.42	2.36
27	.05	4.21	3.35	2.96	2.73	2.57	2.46	2.37	2.31	2.25	2.20	2.06	1.97	1.92	1.88	1.84	1.81
	.01	7.68	5.49	4.60	4.11	3.78	3.56	3.39	3.26	3.15	3.06	2.78	2.63	2.54	2.47	2.38	2.33
28	.05	4.20	3.34	2.95	2.71	2.56	2.45	2.36	2.29	2.24	2.19	2.04	1.96	1.91	1.87	1.82	1.79
	.01	7.64	5.45	4.57	4.07	3.75	3.53	3.36	3.23	3.12	3.03	2.75	2.60	2.51	2.44	2.35	2.30
29	.05	4.18	3.33	2.93	2.70	2.55	2.43	2.35	2.28	2.22	2.18	2.03	1.94	1.89	1.85	1.81	1.77
	.01	7.60	5.42	4.54	4.04	3.73	3.50	3.33	3.20	3.09	3.00	2.73	2.57	2.48	2.41	2.33	2.27
30	.05	4.17	3.32	2.92	2.69	2.53	2.42	2.33	2.27	2.21	2.16	2.01	1.93	1.88	1.84	1.79	1.76
	.01	7.56	5.39	4.51	4.02	3.70	3.47	3.30	3.17	3.07	2.98	2.70	2.55	2.45	2.39	2.30	2.25
35	.05	4.12	3.27	2.87	2.64	2.49	2.37	2.29	2.22	2.16	2.11	1.96	1.88	1.82	1.79	1.74	1.70
	.01	7.42	5.27	4.40	3.91	3.59	3.37	3.20	3.07	2.96	2.88	2.60	2.44	2.35	2.28	2.19	2.14
40	.05	4.08	3.23	2.84	2.61	2.45	2.34	2.25	2.18	2.12	2.08	1.92	1.84	1.78	1.74	1.69	1.66
	.01	7.31	5.18	4.31	3.83	3.51	3.29	3.12	2.99	2.89	2.80	2.52	2.37	2.27	2.20	2.11	2.06
45	.05	4.06	3.20	2.81	2.58	2.42	2.31	2.22	2.15	2.10	2.05	1.89	1.81	1.75	1.71	1.66	1.63
	.01	7.23	5.11	4.25	3.77	3.45	3.23	3.07	2.94	2.83	2.74	2.46	2.31	2.21	2.14	2.05	2.00
50	.05	4.03	3.18	2.79	2.56	2.40	2.29	2.20	2.13	2.07	2.03	1.87	1.78	1.73	1.69	1.63	1.60
	.01	7.17	5.06	4.20	3.72	3.41	3.19	3.02	2.89	2.78	2.70	2.42	2.27	2.17	2.10	2.01	1.95
55	.05	4.02	3.16	2.77	2.54	2.38	2.27	2.18	2.11	2.06	2.01	1.85	1.76	1.71	1.67	1.61	1.58
	.01	7.12	5.01	4.16	3.68	3.37	3.15	2.98	2.85	2.75	2.66	2.38	2.23	2.13	2.06	1.97	1.91
60	.05	4.00	3.15	2.76	2.53	2.37	2.25	2.17	2.10	2.04	1.99	1.84	1.75	1.69	1.65	1.59	1.56
	.01	7.08	4.98	4.13	3.65	3.34	3.12	2.95	2.82	2.72	2.63	2.35	2.20	2.10	2.03	1.94	1.88
100	.05	3.94	3.09	2.70	2.46	2.31	2.19	2.10	2.03	1.97	1.93	1.77	1.68	1.62	1.57	1.52	1.48
	.01	6.90	4.82	3.98	3.51	3.21	2.99	2.82	2.69	2.59	2.50	2.22	2.07	1.97	1.89	1.80	1.74
160	.05	3.90	3.05	2.66	2.43	2.27	2.16	2.07	2.00	1.94	1.89	1.73	1.64	1.57	1.53	1.47	1.43
	.01	6.80	4.74	3.91	3.44	3.13	2.92	2.75	2.62	2.52	2.43	2.15	1.99	1.89	1.82	1.72	1.66
300	.05	3.87	3.03	2.63	2.40	2.24	2.13	2.04	1.97	1.91	1.86	1.70	1.61	1.54	1.50	1.43	1.39
	.01	6.72	4.68	3.85	3.38	3.08	2.86	2.70	2.57	2.47	2.38	2.10	1.94	1.84	1.76	1.66	1.59
∞	.05	3.84	3.00	2.60	2.37	2.21	2.10	2.01	1.94	1.88	1.83	1.67	1.57	1.51	1.46	1.39	1.35
	.01	6.63	4.61	3.78	3.32	3.02	2.80	2.64	2.51	2.41	2.32	2.04	1.88	1.77	1.70	1.59	1.52

Source: The entries in this table were computed by the authors.

TABLE D. Critical Values of the Chi Square Distribution

Degrees of freedom, df	Alpha levels					
	.10	.05	.02	.01	.002	.001
1	2.706	3.841	5.412	6.635	9.550	10.828
2	4.605	5.991	7.824	9.210	12.429	13.816
3	6.251	7.815	9.837	11.345	14.796	16.266
4	7.779	9.488	11.668	13.277	16.924	18.467
5	9.236	11.070	13.388	15.086	18.907	20.515
6	10.645	12.592	15.033	16.812	20.791	22.458
7	12.017	14.067	16.622	18.475	22.601	24.322
8	13.362	15.507	18.168	20.090	24.352	26.124
9	14.684	16.919	19.679	21.666	26.056	27.877
10	15.987	18.307	21.161	23.209	27.722	29.588
11	17.275	19.675	22.618	24.725	29.354	31.264
12	18.549	21.026	24.054	26.217	30.957	32.909
13	19.812	22.362	25.472	27.688	32.535	34.528
14	21.064	23.685	26.873	29.141	34.091	36.123
15	22.307	24.996	28.259	30.578	35.628	37.697
16	23.542	26.296	29.633	32.000	37.146	39.252
17	24.769	27.587	30.995	33.409	38.648	40.790
18	25.989	28.869	32.346	34.805	40.136	42.312
19	27.204	30.144	33.687	36.191	41.610	43.820
20	28.412	31.410	35.020	37.566	43.072	45.315
21	29.615	32.671	36.343	38.932	44.522	46.797
22	30.813	33.924	37.659	40.289	45.962	48.268
23	32.007	35.172	38.968	41.638	47.391	49.728
24	33.196	36.415	40.270	42.980	48.812	51.179
25	34.382	37.652	41.566	44.314	50.223	52.620
26	35.563	38.885	42.856	45.642	51.627	54.052
27	36.741	40.113	44.140	46.963	53.023	55.476
28	37.916	41.337	45.419	48.278	54.411	56.892
29	39.087	42.557	46.693	49.588	55.792	58.301
30	40.256	43.773	47.962	50.892	57.167	59.703
40	51.805	55.758	60.436	63.691	70.618	73.402
60	74.397	79.082	84.580	88.379	96.404	99.607
100	118.498	124.342	131.142	135.807	145.577	149.449
160	183.311	190.516	198.846	204.530	216.358	221.019
400	436.649	447.632	460.211	468.724	486.274	493.132

Source: The entries in this table were computed by the authors.

TABLE E. Critical Values of Hotelling's T^2 Distribution

Degrees of freedom for error, df_E	Number of dependent variables ($\alpha = .05$)					Number of dependent variables ($\alpha = .01$)				
	$p = 1$	$p = 2$	$p = 3$	$p = 4$	$p = 5$	$p = 1$	$p = 2$	$p = 3$	$p = 4$	$p = 5$
2	18.513					98.503				
3	10.128	57.000				34.116	297.000			
4	7.709	25.472	114.986			21.198	82.177	594.997		
5	6.608	17.361	46.383	192.468		16.258	45.000	147.283	992.494	
6	5.987	13.887	29.661	72.937	289.446	13.745	31.857	75.125	229.679	1489.489
7	5.591	12.001	22.720	44.718	105.157	12.246	25.491	50.652	111.839	329.433
8	5.318	10.828	19.028	33.230	62.561	11.259	21.821	39.118	72.908	155.219
9	5.117	10.033	16.766	27.202	45.453	10.561	19.460	32.598	54.890	98.703
10	4.965	9.459	15.248	23.545	36.561	10.044	17.826	28.466	44.838	72.882
11	4.844	9.026	14.163	21.108	31.205	9.646	16.631	25.637	38.533	58.618
12	4.747	8.689	13.350	19.376	27.656	9.330	15.722	23.588	34.251	49.739
13	4.667	8.418	12.719	18.086	25.145	9.074	15.008	22.041	31.171	43.745
14	4.600	8.197	12.216	17.089	23.281	8.862	14.433	20.834	28.857	39.454
15	4.543	8.012	11.806	16.296	21.845	8.683	13.960	19.867	27.060	36.246
16	4.494	7.856	11.465	15.651	20.706	8.531	13.566	19.076	25.626	33.762
17	4.451	7.722	11.177	15.117	19.782	8.400	13.231	18.418	24.458	31.788
18	4.414	7.606	10.931	14.667	19.017	8.285	12.943	17.861	23.487	30.182
19	4.381	7.504	10.719	14.283	18.375	8.185	12.694	17.385	22.670	28.852
20	4.351	7.415	10.533	13.952	17.828	8.096	12.476	16.973	21.972	27.734

(Continued)

451

TABLE E. (*Continued*)

Degrees of freedom for error, df_E	Number of dependent variables					Number of dependent variables				
	$p = 1$	$p = 2$	$p = 3$	$p = 4$	$p = 5$	$p = 1$	$p = 2$	$p = 3$	$p = 4$	$p = 5$
22	4.301	7.264	10.225	13.409	16.945	7.945	12.111	16.296	20.843	25.959
24	4.260	7.142	9.979	12.983	16.265	7.823	11.820	15.763	19.972	24.616
26	4.225	7.041	9.779	12.641	15.726	7.721	11.581	15.334	19.279	23.565
28	4.196	6.957	9.612	12.359	15.287	7.636	11.383	14.980	18.715	22.721
30	4.171	6.885	9.471	12.123	14.924	7.562	11.215	14.683	18.247	22.029
35	4.121	6.744	9.200	11.674	14.240	7.419	10.890	14.117	17.366	20.743
40	4.085	6.642	9.005	11.356	13.762	7.314	10.655	13.715	16.750	19.858
45	4.057	6.564	8.859	11.118	13.409	7.234	10.478	13.414	16.295	19.211
50	4.034	6.503	8.744	10.934	13.138	7.171	10.340	13.181	15.945	18.718
60	4.001	6.413	8.577	10.668	12.748	7.077	10.137	12.843	15.442	18.018
70	3.978	6.350	8.460	10.484	12.482	7.011	9.996	12.611	15.098	17.543
80	3.960	6.303	8.375	10.350	12.289	6.963	9.892	12.440	14.849	17.201
90	3.947	6.267	8.309	10.248	12.142	6.925	9.813	12.310	14.660	16.942
100	3.936	6.239	8.257	10.167	12.027	6.895	9.750	12.208	14.511	16.740
120	3.920	6.196	8.181	10.048	11.858	6.851	9.657	12.057	14.292	16.444
150	3.904	6.155	8.105	9.931	11.693	6.807	9.565	11.909	14.079	16.156
200	3.888	6.113	8.031	9.817	11.531	6.763	9.474	11.764	13.871	15.877
400	3.865	6.052	7.922	9.650	11.297	6.699	9.341	11.551	13.569	15.473
1000	3.851	6.015	7.857	9.552	11.160	6.660	9.262	11.427	13.392	15.239
∞	3.841	5.991	7.815	9.488	11.071	6.635	9.210	11.345	13.277	15.086

Source: The entries in this table were computed by the authors.

TABLE F. Critical Values of Wilks' Lambda Distribution for α = .05

p = 1, one dependent variable

Degrees of freedom for error, df_E	Degrees of freedom for hypothesis, df_H									
	1	2	3	4	5	6	7	8	9	10
1	6.156*	2.500*	1.5436*	1.112*	.868*	.712*	.603*	.523*	.462*	.413*
2	.098	.050	.034	.025	.020	.017	.015	.013	.011	.010
3	.229	.136	.097	.076	.062	.053	.046	.041	.036	.033
4	.342	.224	.168	.135	.113	.098	.086	.076	.069	.063
5	.431	.302	.236	.194	.165	.144	.128	.115	.104	.096
6	.501	.368	.296	.249	.215	.189	.169	.153	.140	.129
7	.556	.425	.349	.298	.261	.232	.209	.190	.175	.161
8	.601	.473	.396	.343	.303	.271	.246	.225	.208	.193
9	.638	.514	.437	.382	.341	.308	.281	.258	.239	.223
10	.668	.549	.473	.418	.376	.341	.313	.289	.269	.251
11	.694	.580	.505	.450	.407	.372	.343	.318	.297	.278
12	.717	.607	.534	.479	.436	.400	.370	.345	.323	.304
13	.736	.631	.560	.506	.462	.426	.396	.370	.347	.327
14	.753	.652	.583	.529	.486	.450	.420	.393	.370	.350
15	.768	.671	.603	.551	.508	.473	.442	.415	.392	.371
16	.781	.688	.622	.571	.529	.493	.462	.436	.412	.391
17	.792	.703	.639	.589	.548	.512	.482	.455	.431	.410
18	.803	.717	.655	.606	.565	.530	.499	.473	.449	.427
19	.813	.730	.669	.621	.581	.546	.516	.490	.466	.444
20	.821	.741	.683	.636	.596	.562	.532	.505	.482	.460
30	.878	.819	.774	.736	.703	.674	.647	.623	.601	.581
40	.907	.861	.824	.793	.766	.741	.718	.696	.677	.658
60	.937	.905	.879	.856	.835	.816	.798	.781	.766	.751
80	.953	.928	.907	.889	.873	.858	.843	.829	.816	.804
100	.962	.942	.925	.910	.897	.884	.872	.860	.849	.838
120	.968	.951	.937	.925	.913	.902	.891	.882	.872	.863
140	.973	.958	.946	.935	.925	.915	.906	.897	.889	.881
160	.976	.963	.952	.943	.934	.925	.917	.909	.902	.894
180	.979	.967	.958	.949	.941	.933	.926	.919	.912	.905
200	.981	.970	.962	.954	.947	.940	.933	.926	.920	.914
250	.985	.976	.969	.963	.957	.951	.946	.941	.935	.930
300	.987	.980	.974	.969	.964	.959	.955	.950	.946	.942
350	.989	.983	.978	.973	.969	.965	.961	.957	.953	.950
400	.990	.985	.981	.977	.973	.969	.966	.962	.959	.956
600	.994	.990	.987	.984	.982	.979	.977	.975	.972	.970
800	.995	.993	.990	.988	.986	.984	.983	.981	.979	.977
1000	.996	.994	.992	.991	.989	.988	.986	.985	.983	.982

*All entries with an asterisk must be divided by 1000.

Source: The entries in this table were computed by the authors using a derivation from equation 6.15 (page 163) of Rencher (2002). For more extensive Wilks' lambda tables see Table A.9 (page 566) of the Rencher book.

TABLE F. *(Continued)*

p = 2, two dependent variables

Degrees of freedom for error, df_E	Degrees of freedom for hypothesis, df_H									
	1	2	3	4	5	6	7	8	9	10
1	.000	.000	.000	.000	.000	.000	.000	.000	.000	.000
2	2.500*	.641*	.287*	.162*	.104*	.072*	.053*	.041*	.032*	.026*
3	.050	.018	9.528*	5.843*	3.950*	2.849*	2.152*	1.683*	1.352*	1.11*
4	.136	.062	.036	.023	.017	.012	9.554*	7.615*	6.213*	5.165*
5	.224	.117	.074	.051	.037	.028	.023	.018	.015	.013
6	.302	.175	.116	.084	.063	.049	.040	.033	.027	.023
7	.368	.230	.160	.119	.092	.074	.060	.050	.042	.036
8	.425	.280	.203	.155	.122	.099	.082	.069	.059	.051
9	.473	.326	.243	.190	.153	.126	.106	.090	.078	.068
10	.514	.367	.281	.223	.183	.152	.129	.111	.097	.085
11	.549	.404	.316	.255	.212	.179	.153	.133	.116	.102
12	.580	.437	.348	.286	.239	.204	.176	.154	.136	.120
13	.607	.467	.378	.314	.266	.229	.199	.175	.155	.138
14	.631	.495	.405	.340	.291	.252	.221	.195	.174	.156
15	.652	.519	.431	.365	.315	.275	.242	.215	.193	.174
16	.671	.542	.454	.389	.337	.296	.263	.235	.211	.191
17	.688	.562	.476	.410	.359	.317	.282	.254	.229	.208
18	.703	.581	.496	.431	.379	.337	.301	.272	.246	.225
19	.717	.598	.515	.450	.398	.355	.320	.289	.263	.241
20	.730	.614	.532	.468	.416	.373	.337	.306	.279	.256
30	.813	.725	.657	.601	.553	.512	.475	.443	.414	.388
40	.858	.786	.730	.682	.639	.602	.568	.537	.509	.484
60	.903	.853	.811	.774	.741	.710	.682	.656	.632	.609
80	.927	.887	.854	.825	.798	.772	.749	.727	.706	.686
100	.941	.909	.882	.857	.834	.813	.793	.774	.755	.738
120	.951	.924	.900	.879	.860	.841	.823	.807	.791	.775
140	.958	.934	.914	.895	.878	.862	.846	.831	.817	.803
160	.963	.942	.924	.908	.893	.878	.864	.851	.838	.825
180	.967	.949	.932	.918	.904	.891	.878	.866	.854	.843
200	.970	.954	.939	.926	.913	.901	.889	.878	.867	.857
250	.976	.963	.951	.940	.930	.920	.910	.901	.892	.883
300	.980	.969	.959	.950	.941	.933	.925	.917	.909	.902
350	.983	.973	.965	.957	.949	.942	.935	.928	.921	.915
400	.985	.977	.969	.962	.955	.949	.943	.937	.931	.925
600	.990	.984	.979	.975	.970	.966	.961	.957	.953	.949
800	.993	.988	.984	.981	.977	.974	.971	.968	.965	.962
1000	.994	.991	.987	.985	.982	.979	.977	.974	.972	.969

*All entries with an asterisk must be divided by 1000.

TABLE F. *(Continued)*

	p = 3		p = 4		p = 5		p = 6		p = 7	
	df_H		df_H		df_H		df_H		df_H	
df_E	1	2	1	2	1	2	1	2	1	2
1	.000	.000	.000	.000	.000	.000	.000	.000	.000	.000
2	.000	.000	.000	.000	.000	.000	.000	.000	.000	.000
3	.002	.000	.000	.000	.000	.000	.000	.000	.000	.000
4	.034	.010	.001	.000	.000	.000	.000	.000	.000	.000
5	.097	.036	.025	.006	.001	.000	.000	.000	.000	.000
6	.168	.074	.076	.023	.020	.004	.001	.000	.000	.000
7	.236	.116	.135	.051	.062	.017	.017	.003	.001	.000
8	.296	.160	.194	.084	.113	.037	.053	.012	.015	.002
9	.349	.203	.249	.119	.165	.063	.098	.028	.046	.010
10	.396	.243	.298	.155	.215	.092	.144	.049	.086	.023
11	.437	.281	.343	.190	.261	.122	.189	.074	.128	.040
12	.473	.316	.382	.223	.303	.153	.232	.099	.169	.060
13	.505	.348	.418	.255	.341	.183	.271	.126	.209	.082
14	.534	.378	.450	.286	.376	.212	.308	.152	.246	.106
15	.560	.405	.479	.314	.407	.239	.341	.179	.281	.129
16	.583	.431	.506	.340	.436	.266	.372	.204	.313	.153
17	.603	.454	.529	.365	.462	.291	.400	.229	.343	.176
18	.622	.476	.551	.389	.486	.315	.426	.252	.370	.199
19	.639	.496	.571	.410	.508	.337	.450	.275	.396	.221
20	.655	.515	.589	.431	.529	.359	.473	.296	.420	.242
30	.760	.648	.712	.580	.668	.519	.626	.464	.586	.413
40	.816	.724	.779	.668	.744	.617	.711	.570	.679	.526
60	.875	.808	.849	.767	.825	.729	.802	.693	.779	.660
80	.905	.853	.885	.821	.867	.791	.849	.762	.832	.735
100	.924	.881	.908	.854	.893	.830	.878	.806	.864	.783
120	.936	.900	.923	.877	.910	.856	.898	.836	.886	.817
140	.945	.913	.934	.894	.923	.876	.912	.858	.902	.841
160	.952	.924	.942	.907	.932	.891	.923	.875	.914	.860
180	.957	.932	.948	.917	.940	.902	.931	.888	.923	.875
200	.961	.939	.953	.925	.945	.912	.938	.899	.931	.887
250	.969	.951	.962	.940	.956	.929	.950	.919	.945	.909
300	.974	.959	.969	.949	.963	.940	.958	.932	.954	.923
350	.978	.965	.973	.956	.969	.949	.964	.941	.960	.934
400	.981	.969	.976	.962	.973	.955	.969	.948	.965	.942
600	.987	.979	.984	.974	.982	.970	.979	.965	.977	.961
800	.990	.984	.988	.981	.986	.977	.984	.974	.982	.971
1000	.992	.987	.991	.985	.989	.982	.987	.979	.986	.977

Note: The number of dependent variables is denoted by p for each set of two columns of lambda values.

NAME INDEX

Multivariate Analysis for the Biobehavioral and Social Sciences: A Graphical Approach,
First Edition. Bruce L. Brown, Suzanne B. Hendrix, Dawson W. Hedges, Timothy B. Smith.
© 2012 John Wiley & Sons, Inc. Published 2012 by John Wiley & Sons, Inc.

SUBJECT INDEX

Multivariate Analysis for the Biobehavioral and Social Sciences: A Graphical Approach,
First Edition. Bruce L. Brown, Suzanne B. Hendrix, Dawson W. Hedges, Timothy B. Smith.
© 2012 John Wiley & Sons, Inc. Published 2012 by John Wiley & Sons, Inc.